Remote Sensing: Monitoring, Modeling and Applications

Edited by William Ramsay

SYRAWOOD
PUBLISHING HOUSE

New York

Published by Syrawood Publishing House,
750 Third Avenue, 9th Floor,
New York, NY 10017, USA
www.syrawoodpublishinghouse.com

Remote Sensing: Monitoring, Modeling and Applications
Edited by William Ramsay

© 2018 Syrawood Publishing House

International Standard Book Number: 978-1-68286-556-9 (Hardback)

Cataloging-in-Publication Data

Remote sensing : monitoring, modeling and applications / edited by William Ramsay.
    p. cm.
Includes bibliographical references and index.
ISBN 978-1-68286-556-9
1. Remote sensing. I. Ramsay, William.
G70.4 .R46 2018
621.367 8--dc23

TABLE OF CONTENTS

PREFACE

This book was inspired by the evolution of our times; to answer the curiosity of inquisitive minds. Many developments have occurred across the globe in the recent past which has transformed the progress in the field.

Acquiring information relating to objects that are geographically inaccessible to collect data is known as remote sensing. It is used in a number of fields such as military intelligence, oceanography, geology, etc. It can be further divided into active or passive remote sensing. This book outlines the processes and applications of remote sensing in detail. It is a vital tool for all those who are researching or studying remote sensing as it gives incredible insights into emerging trends and concepts.

This book was developed from a mere concept to drafts to chapters and finally compiled together as a complete text to benefit the readers across all nations. To ensure the quality of the content we instilled two significant steps in our procedure. The first was to appoint an editorial team that would verify the data and statistics provided in the book and also select the most appropriate and valuable contributions from the plentiful contributions we received from authors worldwide. The next step was to appoint an expert of the topic as the Editor-in-Chief, who would head the project and finally make the necessary amendments and modifications to make the text reader-friendly. I was then commissioned to examine all the material to present the topics in the most comprehensible and productive format.

I would like to take this opportunity to thank all the contributing authors who were supportive enough to contribute their time and knowledge to this project. I also wish to convey my regards to my family who have been extremely supportive during the entire project.

Editor

Polar Bears from Space: Assessing Satellite Imagery as a Tool to Track Arctic Wildlife

Seth Stapleton[1*¤a], Michelle LaRue[3], Nicolas Lecomte[4¤b], Stephen Atkinson[4], David Garshelis[2,5], Claire Porter[3], Todd Atwood[1]

1 United States Geological Survey, Alaska Science Center, Anchorage, Alaska, United States of America, 2 Department of Fisheries, Wildlife and Conservation Biology, University of Minnesota, St. Paul, Minnesota, United States of America, 3 Department of Earth Sciences, University of Minnesota, Minneapolis, Minnesota, United States of America, 4 Department of Environment, Government of Nunavut, Igloolik, Nunavut, Canada, 5 Minnesota Department of Natural Resources, Grand Rapids, Minnesota, United States of America

Abstract

Development of efficient techniques for monitoring wildlife is a priority in the Arctic, where the impacts of climate change are acute and remoteness and logistical constraints hinder access. We evaluated high resolution satellite imagery as a tool to track the distribution and abundance of polar bears. We examined satellite images of a small island in Foxe Basin, Canada, occupied by a high density of bears during the summer ice-free season. Bears were distinguished from other light-colored spots by comparing images collected on different dates. A sample of ground-truthed points demonstrated that we accurately classified bears. Independent observers reviewed images and a population estimate was obtained using mark–recapture models. This estimate ($\hat{N}$: 94; 95% Confidence Interval: 92–105) was remarkably similar to an abundance estimate derived from a line transect aerial survey conducted a few days earlier ($\hat{N}$: 102; 95% CI: 69–152). Our findings suggest that satellite imagery is a promising tool for monitoring polar bears on land, with implications for use with other Arctic wildlife. Large scale applications may require development of automated detection processes to expedite review and analysis. Future research should assess the utility of multi-spectral imagery and examine sites with different environmental characteristics.

Editor: Yan Ropert-Coudert, Institut Pluridisciplinaire Hubert Curien, France

Funding: Funding for this research was provided by the Changing Arctic Ecosystems Initiative of the U.S. Geological Survey, Ecosystems Mission Area, Wildlife Program; the Bureau of Ocean Energy Management; and the Government of Nunavut. The funders had no role in study design, data collection and analysis, decision to publish, or preparation of the manuscript.

Competing Interests: The authors have declared that no competing interests exist.

* Email: stapl078@umn.edu

¤a Current address: Department of Fisheries, Wildlife and Conservation Biology, University of Minnesota, St. Paul, Minnesota, United States of America
¤b Current address: Université de Moncton, Department of Biology, Moncton, New Brunswick, Canada

Introduction

The loss of Arctic sea ice has accelerated during recent years [1–3], with minimum sea ice extent reaching a record low during September, 2012. A nearly ice-free summer is now forecasted to occur as early as 2016 [4,5]. Such large-scale, precipitous environmental changes will be detrimental for many species dependent on sea ice habitats [6].

Despite potentially massive ecological impacts, regimes for monitoring wildlife remain deficient across large portions of the Arctic. For example, marine mammal assessment programs traditionally have used some combination of costly aircraft- or ship-based surveys and/or mark-recapture programs [7,8], but the precision of resulting demographic estimates is often inadequate to detect trends in abundance [9]. Moreover, some areas are simply too inaccessible for routine monitoring. As such, baseline or long-term data are lacking for numerous species, precluding status and trend assessment and hindering management efforts. Walrus (*Odobenus rosmarus*) [10] and ribbon seals (*Histriophoca fasciata*) [11] are among the Arctic marine mammals currently classified as data deficient by the International Union for Conservation of Nature

(IUCN). Likewise, data are insufficient to assess polar bear (*Ursus maritimus*) status across large portions of their range [12]; even in surveyed areas, monitoring intervals are often inadequate [13]. More frequent, systematic and efficient population surveys are needed to match the data needs of resource managers faced with a rapidly changing environment.

Recent advancements in satellite technology (resolutions of 0.5–5 m) have provided new tools for monitoring wildlife. Previous studies used satellite imagery to estimate abundance at Weddell seal (*Leptonychotes weddellii*) haul-outs [14] and emperor penguin (*Aptenodytes forsteri*) colonies in Antarctica [15]. Similarly, Platonov et al. [16] reported that polar bears, walrus and other marine mammals are visible on satellite imagery, but their findings are limited by an absence of ground-truthed data. Remote sensing affords access to vast expanses of otherwise inaccessible sites, at potentially reduced costs, without concerns about human safety and disturbance to wildlife.

Here, our goal was to evaluate the utility of high resolution satellite imagery to monitor Arctic wildlife, using polar bears as a case study. Whereas polar bears rank among the most studied large mammals globally, with capture datasets in some regions

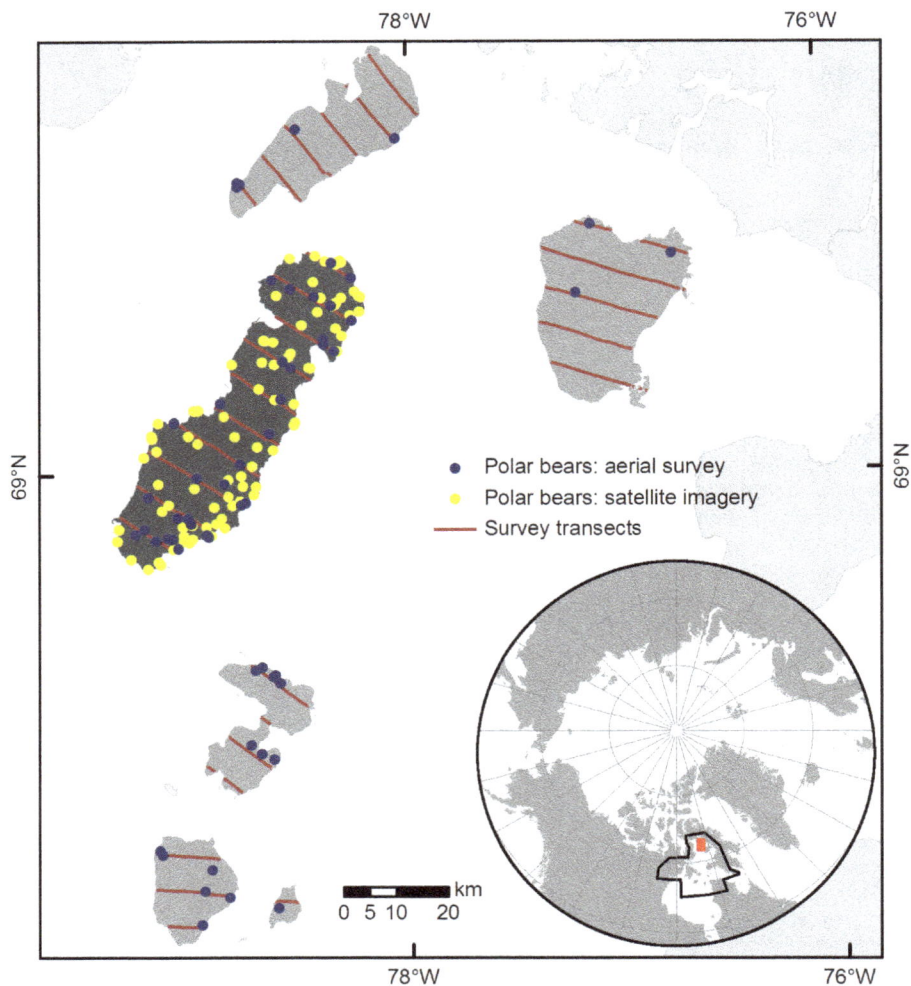

Figure 1. Polar bears detected with high resolution satellite imagery and during the helicopter-based aerial survey. Target imagery was acquired from Rowley Island (dark shade) in northern Foxe Basin, Nunavut with the WorldView-2 and Quickbird satellites on September 3, 2012. Transects were spaced at 7 km intervals during the aerial survey. The Foxe Basin polar bear subpopulation is outlined in black and the study area is shaded red in the inset.

extending >30 years [17], most of that research has focused on a few easily accessible subpopulations. Polar bears are categorized as Vulnerable by the IUCN, largely owing to projected sea ice losses [18], but there is a paucity of population-level data across several broad regions. Additionally, the changing sea ice dynamics have led to shifts in the onshore distribution and abundance of polar bears [19]. Mitigating corresponding increases in human – bear conflicts requires an understanding of and ability to predict these distributional shifts. These issues highlight the need for efficient methods of population assessment that overcome logistical challenges, facilitate regular monitoring, and are consistent with the values of northern communities concerned about disturbance to wildlife.

Methods

Ethics Statement

This research was conducted under Wildlife Research Permit Number 2012-052 (Government of Nunavut). Aerial survey field protocols were approved by the Institutional Animal Care and Use Committee at the University of Minnesota (Permit Number 1207A17284).

Study Area

We conducted our research in Foxe Basin, Nunavut, located in a seasonal ice region of the eastern Canadian Arctic. Recent comprehensive aerial surveys documented high densities of polar bears on relatively small islands (totaling <3,000 km^2) in northern Foxe Basin with low topographic relief and no snow cover during the late summer, ice-free season [20]. As the ice melts across Foxe Basin, bears become stranded on small ice floes and eventually retreat to nearby land masses where they wait for ice to return. Hence, high densities of bears tend to accrue on land adjacent to late-melting ice, especially islands where dispersion is limited. We selected Rowley Island as our study site: its high density of bears during the ice-free season, contrasting dark landscape, and flat terrain provided an ideal setting to evaluate the utility of satellite imagery.

Remote Sensing

We procured target satellite images of Rowley Island ($\sim$1,100 km^2) from DigitalGlobe, Inc. (WorldView-2 satellite, $\sim$0.5 m resolution at nadir; Quickbird, 0.65 m resolution), during early September, 2012. We compared these images to reference images to discriminate non-target objects from bears ($\sim$2-m white objects visible on the target image but not the reference image). Reference imagery was acquired during August, 2009 and 2010 (WorldView-1; 0.5 m resolution) and August, 2012 (Quickbird; all satellite images are available for purchase through DigitalGlobe, Inc., http://www.digitalglobe.com.) We corrected all images for terrain (i.e., orthorectification). To account for any differences in sensor exposure settings and sun irradiance based on time of year and day, we calculated top-of-atmosphere reflectance (following [21]) using relevant metadata from the imagery (per band), earth-sun distance at time of acquisition, and sun elevation angle. We applied an additional histogram stretch to brighten darker, non-ice areas (identical for all images) in order to facilitate image comparison by human analysts. We used a Python script that leverages the open-source Geospatial Data Abstraction Library package for image manipulation and ArcGIS 10.1 (Environmental Systems Research Institute; Redlands, California, USA) to overlay target images on reference imagery.

Two independent observers visually identified potential polar bears on the September, 2012 image and recorded latitude and longitude. Observers initially reviewed imagery at a fixed scale of 1:2,000 to 1:3,000 and subsequently examined potential polar bears at multiple scales (up to $\sim$1:250), and in comparison to reference images to help distinguish likely bears.

Following this independent review, the two observers jointly examined imagery to resolve uncertainties in identification of potential bears. We did not categorize an object as a "presumed bear" unless observers were in agreement and confident in that classification. We thus deleted some points from each observer's initial list of candidate polar bears, but observers did not add points to their respective sightings during this process. We treated each observer's review as an independent sampling period, enabling us to generate capture histories for mark-recapture analysis. We employed a full likelihood-based, closed population model [22], facilitating direct estimation of abundance and detection. We allowed detection probabilities to vary between observers and conducted modeling in Program MARK [23].

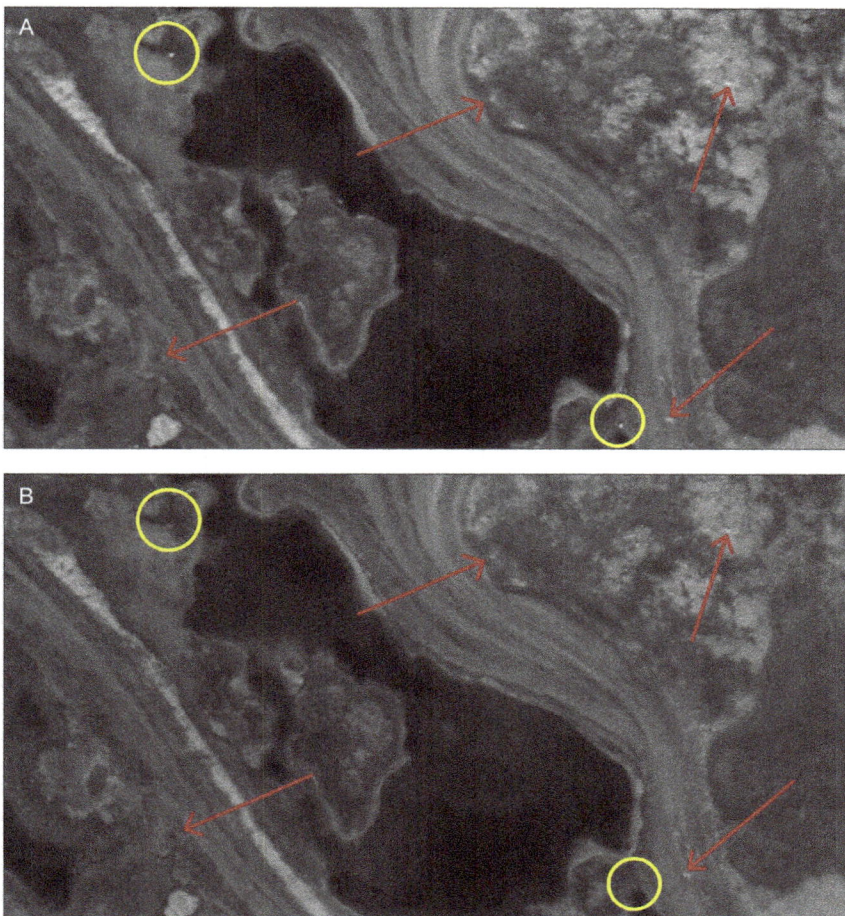

Figure 2. Example of high resolution satellite imagery used to detect polar bears. Imagery was procured from Rowley Island in Foxe Basin, Nunavut during late summer, 2012. The target imagery (a) was searched for polar bears, and the reference imagery (b) was used for comparison. Polar bears are present in the example target image but absent in the reference image (yellow circles). Landscape features that remain consistent between images, including rocks and substrate, are denoted with red arrows. Satellite imagery printed under a CC BY license, with permission from DigitalGlobe ©2013.

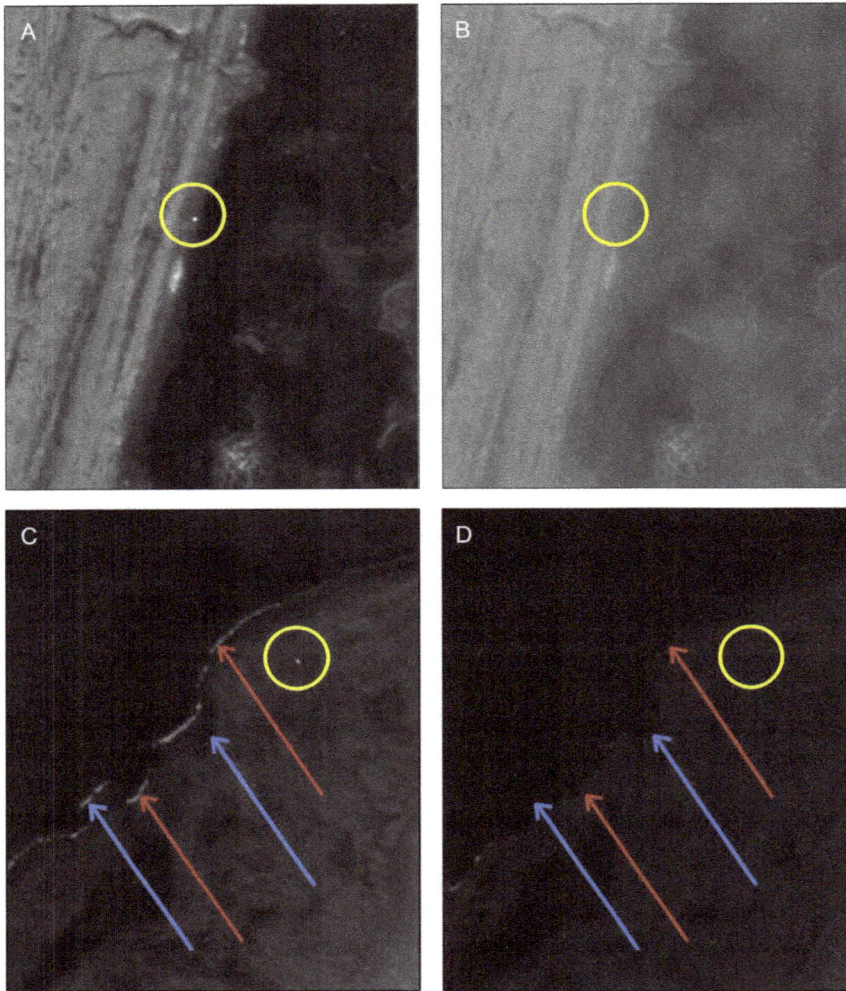

Figure 3. Clouds (top) and water conditions (bottom) are factors that may hamper detection of bears. Bear locations are indicated in target [(a) and (c)] and reference [(b) and (d)] images with yellow circles. Foam accumulating along the edges of water bodies and changes in water levels between target and reference images are indicated in the bottom pair of shots by red and blue arrows, respectively. Satellite imagery printed under a CC BY license, with permission from DigitalGlobe ©2013.

Field Sampling

We used a helicopter (Bell 206L) survey to assess how well we distinguished polar bears from objects of similar size and color on imagery. We categorized 26 points on imagery as either polar bears or non-target, light-colored control points (e.g., rocks, foam on water surface), and we flew to these sites to confirm identity. We assumed that a bear had been present when the site had been photographed if 1) there was no rock or other feature that could be confused with a bear and 2) the site was not prone to ephemeral landscape features (e.g., not downwind of a pond that could have had white foam when the image was collected).

We also conducted a helicopter-based aerial survey to obtain a second population estimate of bears on Rowley Island. We could not directly compare polar bear sightings during this aerial survey (August 30 – September 1, 2012) with points on the target image (September 3) because bears moved in the 2–4 days that elapsed between events. However, we assumed that Rowley Island was a closed population during this short time frame, enabling us to compare abundance estimates derived by the two techniques.

We implemented mark-recapture distance sampling (MRDS) [24] protocols for abundance estimation. MRDS combines distance sampling with a double-observer platform; the double-observer data are incorporated in a mark-recapture modeling framework to explicitly test distance sampling's assumption of perfect detection at distance 0 [25] and inflate density estimates, if necessary. Here, bears observed by the pilot and front seat observer were considered marked, while those observed by rear seat observers were considered recaptured. We surveyed Rowley and nearby islands to obtain a sufficient sample for estimating the detection function. We oriented sampling transects perpendicular to each island's primary axis and extended them across island widths (Fig. 1). Transects were spaced at 7-km intervals, and we sampled at an above ground level altitude of 120 m (400 ft) and target airspeed of 160 km/h (85 knots). Flight parameters were based on previous overland aerial surveys of polar bears in the region [20]. We recorded flight paths and locations of polar bear sightings with a GPS, and we measured distances from transects to observations in a GIS (modified from [26]). We documented group

Table 1. Summary of distance sampling analyses.

Model	ΔAIC$_c$	Estimate (SE)			Goodness of Fit		
		ESW	p	N̂	C-S	K-S	C-vM
Uniform/Cosine	0.00	1234 (65)	0.53 (0.028)	97 (17.8)	0.75	0.79	0.6
Half-normal/None	0.15	1151 (114)	0.50 (0.049)	104 (21.0)	0.75	0.94	0.8
Half-normal/VIS	1.12	1136 (114)	0.49 (0.049)	105 (21.3)	0.66	0.87	0.7
Half-normal/LIGHT	1.38	1112 (115)	0.48 (0.050)	108 (22.1)	0.55	0.91	0.8
Hazard/None	2.62	1201 (168)	0.52 (0.073)	100 (22.5)	0.53	0.89	0.8

Results of distance sampling analyses of a polar bear aerial survey conducted in northern Foxe Basin, Nunavut, Canada during August – September, 2012. Highly supported models (ΔAIC$_c$<3) are presented. In the column Model, the key function is followed by adjustment terms or covariates (VIS = visibility; poor/fair (e.g., glare, light fog or rain) or excellent; LIGHT = light conditions; overcast, mostly cloudy, or partly cloudy/clear). ESW = Effective strip width (meters). p = Detection probability. N̂ = Abundance estimate. Goodness of Fit metrics: C-S = Chi-squared; K-S = Kolmogorov-Smirnov; C-vM = Cramér-von Mises.

size and recorded conditions that may have impacted detection (weather, lighting).

We conducted preliminary double-observer analyses with the Huggins model [27,28], which suggested that detection on and near the transect line was nearly perfect. Hence, we analyzed data in the conventional and multiple covariate distance sampling engines of Program DISTANCE 6.0 [29]. We pooled sightings data from all islands to estimate a common detection function and used encounter rates and group sizes from Rowley Island to obtain an island-specific abundance estimate. We considered models with standard key functions and series expansion terms as well as covariate-based models. Because we could not reliably differentiate family groups on satellite imagery, our aerial survey estimate included only independent bears (i.e., excluded cubs or yearlings with their mother). We used Akaike's Information Criteria, adjusted for small sample sizes (AIC$_c$) [30], for model selection.

Results

Remote Sensing

We detected 92 presumed bears on satellite images of Rowley Island (Figure 1) and documented likely family groups (adult females with cubs) on five occasions. The most highly supported model included separate detection probabilities for the two observers and yielded an abundance estimate of 94 (95% confidence interval: 92–105) independent bears. Individual detection probabilities varied greatly between the two observers (96% [95% CI: 83%–99%] and 42% [95% CI: 32%–52%]). Although it was generally straightforward to distinguish bears from other objects (Figure 2), landscape features and environmental characteristics sometimes complicated detection (Figure 3). About 12% of the reference imagery was obscured by clouds, and strong winds on the date of imagery collection created large expanses of foam along the banks of ponds that initially appeared to be bears, since they were absent from reference imagery. Additionally, some rocks reflected light differently between successive photos, requiring careful scrutiny to differentiate them from bears. However, joint review of imagery enabled us to correctly categorize all points that we ground-truthed via helicopter as presumed bears (n = 13) or inanimate objects (n = 13).

Aerial Survey Abundance Estimation

During the helicopter aerial survey, we sighted 56 polar bear groups totaling 77 individuals along ca. 400 km of transects across all study islands; this included 33 groups (34 independent bears) during ca. 160 km of sampling on Rowley Island (Fig. 1). Despite a small number of detections, our data facilitated estimation of a robust detection function, and abundance estimates were consistent among the most highly supported models (Table 1). Our model-averaged [31] estimate of abundance (including models ΔAIC$_c$<3) yielded 102 independent bears on Rowley Island (95% CI: 69–152).

Discussion

Satellite imagery shows promise as a means to quickly and safely monitor the abundance and distribution of polar bears using onshore habitats. We were able to discriminate among presumed bears and non-targets by comparing high resolution images collected at different points in time. The remarkable consistency between our estimates of abundance derived from imagery and established aerial survey techniques suggests that bear identification using imagery was quite accurate. We believe that the methods employed here (use of reference images, review by multiple observers to build consensus and generate capture histories, and estimation of abundance and detection probabilities via population models) provide a framework for other small-scale studies. However, applications at broader geographic scales may necessitate the development of automated image classification processes to expedite review and analysis. Our initial, independent review of imagery was tedious and required a combined 100 hours; this made it unrealistic to re-examine the images a second time (after our joint examination of points), and also made it difficult to recruit more observers. A reliable, automated process would greatly enhance the applicability of this technique.

Observers differed substantially in their abilities to detect bears with imagery. This finding was an unexpected but important result of this study; this did not diminish the robustness of our results, although precision would improve with higher detection for both observers. In our study, the two observers had vastly different levels of experience: one had several seasons of experience studying polar bears in this landscape during the ice-free season, whereas the other had extensive experience interpreting remote sensing imagery but no direct experience with polar bears. The observer with field experience had better detection of bears on the images, suggesting that familiarity with the study landscape and first-hand knowledge of bear biology and behavior (e.g., variation in color and body outline based on posture) greatly improved detection. Moreover, the observers searched imagery somewhat differently. We found that detection was higher when one regularly compared the target and reference images (one's eye was attracted to white spots on the target image not present on the reference image), rather than using the reference image to simply verify the presence of bears. These experiences suggest that explicit search protocols and a rigorous training program including individuals with relevant, on-the-ground experience with the target species will improve implementation of the technique and ensure appropriate search images.

The two abundance estimation techniques provided significantly different estimates of precision (coefficients of variation for line transect aerial survey: 20.4% versus satellite imagery: 2.5%). Distance sampling incorporates multiple variance components, including detection and encounter rates on sampled transects. Conversely, the satellite imagery modeling only includes a variance component for detection, since we reviewed imagery from the entire island. The very high detection probability of one imagery observer also contributed to this difference. Variance estimated from manual review of imagery would increase in applications in which observers have lower detection probabilities or if images provide less complete coverage of the study site.

Synchronizing collection of satellite imagery with visual surveys is not currently possible, since there is no assurance as to when the satellite's orbit will pass above the study area and if weather will be conducive to shooting imagery or conducting an aerial survey. This reality prohibits directly matching bears identified on photos with bears observed during an aerial survey. As such, absolute confirmation of presumed bears is impossible, and thus some false positives (i.e., inanimate objects classified as bears) or negatives are likely to occur.

Because one observer of the images had a very high detection probability (96%), we deemed it unnecessary to model potential sources of heterogeneity. However, future studies may be compelled to quantify variables potentially impacting detection. We hypothesize that environmental conditions including wind, light, and the presence of clouds and small onshore ice floes may affect detection (Fig. 3). Other prospective covariates may include bear reflectance values, bear size (i.e., pixels), reflectance values and complexity metrics for the surrounding landscape at multiple spatial scales, image exposure, and off-nadir angle at image collection [32].

We presumed that cubs were not consistently identifiable on imagery, given the resolution constraints. Their presence was suggested by multiple white spots of notably different sizes in a cluster (ca.<20 m). We detected only five likely family groups with imagery, whereas the nine family groups sighted on Rowley Island during aerial survey sampling suggest that there were ~28 family groups present island-wide. The inability to reliably discern family groups poses some limits on the utility of imagery for demographic studies. However, the advent of higher resolution imagery (e.g., WorldView-3 platform, set to launch in 2014, will shoot at 0.3 m resolution at nadir) may permit differentiation of cubs, as well as improve detection of smaller species, in the future.

With minimal topographic relief and high densities of polar bears during late summer, our study island provided a model setting to test satellite imagery as a monitoring tool. Conditions elsewhere in the Arctic, however, are less ideal, and further technique development will be required to more broadly apply the technology. Priority research and development areas for polar bears should include assessing onshore sites with lower densities and more variable landscapes (e.g., higher topographic relief) and evaluating sampling intensities necessary to obtain reliable density estimates and distributional information. Additionally, multispectral imagery may better capture unique spectral signatures of the target species, thereby improving manual and automated detection in more challenging onshore environments. Multispectral imagery also may facilitate the detection of polar bears on sea ice, given the apparent spectral differences between bears and snow at short wavelengths (G. LeBlanc, National Research Council Canada and C. Francis, Environment Canada, unpublished data).

The success of this technique with polar bears suggests that satellite imagery would likely provide a useful means to inventory other megafauna as well. In the Arctic, darker species such as musk oxen (*Ovibos moschatus*) and caribou (*Rangifer tarandus*) may be readily detected against a snow-covered, springtime landscape. Whereas satellite imagery does not yield the same detail of information as traditional capture programs and aerial surveys, it has tremendous potential to provide coarse abundance and distribution data from sites otherwise too logistically challenging or costly to routinely access. The technology can open vast, remote regions to regular monitoring, facilitating the collection of data across species' ranges and at global scales. Understanding and predicting shifts in abundance and distribution of wildlife is critical to evaluating the ecological impacts of a rapidly changing climate. With archives dating back nearly a decade, imagery provides the opportunity to establish short-term longitudinal data.

Acknowledgments

Disclaimer: Any use of trade names is for descriptive purposes only and does not constitute endorsement by the U.S. government.

We are very grateful for the logistical support provided by the Department of Environment (Government of Nunavut), the University of Minnesota, and the USGS. We thank Universal Helicopters and our pilot and engineer. This research was conducted under Wildlife Research Permit Number 2012-052 (Government of Nunavut). Aerial survey field protocols were approved by the Institutional Animal Care and Use Committee at the University of Minnesota (Permit Number 1207A17284).

Author Contributions

Conceived and designed the experiments: SS ML NL SA DG CP TA. Performed the experiments: SS ML NL CP. Analyzed the data: SS ML.

Contributed reagents/materials/analysis tools: SS ML CP. Wrote the paper: SS ML NL SA DG CP TA.

References

1. Stroeve J, Holland MM, Meier W, Scambos T, Serreze M (2007) Arctic sea ice decline: Faster than forecast. Geophys Res Lett 34: L09501.
2. Stroeve JC, Serreze MC, Holland MM, Kay JE, Malanik J, et al. (2012) The Arctic's rapidly shrinking sea ice cover: a research synthesis. Clim Change 110: 1005–1027.
3. Comiso JC, Parkinson CL, Gertsen R, Stock L (2008) Accelerated decline in the Arctic sea ice cover. Geophys Res Lett 35: L01703.
4. Maslowski W, Kinney JC, Higgins M, Roberts A (2012) The future of Arctic sea ice. Ann Rev of Earth Planet Sci 40: 625–54.
5. Overland JE, Wang M (2013) When will the summer Arctic be nearly sea ice free? Geophys Res Lett 40: 1–5.
6. Laidre KL, Stirling I, Lowry LF, Wiig Ø, Heide-Jørgensen MP, et al. (2008) Quantifying the sensitivity of arctic marine mammals to climate-induced habitat change. Ecol Appl (Suppl.) 18: S97–S125.
7. Eberhardt LL, Chapman DG, Gilbert JR (1979) A review of marine mammal census methods. Wildl Monogr 63: 3–46.
8. Garner GW, Amstrup SC, Laake JL, Manly BFJ, McDonald LL, et al. (1999) Marine mammal survey and assessment methods. Rotterdam: A.A. Balkema. 291 p.
9. Taylor BL, Martinez M, Gerrodette T, Barlow J, Hrovat YN (2007) Lessons from monitoring trends in abundance of marine mammals. Mar Mamm Sci 23: 157–175.
10. Lowry L, Kovacs K, Burkanov V (IUCN SSC Pinniped Specialist Group) (2008) *Odobenus rosmarus*. In IUCN 2012. IUCN red list of threatened species. Version 2012.2. Available: http://www.iucnredlist.org. Accessed 2013 May 3.
11. Burkanov V, Lowry L (IUCN SSC Pinniped Specialist Group) (2008) *Histriophoca fasciata*. In IUCN 2012. IUCN red list of threatened species. Version 2012.2. Available: http://www.iucnredlist.org. Accessed 2013 May 3.
12. Obbard ME, Thiemann GW, Peacock E, DeBruyn TD (2010) Polar bears: Proceedings of the 15th working meeting of the IUCN/SSC polar bear specialist group, Copenhagen, Denmark, 29 June - 3 July 2009. Gland, Switzerland and Cambridge: IUCN. vii + 235 p.
13. Vongraven D, Aars J, Amstrup SC, Atkinson SN, Belikov S, et al. (2012) A circumpolar monitoring framework for polar bears. Ursus Monogr Ser 5: 1–66.
14. LaRue MA, Rotella JJ, Garrott RA, Siniff DB, Ainley DG., et al. (2011) Satellite imagery can be used to detect variation in abundance of Weddell seals (*Leptonychotes weddelli*) in Erebus Bay, Antarctica. Polar Biol 34: 1727–1737.
15. Fretwell PT, LaRue MA, Morin P, Kooyman GL, Wienecke B, et al. (2012) An emperor penguin population estimate: The first global, synoptic survey of a species from space. PLoS ONE 7: e33751.
16. Platonov NG, Mordvintsev IN, Rozhnov VV (2013) The possibility of using high resolution satellite imagery for detection of marine mammals. Biol Bull 40: 197–205.
17. Regehr EV, Lunn NJ, Amstrup SC, Stirling I (2007) Effects of earlier sea ice breakup on survival and population size of polar bears in Western Hudson Bay. J Wildl Manage 71: 2673–2683.

18. Schliebe S, Wiig Ø, Derocher A, Lunn N (IUCN SSC Polar Bear Specialist Group) (2008) *Ursus maritimus*. In IUCN 2013. IUCN red list of threatened species. Version 2013.1. Available: http://www.iucnredlist.org. Accessed 2013 November 13.
19. Schliebe S, Rode KD, Gleason JS, Wilder J, Proffitt K, et al. (2008) Effects of sea ice extent and food availability on spatial and temporal distribution of polar bears during the fall open-water period in the Southern Beaufort Sea. Polar Biol 31: 999–1010.
20. Stapleton S, Peacock E, Garshelis D, Atkinson S (2012) Foxe Basin polar bear aerial survey, 2009 and 2010: final report. Iqaluit, Nunavut: Government of Nunavut, Department of Environment. Available: http://env.gov.nu.ca/sites/default/files/gn_report_foxe_basin_polar_bears_2012_for_posting.pdf.
21. Updike T, Comp C (2010) Radiometric use of WorldView-2 imagery. Longmont, Colorado: DigitalGlobe. Available: http://www.digitalglobe.com/sites/default/files/Radiometric_Use_of_WorldView-2_Imagery%20(1).pdf.
22. Otis DL, Burnham KP, White GC, Anderson DR (1978) Statistical inference from capture data on closed animal populations. Wildl Monogr 62: 3–135.
23. White GC, Burnham KP (1999) Program MARK: Survival estimation from populations of marked animals. Bird Study 46: S120–S139.
24. Laake J, Borchers D (2004) Methods for incomplete detection at distance zero. In: Buckland ST, Anderson DR, Burnham KP, Laake JL, Borchers DL, et al., editors. Advanced distance sampling. Oxford: Oxford University Press. pp. 108–109.
25. Buckland ST, Anderson DR, Burnham KP, Laake JL, Borchers DL, et al. (2001) Introduction to distance sampling: Estimating abundance of biological populations. Oxford: Oxford University Press. 432 p.
26. Marques TA, Andersen M, Christensen-Dalsgaard S, Belikov S, Boltunov A, et al. (2006) The use of global positioning systems to record distances in a helicopter line-transect survey. Wildl Soc Bull 34: 759–763.
27. Huggins R (1989) On the statistical analysis of capture experiments. Biometrika 76: 133–140.
28. Huggins R (1991) Practical aspects of a conditional likelihood approach to capture experiments. Biometrics 47: 725–732.
29. Thomas L, Buckland ST, Rexstad EA, Laake JL, Strindberg S, et al. (2010) Distance software: design and analysis of distance sampling surveys for estimating population size. J Appl Ecol 47: 5–14.
30. Burnham KP, Anderson DR (2002) Model selection and multimodel inference: A practical information-theoretic approach (2nd edition). New York: Springer. 488 p.
31. Anderson DR, Burnham KP, Thompson WL (2000) Null hypothesis testing: Problems, prevalence and an alternative. J Wildl Manage 64: 912–923.
32. Boltunov A, Evtushenko N, Knijnikov M, Puhova M, Semenova V (2012) Space technology for the marine mammal research and conservation in the Arctic. Results of the pilot project to develop methods of finding walruses on satellite images. Murmansk, Russia: WWF-Russia.

2

Integrating Remote Sensing and GIS for Prediction of Winter Wheat (*Triticum aestivum*) Protein Contents in Linfen (Shanxi), China

Mei-chen Feng[1], Lu-jie Xiao[1], Mei-jun Zhang[1], Wu-de Yang[1]*, Guang-wei Ding[2]*

1 Institute of Dryland Farming Engineer, Shanxi Agricultural University, Taigu, People's Republic of China, 2 Department of Chemistry, Northern State University, Aberdeen, South Dakota, United States of America

Abstract

In this study, relationships between normalized difference vegetation index (NDVI) and plant (winter wheat) nitrogen content (PNC) and between PNC and grain protein content (GPC) were investigated using multi-temporal moderate-resolution imaging spectroradiometer (MODIS) data at the different stages of winter wheat in Linfen (Shanxi, P. R. China). The anticipating model for GPC of winter wheat was also established by the approach of NDVI at the different stages of winter wheat. The results showed that the spectrum models of PNC passed F test. The $NDVI_{4.14}$ regression effect of PNC model of irrigated winter wheat was the best, and that in dry land was $NDVI_{4.30}$. The PNC of irrigated and dry land winter wheat were significantly ($P<0.01$) and positively correlated to GPC. Both of protein spectral anticipating model of irrigated and dry land winter wheat passed a significance test ($P<0.01$). Multiple anticipating models (MAM) were established by NDVI from two periods of irrigated and dry land winter wheat and PNC to link GPC anticipating model. The coefficient of determination R^2 (R) of MAM was greater than that of the other two single-factor models. The relative root mean square error (R_{RMSE}) and relative error (RE) of MAM were lower than those of the other two single-factor models. Therefore, test effects of multiple proteins anticipating model were better than those of single-factor models. The application of multiple anticipating models for predication of protein content (PC) of irrigated and dry land winter wheat was more accurate and reliable. The regionalization analysis of GPC was performed using inverse distance weighted function of GIS, which is likely to provide the scientific basis for the reasonable winter wheat planting in Linfen city, China.

Editor: Wengui Yan, National Rice Research Center, United States of America

Funding: This work was supported by grants from the Key Technologies R&D Program of Shanxi Province, China (20060311140, 20110311038, and 20120311001-2), the Scientific Research Starting Foundation of Shanxi Agricultural University (XB2009016), and the Sci-tech Innovation Foundation of Shanxi Agricultural University (201222). The funders had no role in study design, data collection and analysis, decision to publish, or preparation of the manuscript.

Competing Interests: The authors have declared that no competing interests exist.

* E-mail: sxauywd@126.com (WDY); guangwei.ding@northern.edu (GWD).

Introduction

The winter wheat grain protein content is one of the important standards to evaluate wheat quality [1,2]. However, the traditional grain protein content (GPC) detection methods need to be destructive in sampling, and consuming time, labor, and money [3]. It is difficult to achieve the quality monitoring and forecasting GPC of the large area of winter wheat. With the rapid development of remote sensing technique in recent years [4], the large scale quickly and nondestructive testing approaches become possible to monitor the wheat grain quality [5,6,7].

It is fundamental to realize quantitative remote sensing and precise monitoring methods by building the models of wheat GPC and remote sensing parameters. Apan et al. [8] built the winter wheat GPC estimation model based on spectral vegetation indices by using partial least squares regression method. The model could more accurately predict winter wheat GPC in Australia and the prediction accuracy was 92%. Reyniers et al. [9] monitored the winter wheat GPC using the aerial images and field spectrometer, and the prediction accuracy had achieved 90%. Liu et al. [10] determined the winter wheat GPC using multi-temporal EnviSat-ASAR and Landsat TM satellite images. The model was built

based on the C-HH backscatter and SIPI data and the correlation coefficient was 0.75.

Wright et al. [11] analyzed the nitrogen status of wheat plants and found that nitrogen content of flag leaf could predict GPC at middle growth stage in Minidoka County, Idaho. Huang et al. [12] reported that the GPC could be predicted using nitrogen reflectance index and foliar nitrogen concentration around the anthesis stage. The leaf nitrogen content of winter wheat was significantly ($P<0.01$) correlated with GPC, and spectral vegetation indices significantly correlated to leaf nitrogen content at anthesis stage. Therefore, it was feasible by using remote sensing data to predict GPC at anthesis stage of winter wheat [13].

The normalized difference vegetation index (NDVI) is a most commonly employed vegetation index, which is sensitive to vegetation growth status, productivity, and other biophysical and biochemistry characteristics [14]. It is widely used in the land use cover monitoring [15], vegetation coverage density evaluation [16], crop recognition [17], and crop yield forecast [18,19] and so on. Stone et al. [20] found that the NDVI and wheat plant nitrogen content were highly correlative at several different growth stages. Freeman et al. [21] demonstrated that NDVI was well correlated with straw N uptake and total N uptake at Feekes growth stages 9 and 10.5 in both cropping cycles at Hennessey and Stillwater.

As there were very limited studies in using vegetation index to simulate the models and predict winter wheat GPC in Shanxi, the current research attempted to predict the GPC of winter wheat in Linfen using plant nitrogen content extracted from satellite remote sensing data. Compared to other studies in the literature [8–11,20,21], our current research not only established the single-factor model, but also established multiple-factor anticipating model to predict the winter wheat protein contents (different stages) in the identified area. This study also mapped the regionalization characterization of the winter wheat production in the Linfen region based on the GPC character. In addition, the selection of winter wheat variety and management system (*i.e.*, irrigated or dry land) might also lead to different results and need further consideration. The specific objectives were to: (i) extract the planted area and NDVI of different irrigated type winter wheat, (ii) analyze the relation between PNC and NDVI, PNC and GPC, and (iii) build GPC prediction model and realize the regionalization analysis of winter wheat GPC based GIS.

Materials and Methods

Site description

The study was carried out in the southwest of Shanxi province in China (north latitude 35°23′–36°56′, east longitude 110°23′–112°33′) (Figure 1). This region is situated in Loess Plateau and the middle-downstream of Fen River. It is temperate continental monsoon climate in the region. The winter is frigid and dry and the summer is torrid and rainy. The temperature difference between winter and summer is large, and the precipitation is concentrated in summer (June, July, and August). The mean annual temperature is 10.7°C. The average temperature of January and July is −4°C and 26°C, respectively. Frostless season is 180 days annually. The mean annual precipitation is 550 mm. The primary crops include winter wheat, cotton, mealie, and paddy rice. The winter wheat is primarily distributed in central basin where is one of the key winter wheat production areas in Shanxi province, China.

Sampling and data collection

There were 38 irrigated wheat and 16 dry land winter wheat fields selected from the study areas (Figure 2) in 2006, 2007, and 2009 under the same sowing, fertilization, with a size of at least 5 ha. The irrigated winter wheat main variety was "Linyou 7287," and the average GPC was 14.06%. The dry land winter wheat main variety was "Jinmai 79," and the average GPC was 14.84%. Winter wheat both in irrigated and dry land was sowed in early October and harvested in late May (dry land) and mid-June (irrigated wheat) respectively. At the different growth stage (returning green, joining, booting, heading, filling, and maturing stages) of winter wheat, the wheat plant samples were collected in the selected area and threshed manually for plant nitrogen content analysis. At the harvest of wheat, the grains were collected for GPC content determination. More specifically, in each plot, an area of 1 m×1 m was cropped manually just above ground and brought to the lab for processing. For each field, there were triplicate plots (50 m in length by 50 m in width). The GPS was used in the region for accurate positioning. The Aridisols is the predominant soil type in this area. The data collected in 2006 and 2007 were used for model establishment and the data of 2009 were collected to test the established models. We need to note that the field study was authorized by the Agricultural Bureau of Linfen City in Shanxi Province (P. R. China). In addition, no specific

permissions were required for these locations/activities because the research activities were for the local agricultural service. Furthermore, the field studies did not involve endangered or protected species and this study also did not involve vertebrate species.

All the samples were placed in the oven to dry at 70°C for 24 h. The PNC (%) and GPC (%) of all the samples were determined by using the Kjeldahl-nitrogen method [22]. The PC in grain was calculated as Kjeldahl-nitrogen content multiplied by 5.7 [23].

Table 1 listed the mean, standard deviation, and range of PNC and GPC values of different irrigation type winter wheat. The range of variation was wide within the wheat variables, *i.e.*, 5-fold and 1.5-fold variation in PNC and GPC of irrigated wheat, 2-fold variation in PNC and GPC of dry land wheat. However, the mean values of different irrigation type wheat were close. And the range of GPC in dry land wheat was so wide; it could be caused by uneven regional rainfall. The wide range of wheat variables was more realistic and universal while the relationship between PNC, GPC and reflectance was made.

Data acquisition and treatment of remote sensing data

The remote sensing data in the current study was Landsat TM5 data and MODIS land surface Reflectance (LSR) data synthesized in eight days. The MODIS data to be used were daily level 3 surface reflectance. We downloaded all tiles of h26v05 and h27v05 from the NASA's EOS data gateway, covering the entire extent of Linfen region. Both MODIS and TM data were used in current investigation mainly based on the fact that compared to the TM, MODIS had a higher time resolution and therefore available MODIS was used in time series analysis; and TM had a higher spatial resolution compared to the MODIS, TM could be used for extraction of crop area. Thus, the crop character through a combination of the two datasets could be more effectively monitored.

TM data preprocessing

Different plants have different seasonal rhythm [17]. The winter wheat growth area was extracted by using TM data (mosaicking disposal has been conducted on purchase) on April 8th, 2007 by the Themaic Mapper sensor. The blue (450–520 nm), green (520–600 nm), red (630–690 nm), and near infrared (NIR, 760–900 nm) bands of electromagnetic spectrum were identified at a resolution of 30 m. The planting status of winter wheat in study areas had been identified as the same in recent years and the change of planting area and position could be neglected. Only one scene TM image in 2007 was used to extract planting area, which was conformable with the practical investigation result in 2006.

Preprocessing consisted of:

(1) Atmosphere adjustment. The atmospheric correction module FLAASH (Fast Line-of-sight Atmospheric Analysis of Spectral Hypercubes) was used here to adjust the TM remote sensing image [24].

(2) Geometric correction. Coarse geometric correction and precise geometric correction were completed using 1:10 000 digitized raster map and ground control points. The cubic convolution interpolation was used in the georeferencing process to assure that the error was less than one pixel [25].

(3) Extraction of researching region. The Mahalanobis Distance taxonomy [26] was used for classification and the best classification effects were reached at the threshold of 2.9. For a fraction of leakage and spillage image after classification, the planting area vectograph produced by classification was superimposed by TM remote sensing image in GIS through second visual interpretation to produce area vectograph

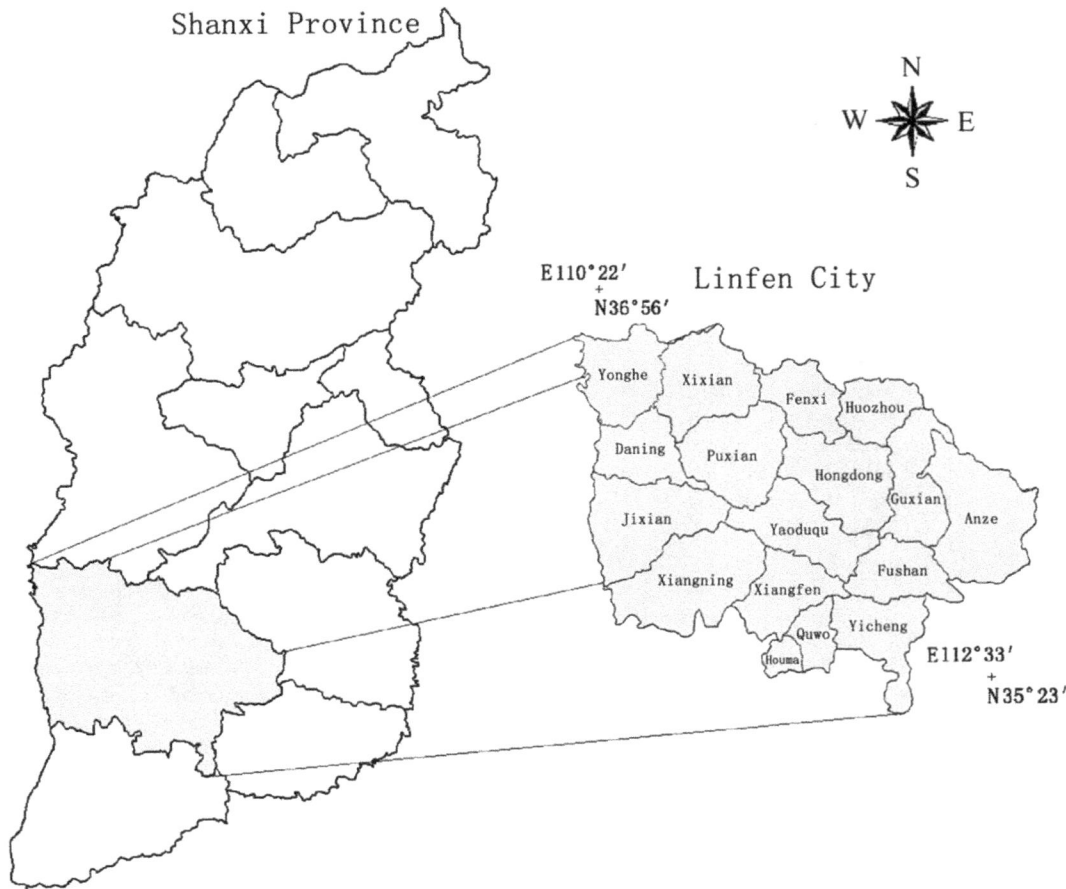

Figure 1. The locations of the study area.

ultimately. Then the mask was made and TM image was cropped, thereby the winter wheat planting area was obtained.

(4) NDVI calculation. The NDVI of TM image was calculated using the following method [27]:

$$NDVI = \frac{\lambda_{NIR} - \lambda_R}{\lambda_{NIR} + \lambda_R}$$

Where λ_{NIR} is the reflectance (%) of the near infrared (NIR) band and λ_R is the reflectance (%) of the red band.

MODIS data processing

The MODIS data were superior to the TM (in time series resolution identification), NOAA/AVHRR [28] in monitoring of crop growth by features such as high time resolution, high spectral resolution as well as moderate spatial resolution. The MODIS LSR data including two bands (Band1 as red band, 620–670 nm; Band2 as NIR band, 841–876 nm) were obtained from LPDAAC and synthesized in eight days at a 250×250 m spatial resolution. And duration of the data collection was from January to July of 2006, 2007, and 2009 in this study.

Processing consisted of:

(1) Image mosaicism. The Mosaicking method [29] localized by geographical coordinates and the Feathering function [30] was used for edge feathering.

(2) Geometric correction. The geographical coordinates locating information carried by the head file of MODIS data were used for geometric correction.

(3) Atmosphere adjustment. The histogram method [4] was used for the atmosphere adjustment of the images.

(4) Extraction of researching region and NDVI calculation. The detailed methods were of the same with TM image.

Flow scheme of winter wheat planted area extraction

As winter wheat has different growing process under different irrigated conditions, in order to improve the monitoring precision, the irrigated and dry land winter wheat should be distinguished in the study of crop quality and growth monitoring. In general, the winter wheat varieties in irrigated and dry land area of Shanxi province were different and the sowing time was almost the same. However, winter wheat of dry land harvested earlier than of the irrigated mainly because of the longer filling time and low levels of drought stress of irrigated winter wheat.

The winter wheat plant area of Linfen region included the irrigated winter wheat area of Linfen basin and dry land winter wheat area of Jinnan hilly region. The irrigated winter wheat area of Linfen basin included hirakawa counties of Houzhou, Hongtong, Linfen, Xiangfen, Quwo, and Houma. The altitude was from 350 to 600 m. The dry land winter wheat area of Jinnan hilly region included most of wheat fields of Fushan and Yicheng

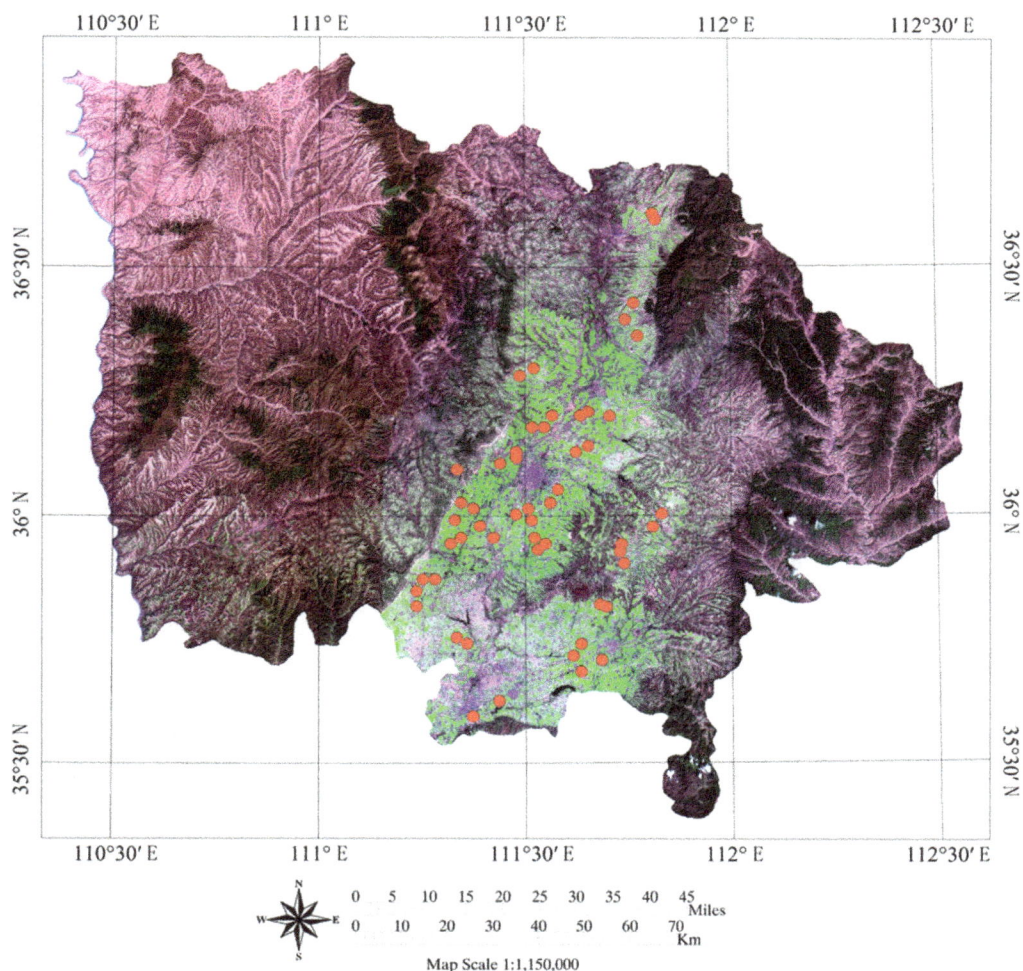

Figure 2. The spatial distribution image of sampling sites.

counties. The altitude (above sea level) was from 475 to 700 m. The slope of Linfen basin was less than 15°, and the slope of the hilly region was in the range of 15°~20°. The 3D model image was shown in Figure 3.

We need to clarify the following: (1). Figure 1 was indeed very difficult to distinguish between basins, but 3D remote sensing image in Figure 3 made it clear that the basin was located, *e.g.*, two mountains between L-type, with 20–25 km wide, 200 km long. The irrigated wheat was concentrated in 350–600 m above sea level, dry land was concentrated in 475–700 m above sea level; partial dry land were included in the irrigated wheat elevations.

(2). To extract the 3D remote sensing images, by using the ArcScene software, it was hard to figure the superscript in the elevation data. (3). Winter wheat growing region in Linfen could be categorized to two areas including basin area of irrigated winter wheat and in hilly area of dry land winter wheat growing. Two major irrigation systems (river irrigation and well irrigation) managed by local farmers were characterized.

Using the above-mentioned discrimination characteristics, irrigated and dry land winter wheat plant area were extracted by building the decision tree (Figure 4) extraction model. The NDVI, elevation, and slope values were obtained from TM image

Table 1. The summary of variables measured different irrigation type winter wheat.

Irrigation Type	Variables	Mean	Standard deviation	Min	Max	Range
irrigated wheat	PNC	1.6690	0.7124	0.7643	3.8479	3.0836
	GPC	12.7617	1.1722	10.4801	15.4102	4.9301
dry land wheat	PNC	1.1928	0.3059	0.8748	1.8144	0.9396
	GPC	12.4966	2.4967	7.1484	14.9925	7.8441

Figure 3. The 3D TM image of remote sensing of Linfen City.

of winter wheat planted area and 3D model, respectively. The decision tree structure image was shown in Figure 4 and Table 2 listed the classification results of irrigated and dry land winter wheat. The actual area came from statistical data of agricultural statistics department of Shanxi Province, China. The accuracy estimation was determined by the following formula:

$$* \ Accuracy(\%) = \left(1 - \left|\frac{extracted \ area - actual \ area}{actual \ area}\right|\right) \times 100\%$$

Data analysis and calculation methods

Excel software was used for data collation, analysis, and mapping. All data were analyzed statistically to use DPS (i.e., statistical analysis, regression analysis, and variance analysis). The DPS (Data Processing System) 7.05 is a kind of statistical analysis software [31].

The simple linear regression models were established using the data of PNC and NDVI values and PNC and GPC. Two prediction models were combined into GPC monitoring models.

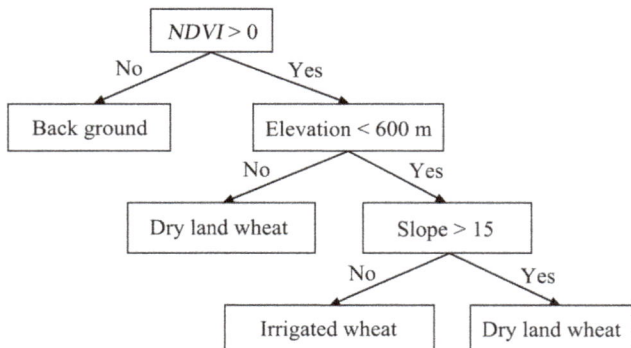

Figure 4. The classification image of decision tree of irrigated and dry land winter wheat.

The GPC model was grain protein content monitoring model. Its direct variable was PNC, and indirect variable was NDVI.

The assessment model was based on multiple correlation coefficients [32], F-test values of significance [7], relative root mean square errors (R_{RMSE}) [33], and relative error (RE) [34].

The relative root mean square error (R_{RMSE}) and relative error (RE) are defined by:

$$R_{RMSE} = \sqrt{\frac{1}{n}\sum_{i=1}^{n}(O_i - S_i)^2} \bigg/ \frac{1}{n}\sum_{i=1}^{n}O_i$$

$$RE = \frac{1}{n}\sum_{i=1}^{n}\left(\frac{|O_i - S_i|}{O_i}\right)$$

Where, O_i and S_i are the observed and simulated values of the sample i and n is the number of samples.

Sampling points monitored by current study were conducted in the state of nature. The research process did not interfere with the farmer's production and operation, in this case the coefficient of determination of the model, which can meet the needs of the model.

$$MCC/R^2 = \frac{\sum(\hat{y} - \bar{y})}{\sum(y - \bar{y})}$$

$$F - test = \frac{v_2 R^2}{v_1(1 - R^2)}$$

Where y, $\bar{y}$ and $\hat{y}$ are the observed value, mean value, and prediction value, degrees of freedom $v_1 = $ m, $v_2 = $ n-m-1.

Results and Discussion

Relation of NDVI and nitrogen content of plants

The correlation between NDVI and nitrogen content of plants was established. The degree of correlation was different for different irrigation type winter wheat at different growth stages

Table 2. The classification result of winter wheat in Linfen City.

Winter wheat	Pixel numbers	Extracting area/ha	Actual area/ha	Accuracy/%
Irrigated wheat	1194314	107488	94400	86.15
Dry land wheat	1414335	127290	147733	86.16
Total area	2608649	234778	242133	96.96

(Figure 5). Through data analysis, there was a significant (P<0.01) negative correlation between $NDVI_{4.14}$, $NDVI_{5.8}$ and nitrogen content of plants of irrigated winter wheat, *i.e.*, at early heading stage (May 8) the correlation coefficient were −0.45 and −0.44, respectively. There was a significant difference (P<0.01) between $NDVI_{4.30}$, $NDVI_{5.8}$, and nitrogen content of plants of dry land winter wheat. The max-relativity also appeared in the May (8th) and the correlation coefficient were −0.75 and −0.67, respectively. Our data might suggest that the NDVI could be used as the indicators to predict nitrogen content of irrigated and dry land winter wheat plants in Shanxi.

According to the relationship between NDVI and nitrogen content of different irrigation type winter wheat plants, the statistical evaluation models of nitrogen content of plants were developed. The simulation models were showed in Table 3. The results of the regression models and significant F-tests were conducted for all models (Table 3). The models of $NDVI_{4.14}$ and $NDVI_{4.30}$ had the best prediction effects on irrigated and dry land, respectively. It was clear that all R^2 of the irrigated wheat models were lower than dry land and the R_{RMSE} and RE were higher than dry land. This result demonstrated that dry land models had a better ability to predict nitrogen content of winter wheat.

Theoretically, the R^2 of model should be closer to 1 for ideal result, and RE, RMSE should be closer to 0 for ideal result. However, in this work, R^2 was only 0.569 for dry land wheat, but the model passed a 0.01 significant level test. The large sample numbers could be one of the reasons. In addition, we did not interfere with farmers' production and management in the whole study process; this might also be the main reason for the emergence of such results. Therefore, it speculated that there might be a possibility by using remote sensing data for large area monitoring winter wheat protein content.

And the multiple anticipating models were also established using two temporal NDVI and nitrogen content of plant, the models were as follows:

Irrigated winter wheat: $N(\%) = 2.9358 - 0.0126NDVI_{4.14} - 0.0119NDVI_{5.8}$, (n = 38, $R^2 = 0.207$, F = 9.4, $R_{RMSE} = 0.302$, RE = 0.254)

Dry land winter wheat: $N(\%) = 3.1581 - 0.0216NDVI_{4.30} - 0.0156NDVI_{5.8}$, (n = 16, $R^2 = 0.527$, F = 15.6, $R_{RMSE} = 0.124$, RE = 0.106)

The above mentioned multiple anticipating models (MAM) showed that they had passed F test and also reached significant level (P<0.01). The R^2 value of MAM was higher than the unifactor models for irrigated winter wheat. The dry land R^2 of MAM value was in the medium between the two unifactor models.

Relation of nitrogen content of plants and GPC

The correlation between PNC and GPC was carried out at different stages. There was a significant (P<0.01) positive correlation between plant nitrogen content and GPC of irrigated and dry land winter wheat in early heading (the correlation coefficient of 0.541 and 0.567, respectively). The difference was not significant (P>0.01) in other stages. And the agronomy models were established to predict GPC; the models were showed in Table 4.

The F-test value of the GPC evaluation models was greater than F-critical value (Table 4). It indicated that the models passed a 0.01 significant level test. The R_{RMSE} and RE of irrigated and dry land wheat models were 0.063 and 0.054, 0.205 and 0.141, respectively. It showed that the two models were more suitable for prediction.

It was also noted that the former researches for constructing model were mainly concentrating on mixing the different irrigation type winter wheat to monitor the crop growth status and estimate crop yield. In fact, the growing process of irrigated and dry land winter wheat as affected by varieties and environmental factor was of difference. For example, the growth stage for irrigated land winter wheat was middle-heading stage on

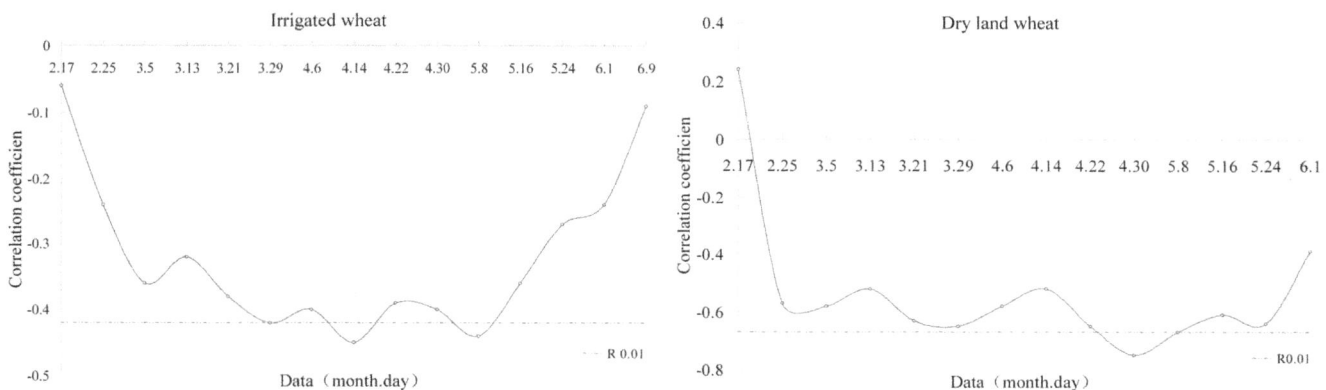

Figure 5. Correlation coefficients between NDVI and plant nitrogen content of irrigated and dry land wheat at different stages.

Table 3. The statistical evaluation models of nitrogen content of different irrigation type winter wheat.

Irrigation Type	Model	R^2	F-test	F-crit	R_{RMSE}	RE
Irrigated wheat	$N(\%) = 2.7597 - 0.0241NDVI_{4.14}$	0.202	9.1	7.4	0.369	0.291
	$N(\%) = 2.9790 - 0.0225NDVI_{5.8}$	0.192	8.6	7.4	0.346	0.278
Dry land wheat	$N(\%) = 3.4194 - 0.0440NDVI_{4.30}$	0.569	18.5	8.5	0.131	0.111
	$N(\%) = 2.7682 - 0.0281NDVI_{5.8}$	0.446	11.3	8.5	0.171	0.146

Table 5. The spectral GPC estimation models of irrigated and dry land winter wheat.

Irrigation Type	Model	R^2	F-test	F-crit	R_{RMSE}	RE
Irrigated wheat	$Pro(\%) = 14.0815 - 0.0292NDVI_{4.14}$	0.235	11.1	7.4	0.083	0.063
	$Pro(\%) = 14.3469 - 0.0272NDVI_{5.8}$	0.228	10.6	7.4	0.082	0.063
Dry land wheat	$Pro(\%) = 26.1742 - 0.2700NDVI_{4.30}$	0.487	13.3	8.5	0.185	0.123
	$Pro(\%) = 22.1738 - 0.1725NDVI_{5.8}$	0.557	17.6	8.5	0.189	0.119

May 8; while late-heading stage for dry land winter wheat was observed on May 8. In addition, the dry land winter wheat matured on June 1; while for irrigated land wheat was on June 9. Thus, it was hard to construct the mixed model based on the specific growth stage. The models were simulated in our investigation only in dry land wheat and irrigated wheat.

Spectral GPC estimation models

Because PNC was strongly associated with GPC, the PNC could indirectly be used to predict GPC though the correlation between PNC and vegetation index. The spectral GPC estimation models were established by using the plant nitrogen content as the connecting point as shown in Table 5.

All the F values of spectral prediction models for GPC in irrigated and dry land were greater than the value of F-critical, revealing that all of the prediction models for GPC of irrigated and dry land winter wheat passed the significance level test of 0.01. The R values of the models constructed by $NDVI_{4.14}$ for irrigated winter wheat were higher than those of the models constructed by $NDVI_{5.8}$, therefore the model in April (14th) was selected as the prediction models for GPC of irrigated winter wheat. The model in May (8th) was selected as the prediction models for GPC of dry land winter wheat.

Simultaneously, hybrid prediction model was constructed by NDVI in two time phases of irrigated and dry land winter wheat as well as nitrogen content in plants to link the prediction model for GPC, the formulas were as following:

Irrigated winter wheat:
$Pro(\%) = 14.2946 - 0.0152NDVI_{4.14} - 0.0145NDVI_{5.8}$, (n = 38, $R^2 = 0.244$, F = 11.6, $R_{RMSE} = 0.081$, RE = 0.062);

Dry land winter wheat:
$Pro(\%) = 24.5688 - 0.1325NDVI_{4.30} - 0.0955NDVI_{5.8}$, (n = 16, $R^2 = 0.632$, F = 24.0, $R_{RMSE} = 0.144$, RE = 0.106).

Table 4. The agriculture statistical evaluation models of GPC of different irrigation type winter wheat.

Irrigation Type	Model	R^2	F-test	F-crit	R_{RMSE}	RE
Irrigated wheat	$Pro\ (\%) = 10.7421 + 1.2101N\ (\%)$	0.541	42.4	7.4	0.063	0.054
Dry land wheat	$Pro\ (\%) = 5.1689 + 6.1431N\ (\%)$	0.567	18.3	8.5	0.205	0.141

Significance level test of correlation coefficient to the formulas above with F test indicated that spectral GPC prediction models for winter wheat in all regions passed the significance level test of 0.01, which showed highly significant relationships between NDVI and GPC. The purpose of this model construction was just for prediction, and then R^2 could be used to test the models [35]. The R^2 values of hybrid prediction models for irrigated and dry land winter wheat were larger than those of the other two unifactorial models. Additionally, R_{RMSE} and RE were smaller than the two unifactorial models, lowered 0.2% (0.1%) and 1.7% (1.3%), respectively.

Furthermore, we constructed a mixed GPC model of irrigated and dry land wheat by using the same growth stage data (8th of May). The formula listed as the following:

$Pro(\%) = 17.5174 - 0.1101NDVI_{5.8} - 0.0004NDVI_{5.8}^2$, (n = 54, $R^2 = 0.174$, F = 6.57, $R_{RMSE} = 0.254$, and RE = 0.340)

Although the GPC mixed prediction model for irrigated and dry land winter wheat passed the significance level test of 0.01, the R^2 value of mixed model was lower than classification models; the R_{RMSE} and RE were higher. Again, the growing process of irrigated land and dry land winter wheat as affected by varieties and environmental factor was of difference. For example, the growth stage for irrigated land winter wheat was middle-heading stage on May 8; while late-heading stage for dry land winter wheat was observed on May 8. In addition, the maturity stage was also different for both irrigated wheat and dry land wheat. Thus, our results demonstrated that the precision of mixed model was far below the classification models using the same stage data.

Additionally, based on our study, liner relationship between predicted and measured values of GPC (independent dataset from 2009) of irrigated and dry land wheat was further explored in Figure 6. The GPC (%) was predicted from $NDVI_{4.14}$ and $NDVI_{5.8}$, resulting in a prediction of GPC [$R^2 = 0.453$, $R_{RMSE} = 0.054$, slope = 0.296 (p<0.01), and intercept = 8.896] for irrigated winter wheat. The combination of $NDVI_{4.30}$ and $NDVI_{5.8}$ was an excellent predictor of GPC content [$R^2 = 0.624$, $R_{RMSE} = 0.118$, slope = 0.711 (p < 0.01), and intercept = 3.295] for dry land wheat. In sum, the test results of hybrid prediction models were better than those of the unifactorial models and the hybrid prediction models would be more reliable in predicting the GPC of irrigated and dry land winter wheat.

The regionalization analysis of irrigated and dry land winter wheat

The spatial distribution maps of irrigated and dry land winter wheat GPC were composited according to the hybrid protein

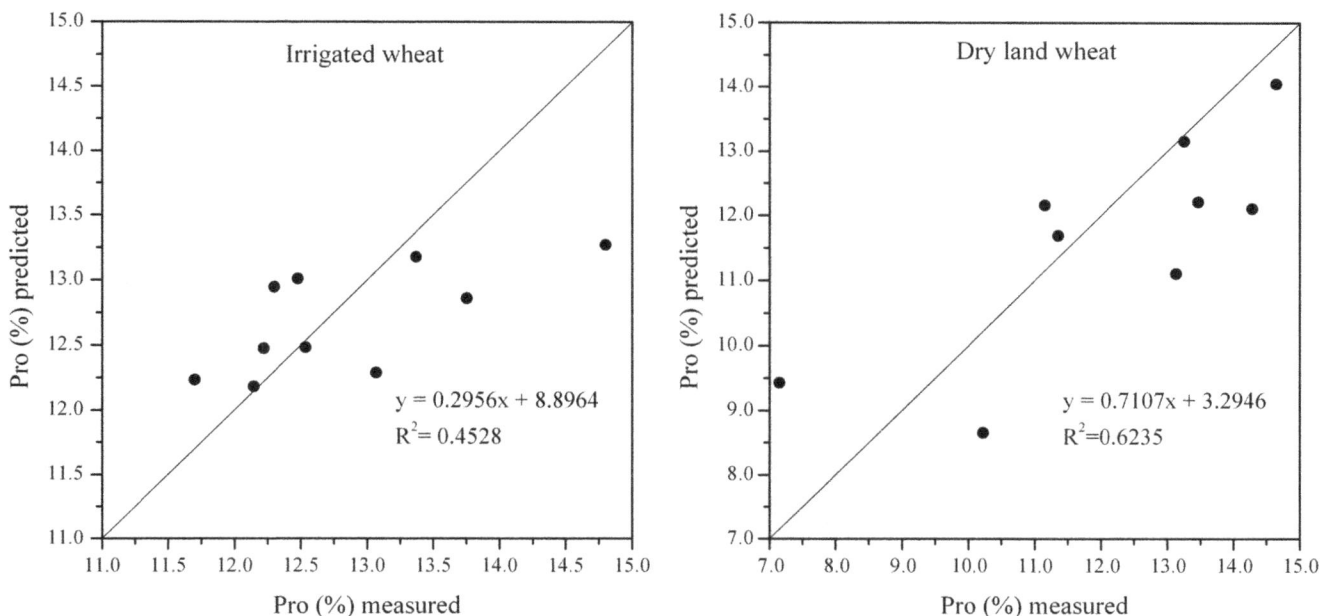

Figure 6. Liner relationships between predicted and measured values of GPC of irrigated and dry land wheat.

prediction models and images operation (Figure 7). From this graph, it demonstrated that irrigated winter wheat with lower GPC than 12.5% was mainly distributed in cities and counties such as Hongtong, Linfen, Xiangfen, and Yicheng. Winter wheat with GPC between 12.5% and 13% was unequally distributed in all counties. Winter wheat with GPC between 13% and 13.5% was mostly distributed in the North and South region. Winter wheat having higher GPC than 13.5% was scattered distributed in all counties. As for in dry land, winter wheat with GPC between 10% and 12% was mainly distributed in cities and counties such as Linfen, Huozhou, and Xiangning; while winter wheat with GPC between 14% and 16% was primarily distributed in the Southeast region.

From Figure 7, it was noted that the spatial distribution of different winter wheat GPC was comparatively scattered in a large scale and GPC in different point had significant difference. Therefore, it was very difficult to play a key role for winter wheat practice with various GPC. This paper focused on the spatial analysis capability by GIS aiming for better regionalization analysis of winter wheat. The outermost dispersion points were connected to form wheat planting region and the Kriging interpolation method [36] was used to realize wheat GPC regionalization (Figure 8). The special distribution images of winter wheat GPC of irrigated and dry land were introduced into ArcGIS, and translated into point layer. Then the peripheral point outline was drawn to form the polygon layer. The regionalization analysis of GPC was performed using inverse distance weighted function, clipped by the polygon layer. The results were showed in Figure 8.

Summary

In this study, taking nitrogen content as the connecting point, a prediction model for GPC in winter wheat based on different breeding periods was established. Compared to the other two unifactor models, the values of R^2 of MAM of winter wheat in irrigated and dry land were greater. Simultaneously, R_{RMSE} and RE in the MAM were lower than those in the unifactor models,

lowering 0.1% (0.1%) and 1.7% (1.3%), respectively. Generally, all of the prediction results in the MAM for GPC were better than those in the unifactor models. Therefore, the MAM for forecasting GPC of winter wheat in irrigated and dry land will be more accurate and reliable. And the analysis of regionalization in GPC of irrigated and dry land winter wheat was preceded in ArcGIS, which could serve wheat research and production well.

The mean, min, and max values of the measured GPC (Table 1) of irrigated wheat were all higher than that of the dry land wheat. However, as shown in Figure 7 and 8, the most of the predicted GPC values of irrigated wheat appeared lower than those of the dry land wheat. This weird observation might be explained by the following. On the one hand, the difference mainly came from the modeling precision (e.g., $R^2 = 0.244$ for irrigated wheat was much less than $R^2 = 0.632$ for dry land wheat). In addition, actual wheat planting area was scattering in the whole dry land wheat distribution region. Therefore, by statistical interpolation statistics method, the dry land wheat planting area was unilateral enlarged.

Most of the previous researchers [2,23] established the hyperspectral model for nitrogen and protein contents in leaf in the flowering period to forecast GPC of winter wheat. However, in the monitoring of large areas, because of difficulty in selection of sampling point, punctuality of sampling, and storage of samples, nitrogen content of different growth periods was selected for calculation in this article. Consequently, this method needs further study and consolidate the applications of established model.

Ideally, once spectral signatures are available it should be able to predict the satisfactorily the GPC under diverse condition. However, in the large area of winter wheat by remote sensing monitoring study, due to the choice of sample sites (related to transportation, information and communications, science and technology and other factors) and timeliness of the sample for the entire region are difficult to achieve synchronous sampling. Furthermore, the inconsistencies in sampling time and the preservation of the samples are also inevitable. This article focused

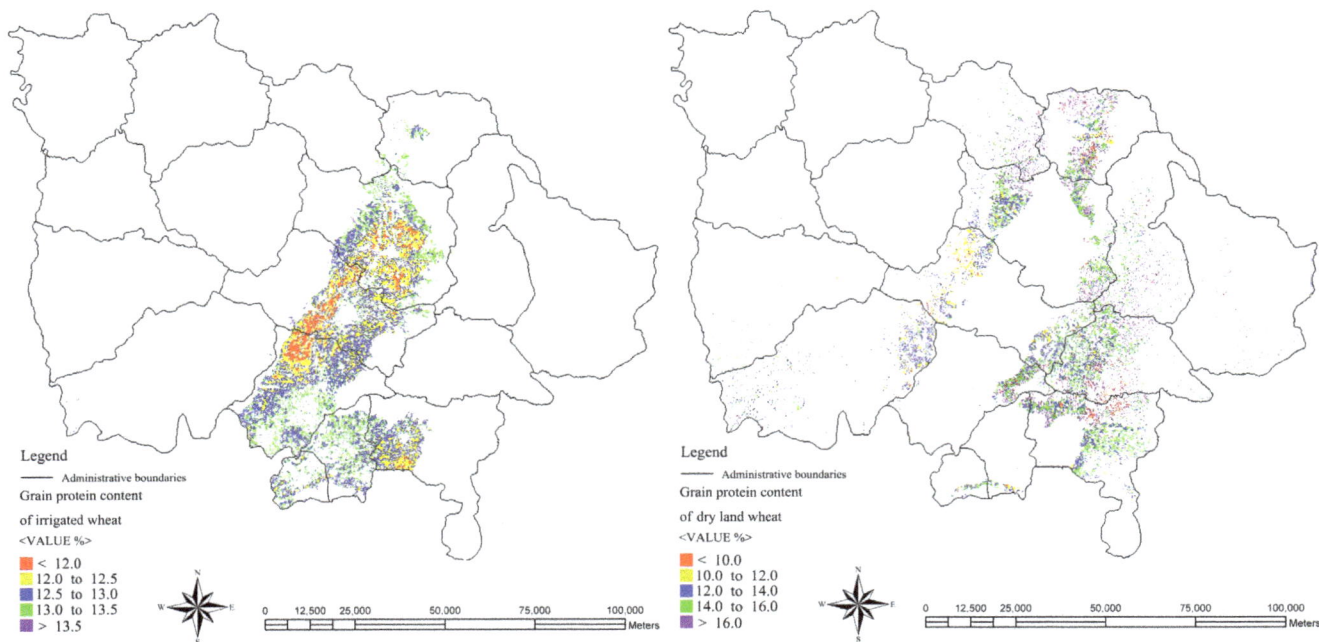

Figure 7. The GPC spatial distribution images of irrigated and dry land wheat.

only on the plant nitrogen content-related research. The forecast accuracy needs to be further study; establishing a research base on the ground may be the key to addressing this problem.

In order to improve the precision of the prediction model and minimize the interference of the mixed pixels, the TM and MODIS data were used in this study through overlay analysis, spatial scaling, adjust MODIS spatial resolution to match TM data. However, due to the terrain surface complexity and the scattered (fragmentation) of the winter wheat sowing area in the whole study region, the mixed pixels interferences were not completely erased and still existed in some degree. Hence, it is necessary to summarize current research to identify (develop) the appropriate technique of partition of mixed pixels for future research.

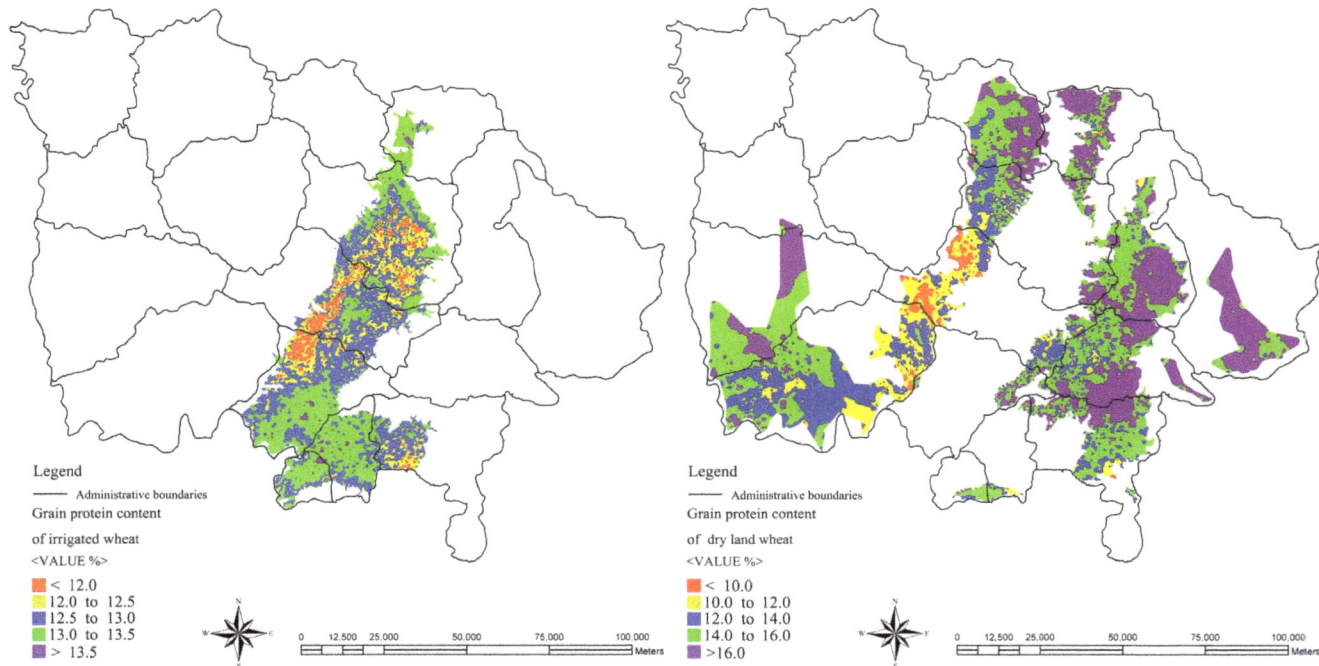

Figure 8. The GPC regionalization images of irrigated and dry land wheat.

Meteorological factors including rainfall, temperature, and sunshine have the important influence on quality of grain. The corresponding correlation analysis of meteorological factors was conducted (unpublished data) and the correlation was not significant. The meteorological factors were not brought into establishment of the model in this study. However it did not mean meteorological factors had no effect on grain quality. The indication of insignificant correlation was probably due to distant sampling points to weather stations. The factors such as cultivars (wheat variety), fertilizers [37,38] also affect grain quality of winter wheat which was not involved in the study. Thereafter, more solid study on the factors affecting quality of winter wheat needs for more consideration. In the end, remote sensing monitoring and regionalizing of GPC of winter wheat was dissected in the study and monitoring and regionalizing of other qualities of winter wheat need further research.

Acknowledgments

We are grateful to Drs. Shu-jin Guo and Yu-xiang Wu for their help. We appreciate Dr. Wengui Yan (academic editor), Dr. BC Kusre, and another anonymous reviewer for their constructive comments to improve our manuscript quality.

Author Contributions

Conceived and designed the experiments: WDY MCF . Performed the experiments: WDY MCF. Analyzed the data: MCF GWD WDY. Contributed reagents/materials/analysis tools: WDY MCF MJZ LJX. Wrote the paper: MCF GWD.

References

1. Shewry PR 2004/Rev. (2006) Improving the protein content and quality of temperate cereals: wheat, barley and rye. In Impacts of agriculture on human health and nutrition [online]. [cit. 2010-04-08]. EOLSS website. Available: http://www.eolss.net/ebooks/Sample%20Chapters/C10/E5-21-04-04.pdf. Accessed 2013 November 9.
2. Matsunak T, Watanabe Y, Miyawaki T, Ichikawa N (1997) Prediction of grain protein content in winter wheat through leaf color measurements using a chlorophyll meter. Journal of Soil Science & Plant Nutrition 43(1): 127–134.
3. Wang ZJ, Wang JH, Liu LY, Huang WJ, Zhao CJ, et al. (2004) Prediction of grain protein content in winter wheat (Triticum aestivum L.) using plant pigment ratio (PPR). Field Crops Research 90: 311–321.
4. Lu D, Mausel P, Brondizio E, Moran E (2004) Change detection techniques. International Journal of Remote Sensing 25: 2365–2407.
5. Gitelson AA, Kaufman YJ (1998) MODIS NDVI optimization to fit the AVHRR data series-Spectral considerations. Remote Sensing of Environment 66: 343–350.
6. Hansen PM, Jorgensen JR, Thomas A (2002) Predicting grain yield and protein content in winter wheat and spring barley using repeated canopy reflectance measurements and partial least squares regression. Journal of Agricultural Science 139: 307–318.
7. Manjunath KR, Potdar MB (2002) Large area operational wheat yield model development and validation based on spectral and meteorological data. International Journal of Remote Sensing 23(15): 2023–3038.
8. Apan A, Kelly RW, Phinn SR, Strong WM, Lester BD, et al. (2006) Predicting grain protein content in wheat using hyperspectral sensing of in-season crop canopies and partial least squares regression. International Journal of Geoinformatics 2(1): 93–108.
9. Reyniers M, Vrindts E, Baerdemaeker JD (2006) Comparison of an aerial-based system and an on the ground continuous measuring device to predict yield of winter wheat. European Journal of Agronomy 24(2): 87–94.
10. Liu LY, Wang JJ, Bao YS, Huang WJ, Ma ZH, et al. (2006) Predicting winter wheat condition, grain yield and protein content using multi-temporal EnviSat-ASAR and Landsat TM satellite images. International Journal of Remote Sensing 27(4): 737–753.
11. Wright DL, Rasmussen VP, Ramsey RD, Ellsworth JW (2004) Canopy reflectance estimation of wheat nitrogen content for grain protein management. GIScience and Remote Sensing 41(4): 287–300.
12. Huang WJ, Wang JH, Song XY, Zhao CJ, Liu LY (2007) Wheat grain quality forecasting by canopy reflected spectrum. International Federation for Information Processing. Volume 259: Computer And Computing Technologies In Agriculture, Volume II: 1299–1301.
13. Zhao CJ, Liu LY, Wang JH, Huang WJ, Song XY, et al. (2005) Predicting grain protein content of winter wheat using remote sensing data based on nitrogen status and water stress, International Journal of Applied Earth Observation and Geoinformation 7(1): 1–9.
14. Boken VK, Shaykewich CF (2002) Improving an operational wheat yield model using phonological phase-based Normalized Difference Vegetation Index. International Journal of Remote Sensing 23(20): 4155–4168.
15. Shih SF (1994) NOAA Polar-Orbiting satellite HRPT data and GIS in vegetation index estimation for the everglades agricultural area. Soil and Crop Science Society of Florida Proceedings 53: 19–24.
16. Zribi M, Hégarat-Mascle L (2003) Derivation of Wild Vegetation Cover Density in Semi-arid Region: ERS2/SAR Evaluation. International Journal of Remote Sensing (24): 1335–1352.
17. Yafit C, Maxim S (2002) A national knowledge-based crop recognition in Mediterranean environment. International Journal of Applied Earth Observation and Geoinformation 4: 75–87.
18. Daughtry CST, Gallo KP, Goward SN, Prince SD, Kustas WP (1992) Spectral estimates of absorbed radiation and phytomass production in corn and soybean canopies. Remote Sensing of Environment 39: 141–152.
19. Serrano L, Filella I, Peñuelas J (2000) Remote sensing of biomass and yield of winter wheat under different nitrogen supplies. Crop Science 40: 723–731.
20. Stone ML, Solie JB, Raun WR, Whitney RW, Taylor SL, et al. (1996) Use of spectral radiance for correcting in-season fertilizer nitrogen deficiencies in winter wheat. Trans ASAE, 39: 1623–1631.
21. Freeman KW, Raun WR, Johnson GV, Mullen RW, Stone ML, et al. (2003) Late-season prediction of wheat grain yield and grain protein. Communications in Soil Science Plant Analysis 34(13): 1837–1852.
22. Sahrawat KL (1995) Fix ammonium and carbon-nitrogen ratios of some semi-arid tropical Indian soils. Geoderma 68: 219–224.
23. Reyniers M, Vrindts E (2006) Measuring wheat nitrogen status from space and ground-based platform. International Journal of Remote Sensing 27(3): 549–567.
24. Agrawal G (2011) Comparision of QUAC and FLAASH atmospheric correction modules on EO-1 Hyperion data of Sanchi. International Journal of Advanced Engineering Sciences and Technologies 4(1): 178–186.
25. Thomlinson JR, Bolstad PV, Cohen WB (1999) Coordinating Methodologies for Scaling Landcover Classifications from Site-Specific to Global. Steps toward Validating Global Map Products, 70(1): 16–28.
26. Maria-Pza D, Christian C, Borja M, Pilar B, Constantino V, et al. (2012) Grapevine yield and leaf area estimation using supervised classification methodology on RGB images taken under field conditions. Sensors 12: 16988–17006.
27. Rouse JW, Haas RH, Schell JA, Deering DW (1974) Monitoring vegetation systems in the great plains with ERTS//NASASP 2351, third ERTS 21 Symposium, Washington, D. C. 309–317.
28. Gitelson AA, Kaufman YJ (1998) MODIS NDVI optimization to fit the AVHRR data series-Spectral considerations. Remote Sensing of Environment 66: 343–350.
29. Shum HY, Szeliski R (2000) Systems and experiment paper: construction of panoramic image mosaics with global and local alignment. International Journal of Computer Vision 36(2): 101–130.
30. Peter JB, Edward HA (1983) A multiresolution spline with application to image mosaics. Acm Transactions on Graphies 2(4): 217–236.
31. Tang QY, Zhang CX (2013) Data Processing System (DPS) software with experimental design, statistical analysis and data mining developed for use in entomological research. Insect Science. 20: 254–260.
32. Abdi H (2007) Multiple correlation coefficient. In N.J. . Salkind (Ed.): Encyclopedia of Measurement and Statistics. Thousand Oaks (CA): Sage, 648–651.
33. Marco G, Richard WM, Detlev H (2006) Influence of three-dimensional cloud effects on satellite derived solar irradiance estimation—First approaches to improve the Heliosat method. Solar Energy 80(9): 1145–1159.
34. White GC, Anderson DR, Burnham KP, Otis DL (1982) Capture-recapture and removal methods for sampling closed populations. Los Alamos National Laboratory, Los Alamos.
35. Guiarati DN (1995) Basic econometrics. 3rd edition (New York: McGraw Hill).
36. Matheron G (1963) Principles of Geostatistics. Economic Geology 58: 1246–1266.
37. Stark J, Souza E, Brown B, Windes J (2001) Irrigation and Nitrogen Management Systems for enhancing Hard Spring Wheat Protein. American Society of Agronomy Annual Meetings. Charlotte, North Carolina, October 24, 2001. Available: http://www.extension.uidaho.edu/swidaho/nutrient%20management/asa%20protein%20presentatns/2001asastark1/index.htm. Accessed: 9 Nov 2013
38. Terman GL, Ramig RE, Dreier AF, Olson RA (1969) Yield-protein relationships in wheat grain as affected by nitrogen and water. Agronomy Journal 611: 755–759.

3

The Augmented Lagrange Multipliers Method for Matrix Completion from Corrupted Samplings with Application to Mixed Gaussian-Impulse Noise Removal

Fan Meng[1,2], Xiaomei Yang[1]*, Chenghu Zhou[1]

1 Institute of Geographic Sciences and Natural Resources Research, Chinese Academy of Sciences, Beijing, China, **2** University of Chinese Academy of Sciences, Beijing, China

Abstract

This paper studies the problem of the restoration of images corrupted by mixed Gaussian-impulse noise. In recent years, low-rank matrix reconstruction has become a research hotspot in many scientific and engineering domains such as machine learning, image processing, computer vision and bioinformatics, which mainly involves the problem of matrix completion and robust principal component analysis, namely recovering a low-rank matrix from an incomplete but accurate sampling subset of its entries and from an observed data matrix with an unknown fraction of its entries being arbitrarily corrupted, respectively. Inspired by these ideas, we consider the problem of recovering a low-rank matrix from an incomplete sampling subset of its entries with an unknown fraction of the samplings contaminated by arbitrary errors, which is defined as the problem of matrix completion from corrupted samplings and modeled as a convex optimization problem that minimizes a combination of the nuclear norm and the l_1-norm in this paper. Meanwhile, we put forward a novel and effective algorithm called augmented Lagrange multipliers to exactly solve the problem. For mixed Gaussian-impulse noise removal, we regard it as the problem of matrix completion from corrupted samplings, and restore the noisy image following an impulse-detecting procedure. Compared with some existing methods for mixed noise removal, the recovery quality performance of our method is dominant if images possess low-rank features such as geometrically regular textures and similar structured contents; especially when the density of impulse noise is relatively high and the variance of Gaussian noise is small, our method can outperform the traditional methods significantly not only in the simultaneous removal of Gaussian noise and impulse noise, and the restoration ability for a low-rank image matrix, but also in the preservation of textures and details in the image.

Editor: Xi-Nian Zuo, Institute of Psychology, Chinese Academy of Sciences, China

Funding: This work was partially supported by grants from the National High Technology Research and Development Program of China (2013AA122901, 2012AA121201) (http://www.863.gov.cn) and the National Natural Science Foundation of China (40971224) (http://www.nsfc.gov.cn). The funders had no role in study design, data collection and analysis, decision to publish, or preparation of the manuscript.

Competing Interests: The authors have declared that no competing interests exist.

* Email: yangxm@lreis.ac.cn

Introduction

Image denoising is highly demanded in the field of image processing, since noise is usually inevitable during the process of image acquisition and transmission, which significantly degrades the visual quality and makes subsequent high-level image analysis and understanding very difficult. There exist two types of most common and extensively studied noise: additive Gaussian noise and impulse noise. Additive Gaussian noise is usually generated during image acquisition and characterized by adding each image pixel a random value from the Gaussian distribution with zero mean and standard deviation σ, while impulse noise is very different in nature from Gaussian noise. Impulse noise can often be introduced in both image acquisition and transmission process by malfunctioning pixels in camera sensors, faulty memory locations in hardware, or transmission in a noisy channel, which mainly includes two models, namely salt-pepper noise and random-valued impulse noise [1]. For the former model, each gray value is replaced with a given probability p by the extreme value g_{min} or g_{max}, leaving remaining pixels unchanged, where $[g_{min}, g_{max}]$ denotes the dynamic range of an image and p determines the level of the salt-pepper noise; as for the latter case, the intensity values of contaminated pixels are taken place by random values identically and uniformly distributed in $[g_{min}, g_{max}]$. In this paper, we will focus on the first model of impulse noise. A fundamental target of image denoising is to effectively remove noises from an image while preserving image details and textures.

For Gaussian noise removal, the non-local means (NL-means) proposed by Buades A. et al. [2] is quite an efficient method in suppressing Gaussian noise while keeping details and structures in the image intact, which better exploits the redundancy of natural images and calculates the weighted average of all the pixels in the image based on similarity of neighborhoods, to achieve a satisfying filtering result. Inspired by nonlocal concept, many more nonlocal methods such as BM3D [3] and K-SVD [4] have been introduced and get state-of-the-art denoising performance for Gaussian noise. In addition, some filters based on wavelet transform [5] and partial differential equations (PDE) [6,7] including nonlinear total variation (TV) are also powerful tools for it. The traditional approaches for impulse noise removal operate locally and nonlinearly on images. Among them, let us mention the classical median filter and its variants like adaptive median filter (AMF), progressive switching median filter (PSMF), and direction weighting median filter (DWMF). This kind of nonlinear filters can remove impulse noise effectively but cannot preserve image details well. In order to better preserve edge structures in images, variational approaches have been developed. In [8], a data-fidelity term of l_1-norm was first introduced to achieve a significant improvement to remove outliers like impulse noise. Some authors [9,10,11] proposed two-phase schemes to better preserve image details: the main idea is to detect the location of noisy pixels corrupted by impulse noise using median-like filters, followed by some variational methods to estimate the gray values for those contaminated pixels.

However, most existing image denoising methods can only deal with a single type of noise, which violates the fact that most of the image noise we encounter in real world can normally be represented by a mixture of Gaussian and impulse noise. Effectively eliminating mixed noise is difficult due to the distinct characteristics of both types of degradation processes. It should be noted that the two-phase approaches [12–16] also show exuberant vitality for mixed noise removal, namely detecting or estimating the outliers before proceeding with the restoration phase. Garnett et al. [12] introduced a universal noise removal algorithm that first detects the impulse corrupted pixels, estimates its local statistics and incorporates them into the bilateral filter, resulting in the trilateral filter. In [13] somewhat similar ideas were used to establish a similarity principle which in turn supplies a simple mathematical justification for the nonlocal means filter in removing Gaussian noises. This is also the case of [15] where outliers are first detected by a median-like filter and then a K-SVD dictionary learning is performed on impulse free pixels to solve a $l_1 - l_0$ minimization problem, and the case of [16] where a median-type filter is used to remove impulse noise first and then NL-means based method is applied to remove the remaining Gaussian noise. Liu et al. [18] proposed a general weighted $l_2 - l_0$ norms energy minimization model to remove mixed noise, which was built upon maximum likelihood estimation framework and sparse representations over a trained dictionary. While in [19], weighted encoding with sparse nonlocal regularization was put forward for mixed noise removal, where there was not an explicit step of impulse pixel detection, and soft impulse pixel detection via weighted encoding was used to deal with impulse and Gaussian noise simultaneously instead. See [17,34,35] for more approaches to mixed noise removal. Although these denoising methods above were proposed specially for mixed Gaussian-impulse removal and indeed can alleviate the impact on image visual effect brought by mixed noises to some extent, most of them will erase details, and cannot better preserve regularly geometrical textures and fine structures.

In recent years, low-rank matrix reconstruction has become a research hotspot in many scientific and engineering domains, with myriad applications ranging from web search to machine learning to computer vision and image analysis, which mainly involves the problem of matrix completion (MC) [20,21,26,27] and robust principal component analysis (RPCA) [22–25,28], namely recovering a low-rank matrix from an incomplete but accurate sampling subset of its entries and from an observed data matrix with an

unknown fraction of its entries being arbitrarily corrupted, respectively. Luckily, there usually exist many regularly geometrical textures and fine structures in both natural image and Remote Sensing (RS) image, due to the self-similarity and redundancy of image, which makes the grayscale matrix of the image possess low-rank features. In light of this, we propose a novel denoising framework for better preserving details and low-rank features in images while removing mixed Gaussian-impulse noise, based on the theory of low-rank matrix reconstruction.

In this paper, we consider the problem of recovering a low-rank matrix from an incomplete sampling subset of its entries with an unknown fraction of the samplings contaminated by arbitrary errors, which is defined as the problem of matrix completion from corrupted samplings (MCCS) and modeled as a convex optimization problem that minimizes a combination of the nuclear norm and the l_1-norm. To exactly solve the problem, we introduce an effective algorithm called augmented Lagrange multipliers (ALM). By detecting impulse noises in a noisy image and treating the impulse free entries of the image matrix as available samplings first, and then regarding the Gaussian noises underlying the samplings as arbitrary errors, we can exploit the proposed algorithm to remove mixed noises and to recover the image matrix with low-rank or approximately low-rank features. To sum up, the main contributions of the paper include modeling the problem of low-rank matrix recovery from incomplete and corrupted samplings of its entries, solving the convex optimization problem via the proposed ALM algorithm and creatively applying it to mixed Gaussian-impulse noise removal.

The rest of the paper is organized as follows. We start in Section 2 by giving some introductions about the theory of low-rank matrix recovery including the problem of MC and Robust PCA. We put forward the ALM algorithm for the problem of MCCS and describe some implementation details and parameter settings for our methods in Section 3, in which we also analyze its powerful performance for exact recovery of low-rank matrices with erasures and errors. In Section 4, we will present in details our full denoising schemes for the impulse detector and mixed Gaussian-impulse noise removal. Experiments and comparisons with recent approaches are demonstrated in Section 5. Finally, we conclude this paper in Section 6.

Low-Rank Matrix Reconstruction

Low-rank matrix plays a central role in large-scale data analysis and dimensionality reduction, since low-rank structure is usually used either to approximate a general matrix, or to recover corrupted or missing data [25]. From the mathematical viewpoint, these practical problems come down to the theory of low-rank matrix reconstruction, which mainly involves the problem of MC and Robust PCA at present.

1 Matrix completion

To describe the problem of MC, suppose to simplify that the unknown matrix $M \in \mathbb{R}^{n \times n}$ is square, and that one has m sampled entries $\{M_{ij} : (i,j) \in \Omega\}$ where Ω is a random subset of cardinality m. The recovery of complete matrix M from these incomplete samplings of its entries is the MC problem.

The issue is of course that this problem is extraordinarily ill-posed because, with fewer samples than entries, there are infinitely many completions. Therefore, it is apparently impossible to identify which of these candidate solutions is indeed the "correct" one without some additional information. In many instances, however, the matrix we wish to recover has low rank or approximately low rank, which may change the property of the problem, and make the search for solutions feasible since the lowest-rank solution now tends to be the right one.

Candes and Recht [20] showed that matrix completion is not as ill-posed as once thought. Indeed, they proved that most low-rank matrices can be recovered exactly from most sets of sampled entries even though these sets have surprisingly small cardinality, and more importantly, they proved that this can be done by solving a simple convex optimization problem

$$\min \|X\|_* \\ s.t. X_{ij} = M_{ij}, (i,j) \in \Omega, \tag{1}$$

provided that the number of samples obeys

$$m \geq C n^{6/5} r \log n \tag{2}$$

for some positive numerical constant C, where r is the rank of matrix M. In (1), the notation $\|\bullet\|_*$ denotes the nuclear norm of a matrix, which is the sum of its singular values. The optimization problem (1) is convex and can be recast as a semidefinite program. In some sense, this is the tightest convex relaxation of the NP-hard rank minimization problem

$$\min rank(X) \\ s.t. X_{ij} = M_{ij}, (i,j) \in \Omega. \tag{3}$$

Another interpretation of Candes and Recht's result is that under suitable conditions, the rank minimization program (3) and the convex program (1) are formally equivalent in the sense that they have exactly the same unique solution. The state-of-the-art algorithms to solve the MC problem include the accelerated proximal gradient (APG) approach and the singular value thresholding (SVT) method [21].

2 Robust principal component analysis

Principal component analysis (PCA), as a popular tool for high-dimensional data processing, analysis, compression and visualization, has wide applications in scientific and engineering fields. It assumes that the given high-dimensional data lie near a much lower-dimensional linear subspace. To large extent, the goal of PCA is to efficiently and accurately estimate this low-dimensional subspace.

Suppose that the given data are arranged as the columns of a large matrix $D \in \mathbb{R}^{m \times n}$. The mathematical model for estimating the low-dimensional subspace is to find a low-rank matrix $A^{m \times n}$, such that the discrepancy between A and D is minimized, leading to the following constrained optimization:

$$\min_{A,E} \|E\|_F, s.t. rank(A) \leq r, D = A + E, \tag{4}$$

where $r \ll \min(m,n)$ is the target dimension of the subspace and $\|\bullet\|_F$ is the Frobenius norm, which corresponds to assuming that the data are corrupted by i.i.d. Gaussian noise. This problem can be conveniently solved by first computing the Singular Value Decomposition (SVD) of D, and then projecting the columns of D onto the subspace spanned by the r principal left singular vectors of D.

As PCA gives the optimal estimate when the corruption is caused by additive i.i.d. Gaussian noise, it works well in practice as long as the magnitude of noise is small. However, it breaks down under large corruption, even if that corruption only affects very few of the observations. Therefore, it is necessary to study whether a low-rank matrix A can still be efficiently and accurately recovered from a corrupted data matrix $D = A + E$, where some entries of the additive error matrix E may be arbitrarily large.

Recently, Wright et al. [23] have shown that under rather broad conditions the answer is affirmative: provided the error matrix E is sufficiently sparse, one can exactly recover the low-rank matrix A from $D = A + E$ by solving the following convex optimization problem:

$$\min_{A,E} \|A\|_* + \lambda |E|_1, s.t. D = A + E, \tag{5}$$

where $|\bullet|_1$ denotes the sum of the absolute values of matrix entries, and λ is a positive weighting parameter. Due to the ability to exactly recover underlying low-rank structure in the data, even in the presence of large errors or outliers, this optimization is referred to as Robust PCA. So far, RPCA has been successfully introduced into several applications such as background modeling and removing shadows and specularities from face images. Ganesh et al. [31] proposed two new algorithms for solving the problem (5), which are in some sense complementary to each other. The first one is an accelerated proximal gradient algorithm applied to the primal, which is a direct application of the FISTA framework introduced by [30], coupled with a fast continuation technique; the other one is a gradient-ascent approach applied to the dual of the problem (5). Recently, the augmented Lagrange multiplier method was introduced by Lin et al. for exact recovery of corrupted low-rank matrices, which shows rather promising performance when dealing with the problem (5).

Problem of MCCS and the ALM Algorithm

1 MCCS and its optimization model

Since the MC problem is closely connected to the RPCA problem, we may formulate the MC problem in the same way as RPCA

$$\min_A \|A\|_*, s.t. A + E = D, \pi_\Omega(E) = 0, \tag{6}$$

where $\pi_\Omega : \mathbb{R}^{m \times n} \to \mathbb{R}^{m \times n}$ is a linear operator that keeps the entries in Ω unchanged and sets those outside Ω (i.e., in $\bar{\Omega}$) zeros. As E will compensate for the unknown entries of D, the unknown entries of D are simply set as zeros. Then the partial augmented Lagrangian function of (6) is

$$L(A,E,Y,\mu) = \|A\|_* + \langle Y, D - A - E \rangle + (\mu/2)\|D - A - E\|_F^2, \tag{7}$$

where μ is a positive scalar. For updating E, the constraint $\pi_\Omega(E) = 0$ should be enforced when minimizing $L(A,E,Y,\mu)$.

The problem of restoring matrix from corrupted entries (i.e., RPCA) is less studied than the simpler MC problem when the available entries are not corrupted. However, in many practical applications, there usually exists the case when missing entries and corrupted entries simultaneously concur in an observed data matrix, which urgently drives researchers to pay more and more attention to the problem of recovering low-rank matrix from its observed data matrix with erasures and errors.

For observed matrix $D = \pi_\Omega(A^* + E^*)$ where the ordered pair (A^*, E^*) denotes the true solutions to low-rank matrix and sparse error matrix respectively, suppose that one can only know the entries of D on subset Ω (the set of indices of known entries), and that one aims to recover low-rank matrix A^* from the observed matrix D with erasures and errors. In this paper, we define the problem as matrix completion from corrupted samplings or robust matrix completion, which is the problem of restoring low-rank matrix from incomplete and corrupted samplings.

Similarly, as the entries of matrix E on set $\bar{\Omega}$ will compensate for the unknown entries of D and the entries of E on set Ω are sparse, we can split E into $E = E_\Omega + E_{\bar{\Omega}}$, where $E_\Omega = \pi_\Omega(E), E_{\bar{\Omega}} = \pi_{\bar{\Omega}}(E)$ and E_Ω denotes sparse error components for the known entries. Now we can formulate the problem of MCCS as follows:

$$\begin{cases} \min_{A,E} \|A\|_* + \lambda |E_\Omega|_1 \\ s.t. A + E = D = \pi_\Omega(A^* + E^*) \end{cases} \quad (8)$$

To recover the low-rank matrix for the MCCS problem (8), we propose a novel and effective algorithm based on the ALM method, whose objective function is as follows:

$$\begin{aligned} L(A, E_\Omega, E_{\bar{\Omega}}, Y, \mu) :&= \|A\|_* + \lambda |E_\Omega|_1 + \langle Y, D - A - E \rangle + \frac{\mu}{2} \|D - A - E\|_F^2 \\ &= \|A\|_* + \lambda |E_\Omega|_1 + \langle Y, D - A - E_\Omega - E_{\bar{\Omega}} \rangle + \frac{\mu}{2} \|D - A - E_\Omega - E_{\bar{\Omega}}\|_F^2 \\ &= \|A\|_* + \lambda |E_\Omega|_1 + \langle Y_\Omega, D_\Omega - A_\Omega - E_\Omega \rangle + \frac{\mu}{2} \|D_\Omega - A_\Omega - E_\Omega\|_F^2 \\ &+ \langle Y_{\bar{\Omega}}, D_{\bar{\Omega}} - A_{\bar{\Omega}} - E_{\bar{\Omega}} \rangle + \frac{\mu}{2} \|D_{\bar{\Omega}} - A_{\bar{\Omega}} - E_{\bar{\Omega}}\|_F^2 \end{aligned} \quad (9)$$

In (9), as E_Ω and $E_{\bar{\Omega}}$ are complementary to each other, we can ignore one when solving another one, and finally combine the two to obtain E. Unlike the technique presented in [29], we separate E into two parts in the objective function, which leads to the splitting of Y, D, A afterwards, and aims to achieve subsection optimization.

2 The proposed ALM algorithm and parameter settings

The ALM algorithm for the Robust MC problem proposed in the paper is described in Algorithm 1. It is apparent that computing the full SVD for the Robust MC problem is unnecessary, so we only need those singular values that are larger than a particular thresholding and their corresponding singular vectors. Firstly, we should predict the dimension of principal singular space, and in the paper, we exploit the following prediction rule:

$$sv_{k+1} = \begin{cases} svp_k + 1, & if svp_k < sv_k \\ \min(svp_k + round(0.05d), d), & if svp_k = sv_k \end{cases} \quad (10)$$

where $d = \min(m, n)$, sv_k is the predicted dimension and svp_k denotes the number of singular values in the sv_k singular values that are larger than μ_k^{-1}, and $sv_0 = 5$. In the paper, we choose $\lambda = 1/sqrt(p * \max(m, n)), p = |\Omega|/(mn)$, and set $\mu_0 = 0.5/\|D\|_2$. The following conditions are chosen as the stopping criterion:

$$\frac{\|D - A^k - E^k\|_F}{\|D\|_F} < \varepsilon_1 \ and \ \frac{\mu_{k-1}\|E^k - E^{k-1}\|_F}{\|D\|_F} < \varepsilon_2 \quad (11)$$

For convenience, we introduce the following soft-thresholding (shrinkage) operator:

$$\mathcal{S}_\varepsilon[x] \doteq \begin{cases} x - \varepsilon, & if x > \varepsilon \\ x + \varepsilon, & if x < -\varepsilon \\ 0, & otherwise \end{cases} \quad (12)$$

where $x \in \mathbb{R}$ and $\varepsilon > 0$. This operator can be extended to vectors and matrices by applying it element-wise. In Algorithm 1, $J(D) = \max(\|D\|_2, \lambda^{-1}\|D\|_\infty)$, where $\|\bullet\|_\infty$ is the maximum absolute value of the matrix entries, and $\|\bullet\|_2$ denotes the maximum singular value of a matrix.

Algorithm 1(MC from Corrupted Samplings via the ALM Method)

Input : Observation matrix $D \in \mathbb{R}^{m \times n}$, Ω, λ.

$Y_\Omega^0 = D_\Omega/J(D), Y_{\bar{\Omega}}^0 = 0, A^0 = E^0 = 0, \mu_0 > 0, \rho > 1, k = 0.$

While not converged do

$//$Solve$E^{k+1} = \arg\min_E L(A^k, E, Y^k, \mu_k).$

$E_\Omega^{k+1} = \mathcal{S}_{\lambda\mu_k^{-1}}[D_\Omega - A_\Omega^k + \mu_k^{-1} Y_\Omega^k];$

$E_{\bar{\Omega}}^{k+1} = D_{\bar{\Omega}} - A_{\bar{\Omega}}^k + \mu_k^{-1} Y_{\bar{\Omega}}^k;$

$//$Solve$A^{k+1} = \arg\min_A L(A, E^{k+1}, Y^k, \mu_k).$

$(U, S, V) = svd(D - E^{k+1} + \mu_k^{-1} Y^k);$

$A^{k+1} = U \mathcal{S}_{\mu_k^{-1}}[S] V^T;$

$Y^{k+1} = Y^k + \mu_k(D - A^{k+1} - E^{k+1});$

Update$\mu_k$to$\mu_{k+1} = \min(\rho\mu_k, \bar{\mu});$

$k \leftarrow k + 1.$

End while

Output : $(A^k, E^k).$

---

3 Performance analysis of the proposed algorithm

In this subsection, we first demonstrate the recovery accuracy of the proposed ALM algorithm by simulation experiments; and then for clarifying how the erasure rate (the fraction of the unknown entries in observed matrix D, i.e., $1 - d(\Omega)$, where $d(\Omega) = |\Omega|/(m * n)$) and the error probability (the fraction of randomly corrupted entries in the known entries available for restoration, i.e., $\|E_\Omega\|_0/|\Omega|$, where $\|\bullet\|_0$ denotes the sparsity of a matrix) affect the performance of our method, we carry out the phase transition analysis for the proposed algorithm. We generate the rank-r matrix A^* as a product LR^T, where L and R are independent $m \times r$ matrices whose elements are i.i.d. Gaussian random variables with zero mean and unit variance, and set a fraction of entries chosen randomly in A^* as zeros; and then, treat a fraction of non-zero entries also chosen randomly in A^* as corrupted entries after adding arbitrarily large errors E^* to them; finally, we get the observed data matrix $D^{m \times m}$ with erasures valued zeros.

We exploit different data matrices and various combinations (ρ_Ω, ρ_s) to do the experiments, where $\rho_\Omega = d(\Omega)$ and ρ_s stands for the error probability, and the reconstruction results are presented in Table 1. We observe that the proposed algorithm can accurately recover the low-rank matrices and sparse errors even when erasure rate reaches 30% and 20% of the available entries are corrupted.

Table 1. Recovery results of low-rank matrices with different (ρ_Ω, ρ_s) using the proposed ALM.

$\rho_r = rank(A^*)/m$	$\rho_\Omega = |\Omega|/m^2$	$\rho_s = \|E^*\|_0/|\Omega|$	$\|A^* - \hat{A}\|_F / \|A^*\|_F$	$rank(\hat{A})$	$\|\hat{E}_\Omega\|_0$
$20/500$	$\frac{225000}{250000}$	0.1	1.20e-6	20	22500
		0.2	2.47e-6	20	45000
	$\frac{200000}{250000}$	0.1	1.38e-6	20	20000
		0.2	2.22e-6	20	39999
	$\frac{175000}{250000}$	0.1	1.27e-6	20	17500
		0.2	2.54e-6	20	35109
$60/1000$	$\frac{9 \times 10^5}{10^6}$	0.1	6.20e-7	60	90000
		0.2	4e-4	60	181894
	$\frac{8 \times 10^5}{10^6}$	0.1	6.82e-7	60	80000
		0.2	2e-4	60	163523
	$\frac{7 \times 10^5}{10^6}$	0.1	4.84e-7	60	70233
		0.2	7.47e-4	60	154281

As for the experiments of phase transition analysis, we fix $m = 512$, set $r = 8, 16, 64$ respectively, and vary ρ_Ω and ρ_s between 0 and 1. For each r and each (ρ_Ω, ρ_s) pair, we generate 5 pairs (A^*, E^*) to obtain 5 observed matrices as described above. We deem the recovery successful if the recovered $\hat{A}$ satisfies $\|\hat{A} - A^*\|_F / \|A^*\|_F < 0.01$ for all the 5 observed data matrices. The curves colored red, green and blue in Figure 1 define "phase transition" bounds for the case of $r = 8, 16, 64$, respectively. In Figure 1, the horizontal coordinate indicates ρ_Ω, while the vertical coordinate denotes ρ_s. At the points of these curves, all 5 trials were accomplished with success, whereas for the points above the curves, at least one attempt failed. It is assumed that the regions under the curves are "regions of success". Our experiments show that for fixed m, the smaller rank r is, the larger area the region of

success will have. In fact, when other factors keep unchanged, the recovery performance of the algorithm is inversely proportional to $rank(A^*)/m$. In addition, we can also observe that erasure rate and error probability keep the relationship of interacting.

Our Denoising Scheme for Mixed Gaussian-Impulse Noise

Image denoising is a research hotspot in the area of image processing all the way. However, in real world, images are typically contaminated by more than one type of noise during image acquisition and transmission process. As a matter of fact, we often encounter the case where an image is corrupted by both Gaussian and impulse noise. Such mixed noise could occur when an image that has already been stained by Gaussian noise in the procedure of image acquisition with faulty equipment suffers impulsive corruption during its transmission over noisy channels successively.

As we know, there usually exist many regularly geometrical textures and similar structures in both natural image and Remote Sensing image, due to the self-similarity and redundancy of image, which makes the grayscale matrix of the image possess low-rank features. Matrix completion can accurately reconstruct low-rank information by exploiting reliable pixels after detecting the isolated noises, which is quite suitable for impulse noise (especially for salt-pepper noise) removal. Robust PCA can recover low-rank matrix from observed matrix with sparse errors, which may deal with the problem of Gaussian noise removal to some extent, since for Gaussian noise, corruptions with large magnitude are rarely distributed while most of the corruptions are located near zero. Luckily, the proposed ALM algorithm for the Robust MC problem, also supplies a powerful technique for removing mixed Gaussian-impulse noise from images with low-rank features.

In this paper, we adopt the two-phase scheme for mixed noise removal: By detecting impulse noises in a noisy image and treating the impulse free entries of the image matrix as available samplings first, and then regarding the Gaussian noises underlying the samplings as arbitrary errors, we can exploit the proposed algorithm to remove mixed noises and to recover the grayscale matrix of the image with low-rank or approximately low-rank features. It should be noted that, we cannot simply use MC for

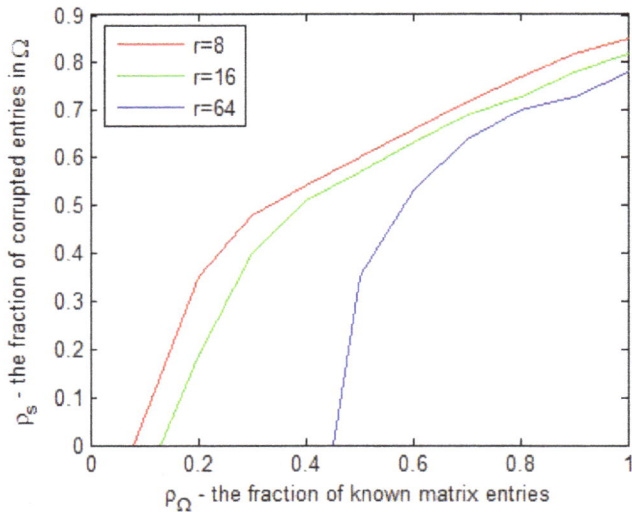

Figure 1. Phase transition with regard to (ρ_Ω, ρ_s) of the proposed ALM algorithm. The curves colored red, green and blue define "phase transition" bounds for the case of $r = 8, 16, 64$, respectively.

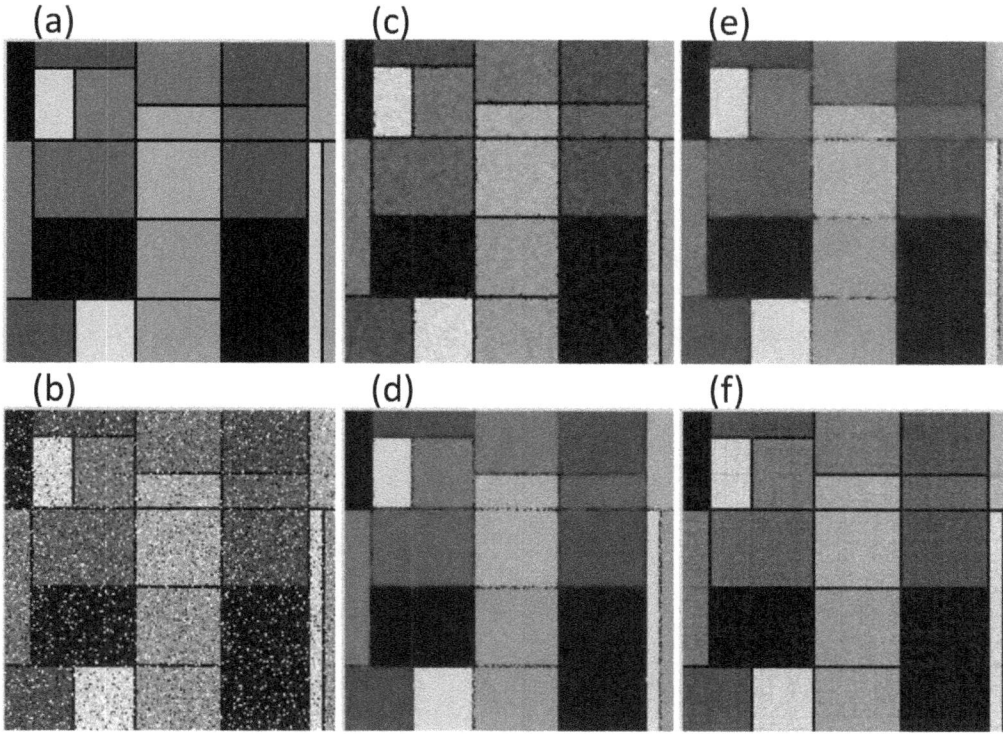

Figure 2. Comparative results on Mondrian image with $p=0.1, \sigma=15/255$**.** The PSNR (dB) and SSIM for (b) ~ (f) are (13.681, 0.883), (27.704, 0.981), (23.542, 0.966), (23.376, 0.966), (**30.158, 0.985**), respectively. Figure (a): Original Image; (b): Noisy Image; (c): Trilateral Filter [12]; (d): Xiao et al. [15]; (e): Xiong and Yin's [16]; (f): Proposed ALM.

removing impulse noises after detecting them, since the remaining Gaussian noises no longer make the image possess low-rank features; in addition, neither can we directly employ Robust PCA to restore the noisy image at first, because impulse noises destroy the sparsity of large errors, which may greatly decrease the accuracy of the restoration.

The success of two-phase approaches for noise removal relies on the accurate detection of impulse noise. Many impulse noise detectors are proposed in the literatures, e.g. AMF is used to detect salt-pepper noise, while adaptive center-weighted median filter (ACWMF) and rank-ordered logarithmic difference (ROLD) [32] are utilized to detect random-valued impulse noise. In this paper, study on detectors for random-valued impulse noise is beyond the scope of the topic.

For salt-pepper noise, we employ the following strategy to detect it after considering its features taken on in the image: (1) Grayscale thresholding. Specifically, set a fixed thresholding α, and regard those pixels whose gray-levels are distributed in the interval $[255-\alpha, 255]$ or $[0, \alpha]$ as candidate noises. (2) Median detecting. For current candidate noise, we consider the absolute difference between its gray-level and the median gray-level of its neighborhood pixels. If the absolute difference is larger than another given thresholding β, we then treat current candidate noise as true noise. Empirically, when the density of salt-pepper noise namely noise level is high, the window size of the neighborhood for median detecting should be large.

Experiments and Discussion

This section is devoted to the experimental analysis and discussion of the denoising scheme introduced in the previous section. We compare our ALM algorithm with some recently proposed approaches [12,15,16] for dealing with the mixture of impulse and Gaussian noise. Comparison results are presented both under the form of statistical index tables and that of visual effects. To evaluate the quality of the restoration results, peak signal to noise ratio (PSNR) and structural similarity (SSIM) are employed for objective evaluation and for subjective visual evaluation, respectively.

Given an image $x \in [0,255]^{m \times n}$, the PSNR of the restored image $\hat{x}$ is defined as follows:

$$PSNR(x,\hat{x}) = 10\log_{10}\frac{255^2}{MSE}, \quad MSE = \|x-\hat{x}\|_F^2 / (m \times n). \quad (13)$$

The larger the value of PSNR is, the better the quality of restoration will be. We can formulate the SSIM between original image and the recovered image as

$$SSIM(x,\hat{x}) = \frac{(2\mu_x\mu_{\hat{x}} + C_1)(2\sigma_{x\hat{x}} + C_2)}{(\mu_x^2 + \mu_{\hat{x}}^2 + C_1)(\sigma_x^2 + \sigma_{\hat{x}}^2 + C_2)} \quad (14)$$

(see [33] for more details). The dynamic range of SSIM is $[0,1]$, which means a better recovery performance with the SSIM closer to the value 1.

As the proposed algorithm is suitable for restoring images with low-rank information, we first choose the classical Mondrian and Barbara image in the field of image processing to do the experiments. In all experiments, parameters of each method have been tuned to yield the best results. On image Mondrian, we

Table 2. Comparisons of statistical indices under different(p,σ) on Barbara Image.

(p,σ)	Statistical Index	Noisy Image	ROAD-Trilateral [12]	Xiao et al. [15]	Xiong and Yin's [16]	Proposed ALM
$(0.1,\frac{5}{255})$	PSNR(dB)	15.132	21.528	27.027	26.472	**27.163**
	SSIM	0.956	0.962	0.983	0.984	**0.986**
$(0.2,\frac{5}{255})$	PSNR(dB)	12.184	19.712	26.325	25.713	**26.386**
	SSIM	0.913	0.949	0.980	0.981	**0.982**
$(0.3,\frac{5}{255})$	PSNR(dB)	10.407	17.652	25.217	24.688	**25.287**
	SSIM	0.870	0.933	0.975	0.976	**0.977**
$(0.4,\frac{5}{255})$	PSNR(dB)	9.179	14.854	24.017	23.438	**24.237**
	SSIM	0.827	0.906	0.969	0.970	**0.971**
$(0.5,\frac{5}{255})$	PSNR(dB)	8.223	12.096	22.977	22.289	**23.315**
	SSIM	0.785	0.857	0.964	0.962	**0.965**
$(0.1,\frac{15}{255})$	PSNR(dB)	14.910	20.726	**26.041**	25.892	25.385
	SSIM	0.938	0.954	**0.978**	0.978	0.975
$(0.2,\frac{15}{255})$	PSNR(dB)	12.044	19.432	**25.542**	25.280	25.067
	SSIM	0.896	0.945	0.975	0.976	0.973
$(0.3,\frac{15}{255})$	PSNR(dB)	10.349	16.838	**24.574**	24.256	24.182
	SSIM	0.855	0.928	**0.971**	0.971	0.970
$(0.4,\frac{15}{255})$	PSNR(dB)	9.131	13.945	**23.555**	23.076	23.307
	SSIM	0.814	0.893	0.965	0.965	**0.965**
$(0.5,\frac{15}{255})$	PSNR(dB)	8.206	11.604	22.492	21.955	**22.684**
	SSIM	0.776	0.846	0.961	0.958	**0.962**

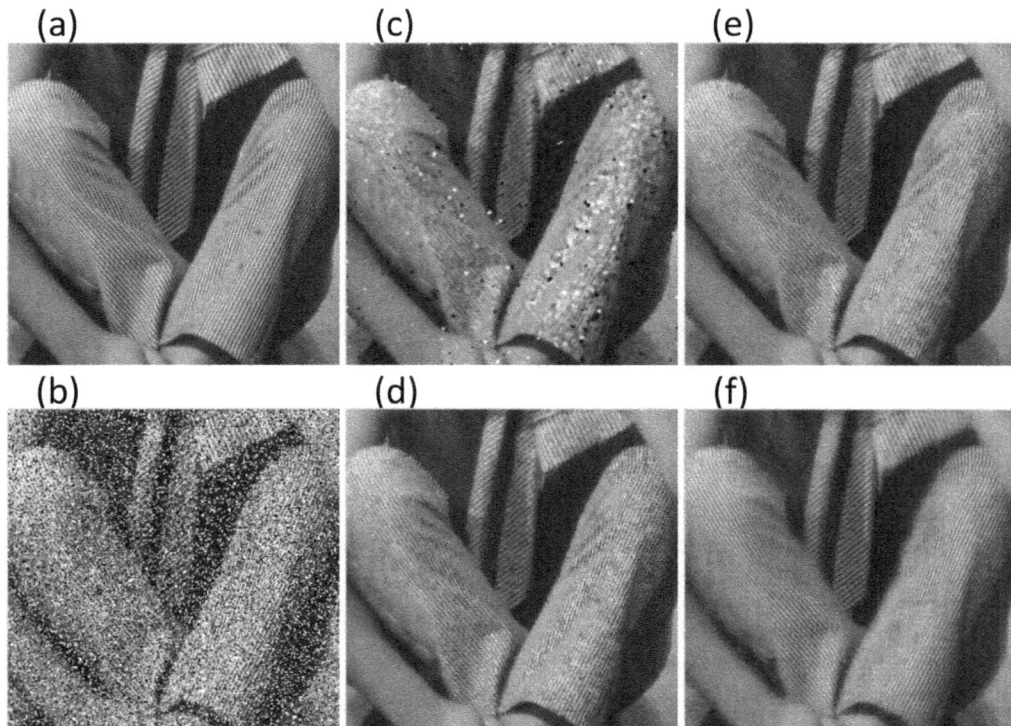

Figure 3. Comparison of visual effect on Barbara image when $p = 0.3, \sigma = 5/255$**.** Figure (a): Original Image; (b): Noisy Image; (c): Trilateral Filter [12]; (d): Xiao et al. [15]; (e): Xiong and Yin's [16]; (f): Proposed ALM.

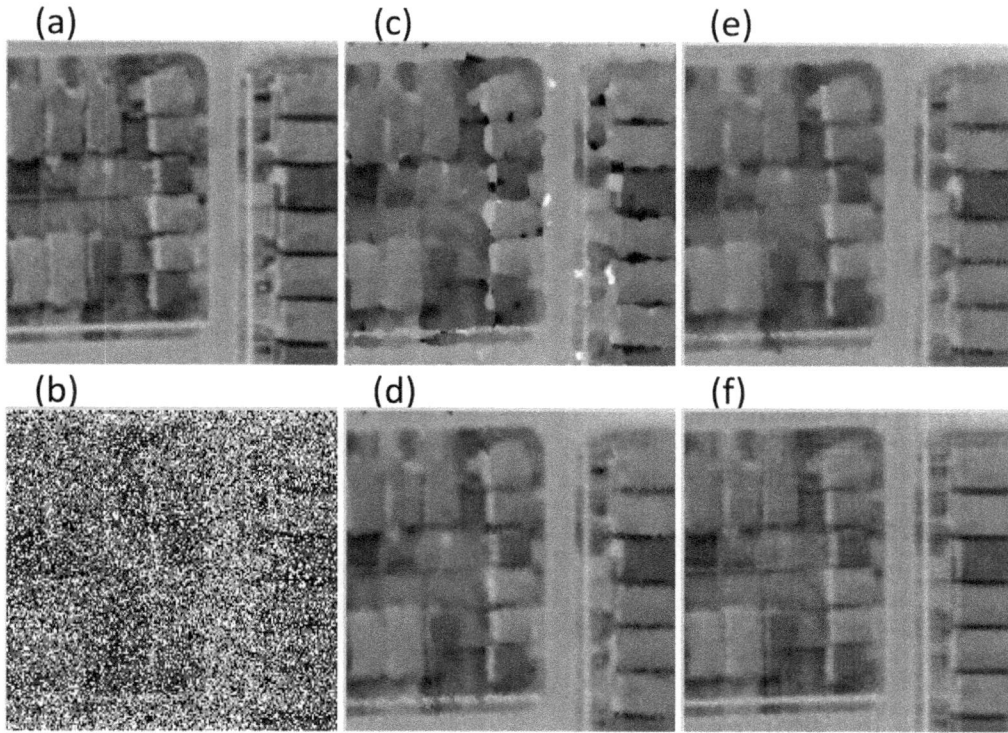

Figure 4. Comparison of visual effect on RS image when $p = 0.5, \sigma = 5/255$**.** Figure (a): Original Image; (b): Noisy Image; (c): Trilateral Filter [12]; (d): Xiao et al. [15]; (e): Xiong and Yin's [16]; (f): Proposed ALM.

Table 3. Comparisons of statistical indices under different (p, σ) on RS Image.

(p, σ)	Statistical Index	Noisy Image	ROAD-Trilateral [12]	Xiao et al. [15]	Xiong and Yin's [16]	Proposed ALM
$(0.1, \frac{5}{255})$	PSNR(dB)	15.620	33.976	34.459	36.079	**36.246**
	SSIM	0.954	0.994	0.994	0.996	**0.996**
$(0.2, \frac{5}{255})$	PSNR(dB)	12.509	30.145	33.993	34.918	**35.170**
	SSIM	0.910	0.989	0.993	0.995	**0.995**
$(0.3, \frac{5}{255})$	PSNR(dB)	10.691	27.797	32.550	33.037	**33.691**
	SSIM	0.861	0.983	0.992	0.992	**0.993**
$(0.4, \frac{5}{255})$	PSNR(dB)	9.559	24.160	30.941	31.603	**32.317**
	SSIM	0.828	0.976	0.988	0.990	**0.991**
$(0.5, \frac{5}{255})$	PSNR(dB)	8.590	19.342	29.151	29.097	**30.892**
	SSIM	0.787	0.948	0.984	0.985	**0.988**
$(0.1, \frac{15}{255})$	PSNR(dB)	15.144	29.618	29.893	**31.581**	28.385
	SSIM	0.932	0.985	0.986	**0.989**	0.981
$(0.2, \frac{15}{255})$	PSNR(dB)	12.393	27.657	29.809	**30.978**	27.943
	SSIM	0.892	0.981	0.985	**0.988**	0.979
$(0.3, \frac{15}{255})$	PSNR(dB)	10.739	26.382	29.277	**30.168**	27.481
	SSIM	0.851	0.976	0.984	**0.987**	0.977
$(0.4, \frac{15}{255})$	PSNR(dB)	9.563	23.601	28.385	**29.103**	26.995
	SSIM	0.829	0.969	0.980	**0.985**	0.975
$(0.5, \frac{15}{255})$	PSNR(dB)	8.543	16.873	26.993	**27.606**	26.617
	SSIM	0.774	0.937	0.975	**0.977**	0.974

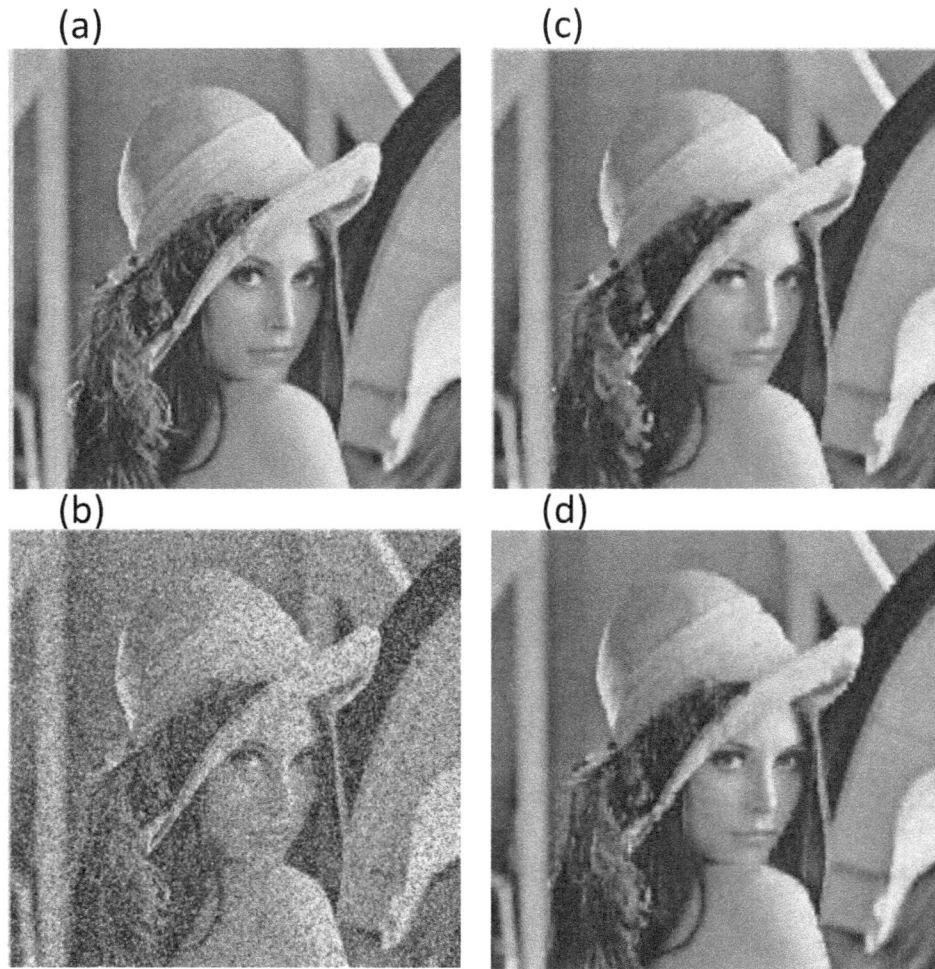

Figure 5. Comparative results on Lena image with $p = 0.3, \sigma = 5/255$. The PSNR (dB) and SSIM for (b) ~ (d) are (10.669, 0.887), (27.237, 0.987), **(28.070, 0.989)**, respectively. Figure (a): Original Lena; (b): Noisy Lena; (c): Trilateral Filter; (d): Proposed ALM.

successively add Gaussian noise with $\sigma = 15/255$ and salt-pepper noise with $p = 0.1$. The visual comparison of the four approaches is presented in Figure 2, from where we will see that our method can simultaneously remove Gaussian noise and salt-pepper noise in the restoration phase, and well reconstruct the original image with low-rank structures; above all, it can better preserve details and edges in the image. By contrast, the other three approaches either cannot thoroughly remove the mixed noises or will destroy fine structures. The values of PSNR and SSIM also demonstrate that the ALM-based approach outperforms other methods significantly not only in objective index, but also in subjective visual effect.

In the following experiment, image Barbara abundant in geometrical textures, is corrupted by Gaussian noise with zero mean and different standard deviations $\sigma = 5/255, 15/255$ first, and then we add salt-pepper noise with different levels $p = 0.1, 0.2, 0.3, 0.4, 0.5$ on the image. Comparisons of statistical indices under different combinations of (p, σ) are provided in Table 2, and the comparison of visual effect when $p = 0.3, \sigma = 5/255$ is also shown in Figure 3. From Figure 3, we also observe that our method can preserve regularly geometrical textures perfectly while removing the mixed Gaussian-impulse noises from the image. However, other approaches will more or less blur the textured information consisting in the image.

Meanwhile, we can see that the differences of recovery performance between the proposed ALM algorithm and the other three methods will be more remarkable when the level of salt-pepper noise is relatively high while the variance of Gaussian noise is small.

To better demonstrate the excellent performance of our method, we do a further comparison experiment of the four approaches in removing mixed Gaussian-impulse noise from Remote Sensing image with low-rank features. As in the previous experiment, we add Gaussian and salt-pepper noise with different values of p and σ on RS image. The experimental results are presented in Figure 4 and Table 3. Again, it shows that our method can better remove the mixed noise and restore the image especially when the level of impulse noise is relatively high while Gaussian noise is small.

The above experiments mainly involve images with significant low-rank features such as similar structures and regular textures, which violates the fact that most of images in the real world cannot possess globally low-rank features. However, most local parts in the image will meet low-rank or approximately low-rank condition, due to the self-similarity and spatial correlation of images. Consequently, we can utilize the proposed algorithm for image de-noising via block processing-based technique. The

Figure 6. Comparative results on Boat image with $p = 0.3, \sigma = 5/255$**.** The PSNR (dB) and SSIM for (b) ~ (d) are (10.696, 0.909), (23.737, 0.973), **(27.192, 0.985)**, respectively. Figure (a): Original Boat; (b): Noisy Boat; (c): Trilateral Filter; (d): Proposed ALM.

following experiments demonstrate the results of removing mixed Gaussian-impulse noise from Lena and Boat images, compared to trilateral filter method. Both of the noisy images are of size 512×512 pixels, either divided into many image blocks of size 16×16 pixels. The de-noising results for Lena and Boat are presented in Figure 5 and Figure 6 respectively, which also show the better performance of our method over trilateral filter. We can see that the details of hairs are not preserved well in Figure 5c, while in Figure 6c, some structures of the masts on the boats are lost.

Conclusion

In this paper, regarding the problem of low-rank matrix recovery from incomplete and corrupted samplings of its entries namely the Robust MC problem, we construct a mathematical model based on convex optimization, and put forward a novel and effective ALM algorithm, to solve this kind of optimization

problem which can be considered as the extension of the Robust PCA and the MC problem. Experiments on performance analysis and mixed Gaussian-impulse noise removal demonstrate the reliability and practicability of the proposed algorithm, which also show that our method can well preserve details and textures and keep the consistency of structures, while simultaneously removing mixed noises from images with low-rank features. By virtue of the novelty and powerful advantages, the approach will bring promising application value in the fields of data mining, image processing, machine learning, RS information processing and so forth.

Author Contributions

Conceived and designed the experiments: FM CZ XY. Performed the experiments: FM. Analyzed the data: FM XY. Contributed reagents/ materials/analysis tools: FM XY. Wrote the paper: FM.

References

1. Pok G, Liu JC, Nair AS (2003) Selective removal of impulse noise based on homogeneity level information. IEEE Transactions on Image Processing 12(1): 85–92.

2. Buades A, Coll B, Morel JM (2005) A non-local algorithm for image denoising. IEEE International Conference on Computer Vision and Pattern Recognition: 60–65.

3. Dabov K, Foi A, Katkovnik V, Egiazarian K (2007) Image denoising by sparse 3-D transform-domain collaborative filtering. IEEE Transactions on Image Processing 16(8): 2080–2095.

4. Elad M, Aharon M (2006) Image denoising via sparse and redundant representations over learned dictionaries. IEEE Transactions on Image Processing 15(12): 3736–3745.

5. Xu J, Osher S (2007) Iterative regularization and nonlinear inverse scale space applied to wavelet-based denoising. IEEE Transactions on Image Processing 16(2): 534–544.

6. You YL, Kaveh M (2000) Fourth-order partial differential equations for noise removal. IEEE Transactions on Image Processing 9(10): 1723–1729.

7. Rudin L, Osher S, Fatemi E (1992) Nonlinear total variation based noise removal algorithms. Physisca D 60: 259–268.

8. Nikolova M (2004) A variational approach to remove outliers and impulse noise. Journal of Mathematical Imaging and Vision 20(1–2): 99–120.

9. Chan R, Dong Y, Hintermuller M (2010) An efficient two-phase L1-TV method for restoring blurred images with impulse noise. IEEE Transactions on Image Processing 19(7): 1731–1739.

10. Cai JF, Chan R, Nikolova M (2010) Fast two-phase image deblurring under impulse noise. Journal of Mathematical Imaging and Vision 36(1): 46–53.

11. Chan R, Hu C, Nikolova M (2004) An iterative procedure for removing random-valued impulse noise. IEEE Signal Processing Letters 11(12): 921–924.

12. Garnett R, Huegerich T, Chui C, He W (2005) A universal noise removal algorithm with an impulse detector. IEEE Transactions on Image Processing 14(11): 1747–1754.

13. Li B, Liu Q, Xu J, Luo X (2011) A new method for removing mixed noises. Science China Information Sciences 54(1): 51–59.

14. Cai JF, Chan R, Nikolova M (2008) Two-phase approach for deblurring images corrupted by impulse plus Gaussian noise. Inverse Problems and Imaging 2(2): 187–204.

15. Xiao Y, Zeng T, Yu J, Ng MK (2011) Restoration of images corrupted by mixed Gaussian-impulse noise via $l_1 - l_0$ minimization. Pattern Recognition 44(8): 1708–1720.

16. Xiong B, Yin Z (2012) A universal denoising framework with a new impulse detector and nonlocal means. IEEE Transactions on Image Processing 21(4): 1663–1675.

17. Delon J, Desolneux A (2013) A patch-based approach for removing impulse or mixed Gaussian-impulse noise. SIAM Journal on Imaging Sciences 6(2): 1140–1174.

18. Liu J, Tai XC, Huang H, Huan Z (2013) A weighted dictionary learning model for denoising images corrupted by mixed noise. IEEE Transactions on Image Processing 22(3): 1108–1120.

19. Jiang J, Zhang L, Yang J (2014) Mixed noise removal by weighted encoding with sparse nonlocal regularization. IEEE Transactions on Image Processing 23(6): 2651–2662.

20. Candes EJ, Recht B (2009) Exact matrix completion via convex optimization. Foundations of Computational Mathematics 9(6): 717–772.

21. Cai JF, Candes EJ, Shen Z (2010) A singular value thresholding algorithm for matrix completion. SIAM Journal on Optimization 20(4): 1956–1982.

22. Candes EJ, Li X, Ma Y, Wright J (2011) Robust principal component analysis? Journal of the ACM. doi: 10.1145/1970392.1970395.

23. Wright J, Ganesh A, Rao S, Ma Y (2009) Robust principal component analysis: exact recovery of corrupted low-rank matrices via convex optimization. Proceedings of Advances in Neural Information Processing Systems: 2080–2088.

24. Xu H, Caramanis C, Sanghavi S (2012) Robust PCA via outlier pursuit. IEEE Transactions on Information Theory 58(5): 3047–3064.

25. Chen Y, Jalali A, Sanghavi S, Caramanis C (2013) Low-rank matrix recovery from errors and erasures. IEEE Transactions on Information Theory 59(7): 4324–4337.

26. Liu Z, Vandenberghe L (2009) Interior-point method for nuclear norm approximation with application to system identification. SIAM Journal on Matrix Analysis and Applications 31(3): 1235–1256.

27. Ji H, Liu C, Shen Z, Xu Y (2010) Robust video denoising using low rank matrix completion. IEEE International Conference on Computer Vision and Pattern Recognition: 1791–1798.

28. Tao M, Yuan X (2011) Recovering low-rank and sparse components of matrices from incomplete and noisy observations. SIAM Journal on Optimization 21(1): 57–81.

29. Liang X, Ren X, Zhang Z, Ma Y (2012) Repairing sparse low-rank texture. Computer Vision-ECCV: 482–495.

30. Beck A, Teboulle M (2009) A fast iterative shrinkage-thresholding algorithm for linear inverse problems. SIAM Journal on Imaging Sciences 2(1): 183–202.

31. Ganesh A, Lin Z, Wright J, Wu L, Chen M, et al. (2009) Fast algorithms for recovering a corrupted low-rank matrix. 3rd IEEE International Workshop on Computational Advances in Multi-Sensor Adaptive Processing (CAMSAP): 213–216.

32. Dong Y, Chan RH, Xu S (2007) A detection statistic for random-valued impulse noise. IEEE Transactions on Image Processing 16(4): 1112–1120.

33. Wang Z, Bovik AC, Sheikh HR, Simoncelli EP (2004) Image quality assessment: from error visibility to structural similarity. IEEE Transactions on Image Processing 13(4): 600–612.

34. Li YR, Shen L, Dai DQ, Suter BW (2011) Framelet algorithms for de-blurring images corrupted by impulse plus Gaussian noise. IEEE Transactions on Image Processing 20(7): 1822–1837.

35. Lopez-Rubio E (2010) Restoration of images corrupted by Gaussian and uniform impulsive noise. Pattern Recognition 43(5): 1835–1846.

4

A Simple Semi-Automatic Approach for Land Cover Classification from Multispectral Remote Sensing Imagery

Dong Jiang, Yaohuan Huang*, Dafang Zhuang, Yunqiang Zhu, Xinliang Xu, Hongyan Ren

State Key Lab of Resources and Environmental Information System, Institute of Geographical Sciences and Natural Resources Research, Chinese Academy of Sciences, Beijing, China

Abstract

Land cover data represent a fundamental data source for various types of scientific research. The classification of land cover based on satellite data is a challenging task, and an efficient classification method is needed. In this study, an automatic scheme is proposed for the classification of land use using multispectral remote sensing images based on change detection and a semi-supervised classifier. The satellite image can be automatically classified using only the prior land cover map and existing images; therefore human involvement is reduced to a minimum, ensuring the operability of the method. The method was tested in the Qingpu District of Shanghai, China. Using Environment Satellite 1(HJ-1) images of 2009 with 30 m spatial resolution, the areas were classified into five main types of land cover based on previous land cover data and spectral features. The results agreed on validation of land cover maps well with a Kappa value of 0.79 and statistical area biases in proportion less than 6%. This study proposed a simple semi-automatic approach for land cover classification by using prior maps with satisfied accuracy, which integrated the accuracy of visual interpretation and performance of automatic classification methods. The method can be used for land cover mapping in areas lacking ground reference information or identifying rapid variation of land cover regions (such as rapid urbanization) with convenience.

Editor: Juan A. Añel, University of Oxford, United Kingdom

Funding: This work was financially supported by the Chinese Academy of Sciences (grant KZZD-EW-08), and the China Postdoctoral Science Foundation (20100480437, 201104133). The funders had no role in study design, data collection and analysis, decision to publish, or preparation of the manuscript.

Competing Interests: The authors have declared that no competing interests exist.

* E-mail: huangyh@lreis.ac.cn

Introduction

Land use and land cover change (LUCC) has increasingly become a central component of global environmental change research [1,2]. Land cover data are the basis of this research and are key input to earth surface processes, including bio-chemical cycles, spatial and temporal distribution of plant biomass and respiration, and coupling between the atmosphere and biosphere [3]. Satellite imagery has been the primary tool for large-scale time-dependent land cover mapping [4]. Numerous classification algorithms utilize pattern recognition techniques, spectral features of the images and empirical geographical knowledge for land use mapping. Recent efforts to map land cover using remote sensing data have taken a variety of approaches, including visual interpretation classification [5], unsupervised clustering coupled with extensive ancillary data and manual labeling of clusters [6], supervised classification [7], expert system classification [8], artificial intelligence neural network classification [9], and decision tree classification [10,11]. Traditional unsupervised classification algorithms, such as maximum likelihood classification, use clustering techniques to identify spectrally distinct groups of data and are the earliest approach of land cover automatic classification that has employed pattern recognition techniques. The drawback of these algorithms is that the accuracy of land cover classification is not guaranteed and the land cover classifications are arbitrary. Supervised classification methods require substantial expertise and

human participation for selecting training samples. Therefore, the result of land cover classification is influenced greatly by classification participants, and it is impossible to classify land cover automatically with these methods. Furthermore, the algorithms such as neural network classification and fuzzy logic classification are highly complicated in their algorithm basis which makes them difficult to understand and apply widely. Decision tree classification methods are widely used in large areas, such as global land cover mapping (e.g. land cover type Yearly 1 km GRID of EOS/MODIS [12]). The main problem presented by decision tree classification is the construction of the decision tree and the assignment of thresholds for each sub nodes, which heavily depends on human experience and varies spatially and temporally.

To improve the accuracy of classification, images taken at multiple times were adopted for change detection. Numerous algorithms have been proposed for identifying and analyzing pixel-by-pixel differences between images of the same area acquired at different times [13], most of which produce reliable change detection results. Although the understanding of the causes of land use change have improved over the last several decades [14], accurate identification of the land cover types of changed pixels based on comparing distinction between two images without validation data is notably difficult. It is because the two images under consideration are often from slightly different times of the year, which may cause errors in change detection. To overcome problems of change detection, Fortier *et al.* proposed a new technique for land classification of imagery lack of ground data [15], which used temporally invariant ground features as calibration and validation data. It provided an innovative and accurate methodology for land cover classification using remote sensing imagery for which ground data are not available. However, manual steps of invariant features identification and classification tree algorithm (CTA) used for image classification made the methodology require human participation and hard to classify images automatically.

In light of the afore-mentioned problems, we propose a simple but robust method for land cover classification using satellite images. Prior accurate land cover data are adopted as important background information. Using only the prior land cover map, a recent image was classified. Our approach can be simply described to two steps: (1) semi-automatically detecting land cover changed pixels from satellite images compared with prior land cover map; (2) semi-automatically classifying the land cover of changed pixels based on pattern recognition and changed rules. This method was evaluated for the Qingpu District in Shanghai, China with images from HJ-1 Environment Satellites (HJ-1), which is a new type of satellite developed by China and launched in 2008, being used for environment and disaster monitoring.

Methods

The main phases for implementation of the new method are shown in Fig. 1.

Automatic Collection of Training Samples

Similar to supervised classification, the first step of our method is obtaining training samples with the purpose of obtaining pure pixels for various land cover classes. Previously, accurate sampling depended entirely on human participation and interpretation, which reduces the automation of land cover classification. In this paper, we propose a new method to extract pure pixels of land cover automatically with an accurate previous land cover dataset as prior knowledge. The 1:100,000 vector land cover maps of China produced by the Chinese Academy of Sciences were used in our study. This database has been validated by intensive field surveys, including a total survey length of 75,271 km across China. The overall accuracy of the land use map is 95% for all land use classes, which is the best map available at the national scale for China [16–19].

Ecologically, the junctions of different land cover classes are fragile areas, where are the main areas of land cover changing. We assumed that the interiors of individual land cover areas are relative ecologically stable areas and that larger patches are always more ecologically stable. Based on this assumption, the patches of land covers were sorted in descending order of their area. Samples of different land covers were selected based on their accumulation area threshold (Pa), the condition of land cover classification that can be defined as follows:

$$Pa_i \leq \frac{\sum_{j=0}^{x} Ac_{ij}}{As_i} \tag{1}$$

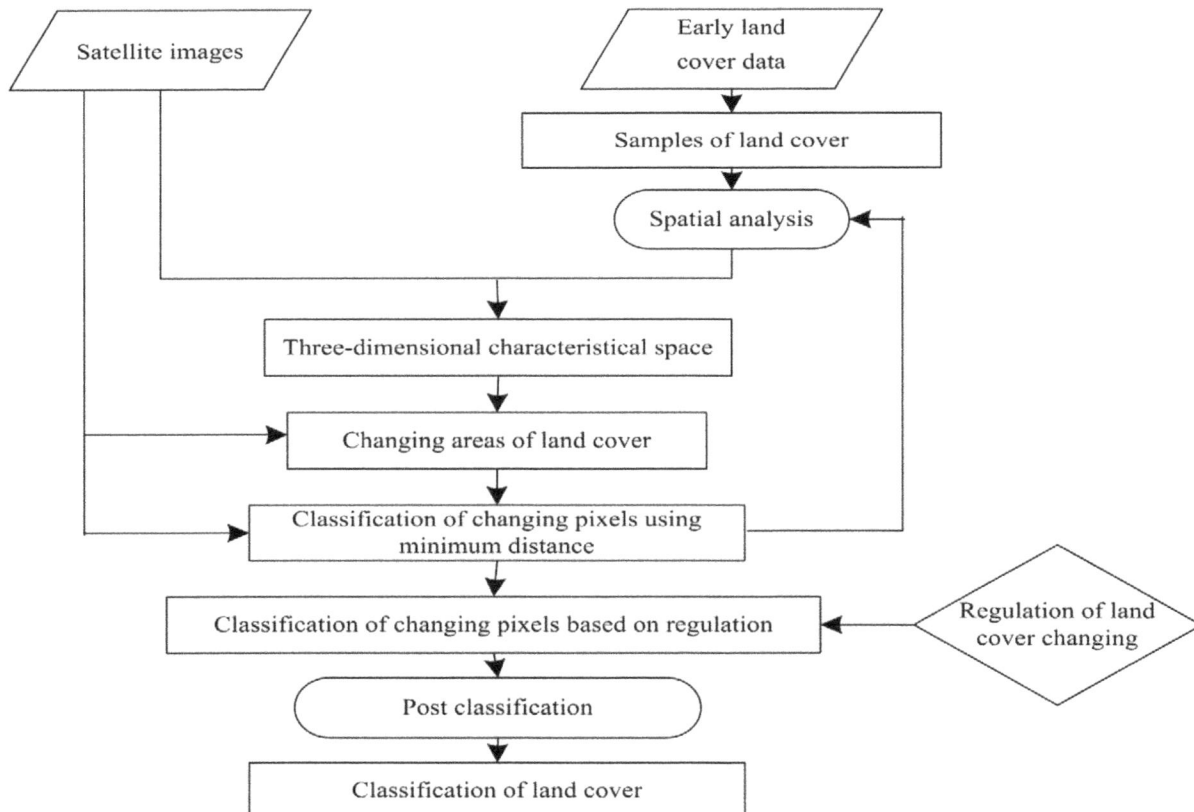

Figure 1. Flowchart of proposed land cover classification method.

where Pa_i is the accumulation area threshold of the ith class of land cover, Ac_{ij} is the area of patches of the ith class of land cover sorted in descending order, x is the number of the ith class of land cover, and As_i is the total area of the ith class of land cover.

We discarded the joint region of different land cover with spatial buffer analysis. The patch areas vary for land cover data; therefore, the buffer area cannot be obtained using the same distance to all patches. The distance of buffer analysis is

$$P_{buffer} = \frac{Ab_d}{A}, d < 0, \qquad (2)$$

where P_{buffer} is area threshold for buffer analysis, Ab_d is the buffer area of the patch with a distance of d, with d negative, and A is the area of the patch. The buffer regions were collected as training samples, and the automatic collection of training samples is illustrated in Fig. 2.

To keep the precision and efficiency of the method we proposed, it is necessary to determine an optimal value of Pa_i and P_{buffer}. Through a series of experiments using data from 4 test areas, including the area of Qingpu which will be used as evaluating area, we determined the (Pa_i, P_{buffer}) combination of (60%, 50% to 60%) to be optimal for samples of land cover classification selection.

Establishment of Three-dimensional Feature Space

The satellite image for classification was first overlaid with a previous land cover map. Multispectral values of each pixel were extracted based on the prior land cover type. Principal component statistical analysis was processed for the data in all spectral bands of each land cover class extracted from the region of interest. Based on previous studies [20–22], we selected the first three principal components for orthogonal decomposition to construct the three-dimensional feature space of different land cover classes:

$$\sum_{j=1}^{3} \frac{(Pji - MPji)^2}{\sigma ji^2} < c_i^2 \qquad (3)$$

where i is the ith class of land cover, Pji is the jth principal component of the ith class of land cover, $MPji$ is the mean value of the jth principal component of the ith class of land cover, σji is the

standard deviation of the jth principal component of the ith class of land cover, and c_i is the ellipsoid radius of the three-dimensional feature space of the ith class of land cover.

The three-dimensional feature space of different land cover classes is later used for the detection and classification of changed land cover pixels in the following steps.

Change Detection

The detection of changed areas of land cover is a key step in our method. During this step, the satellite images were overlaid with early land cover maps, and the spectral data of the images were extracted according to different land cover classes of early land cover maps. For each land cover class, all extracted cell spectral data of that class were applied to equation 2 to calculate the values of the corresponding feature space. The pixels outside of the corresponding three-dimensional feature space were considered to be the changed areas, which is similar to assignment of thresholds for sub nodes in decision tree classification methods.

In equation 2, c_i is determined in this step, for the changed rates of different land cover classes are different. The pixels were calculated iteratively until the proportion of changed areas to the total area of the corresponding land cover class was within a reasonable range. The proportional area of each changed land use cover class was determined with the aid of expert knowledge of the census and other monitor value change in the study area, such as population, gross domestic product (GDP), crops yield and so on. The ellipsoid radius of the three-dimensional feature space of c_i increased with the iterative calculation until this requirement was met.

Classification of Changed Pixels

After obtaining the changed land cover pixels, the satellites images and three-dimensional feature space were used to classify them based on pattern recognition and changed rules.

The initial classification of the changed area of land cover was determined by calculating the minimum spectral distance based on the three-dimensional feature space. For each changed pixel, the spectral data of all bands were input to the formula for all land cover classes in three-dimensional feature space to calculate the minimum spectral distance (d_min):

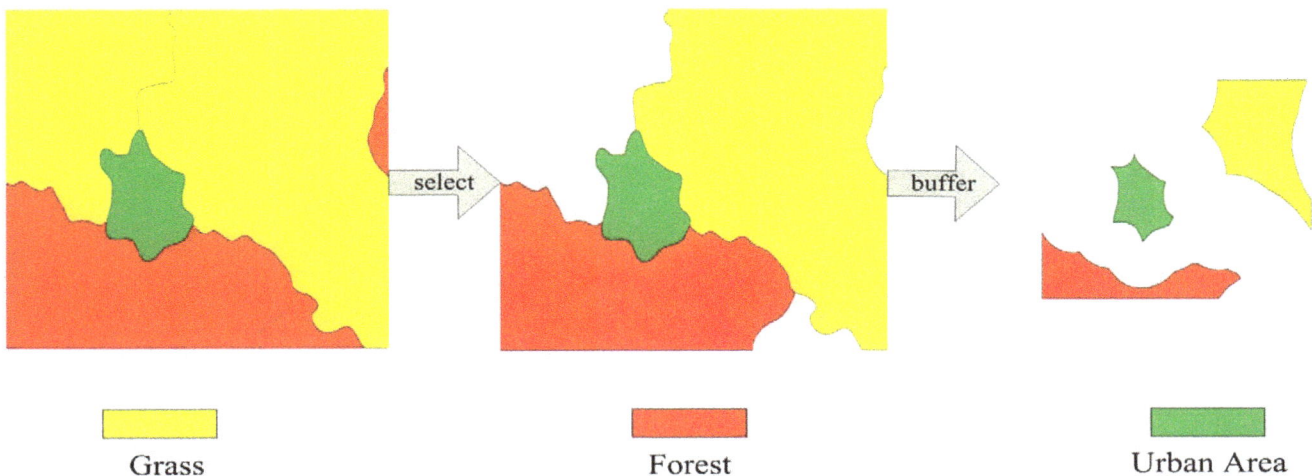

Grass   Forest   Urban Area

Figure 2. Sketch map of automatic collection of training samples.

$$\begin{cases} dmi = (\sum_{j=1}^{3} \dfrac{(Pjm - MPji)^2}{\sigma ji^2})/ci^2, \\ d\_min = \min(dmi) \end{cases} \qquad (4)$$

where dmi is the three-dimensional spectral Euclidean distance of the mth changed pixel to the ith class of land cover based on the feature space, Pjm is the jth principal component of the mth changed pixel in corresponding feature space of the ith land cover class, $MPji$ is the mean value of the jth principal component of the ith land cover class, and ci is the ellipsoidal radius of the three-dimensional feature space of the ith class of land cover. The initial classification of changed pixels was determined by the minimum value of the distance, dmi.

The changing rule of land cover classification was adopted to revise the initial classification result. Previous knowledge of the land cover change could be used in this phase. In several agricultural regions, farmland is protected by a national policy that aims to ensure at least 120,000 km^2 of farmland in China. In other regions with pronounced urbanization, urban areas are expanding rapidly and are not changed into other land cover classes, such as farmland or woodland. In this study, a drag coefficient of changed land cover (ρ) is introduced to express the rules of changed land cover. For all n types of land cover classes, ρ is an n-by-n matrix.

The final land cover classification of changed pixels were determined by combination of dmi and ρ. The minimum distance of the land cover classification based on changed rules is defined as:

$$dLmi = \rho ij * dmi \qquad (5)$$

where $dLmi$ is the minimum distance of the mth pixel to the ith land cover class based on the changed rules and ρij is the drag coefficient of the ith land cover class that changed to the jth land cover class.

Modification of Post-classification Results

Classified images often suffer from a lack of spatial coherency (spotting or holes in classified areas). Low-pass filtering could be used to smooth these images, but the class information would be contaminated by adjacent class codes. This problem is solved by modification of the post-classification results. The land cover classes are clumped together by first performing a dilation operation and later an erosion operation on the classified image [23,24]. In this study, any group of pixels, which was less than 3 pixels in size, was identified as noise in dilation operation. A 3-by-3-pixel moving window was used to eliminate noise and remove salt-and-pepper effects.

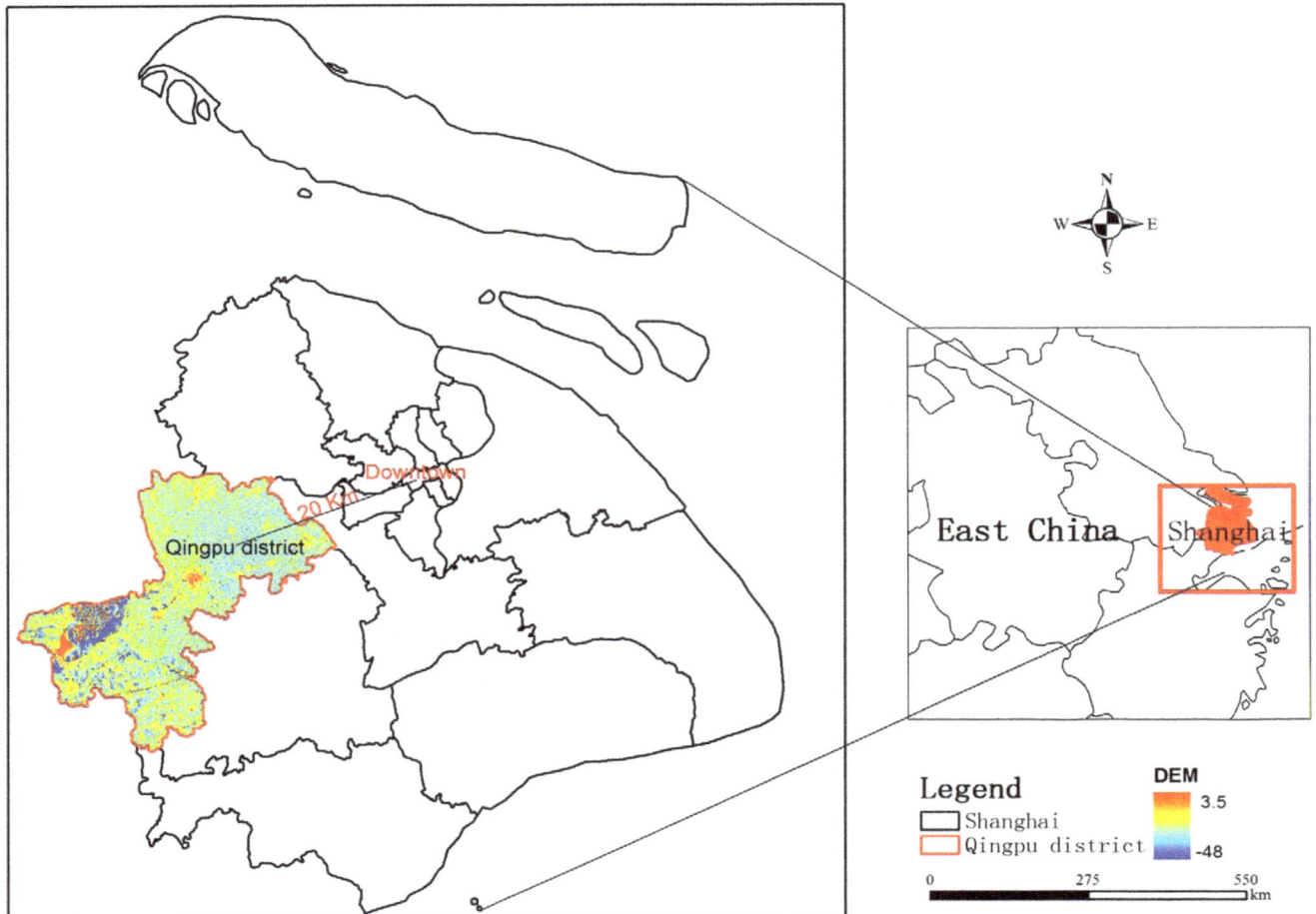

Figure 3. Location of the study area: Qingpu District, Shanghai, China.

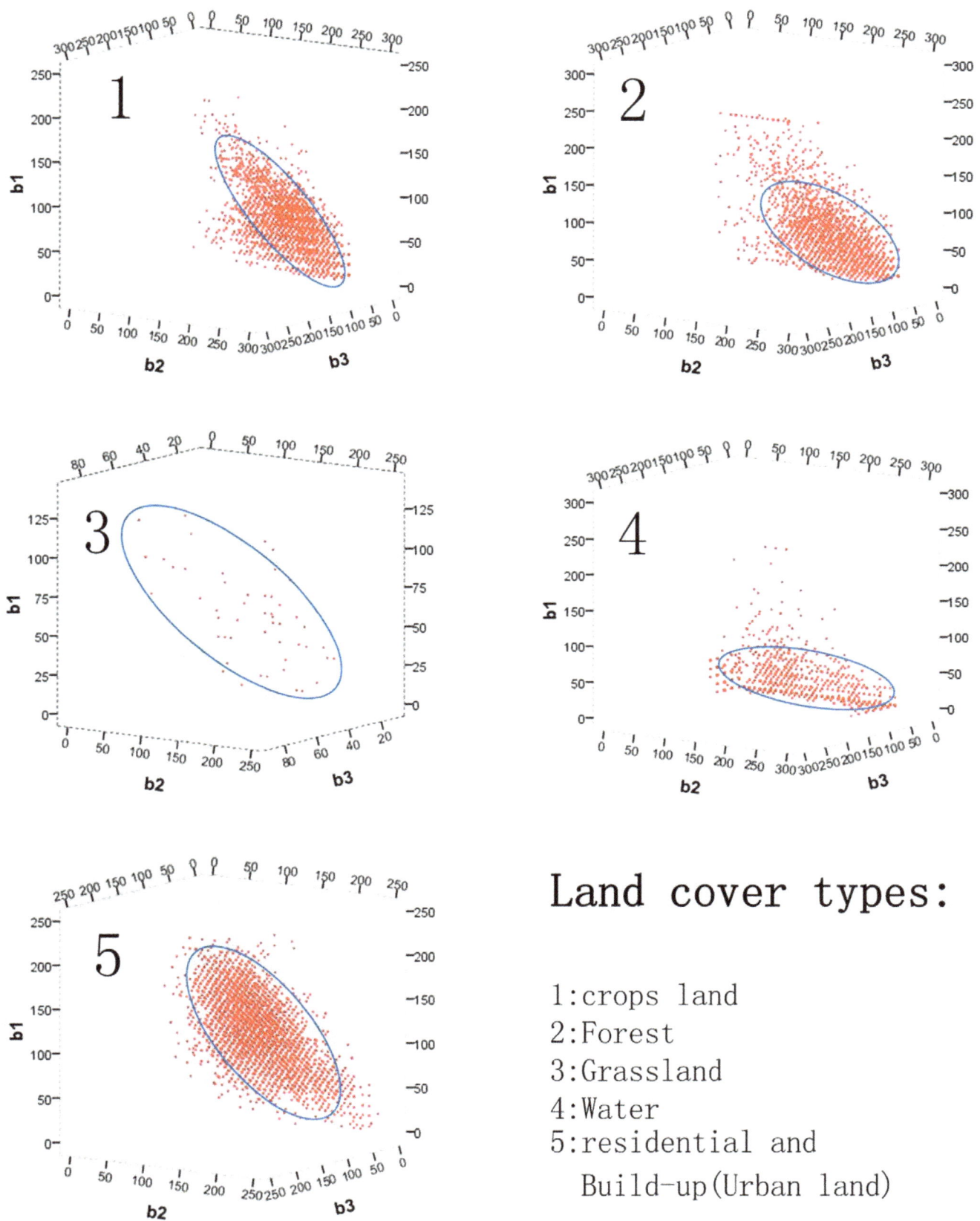

Land cover types:

1:crops land
2:Forest
3:Grassland
4:Water
5:residential and
  Build-up(Urban land)

Figure 4. Three-dimensional scatter plots and feature space of five land cover types.

Results and Discussion

Study Area and Data

Study area. The study area of Qingpu district of Shanghai, China covers 676 km^2 and encompasses over 10% of Shanghai city (Fig. 3). Approximately 60% of study area is dominated by annually double crops land, mainly growing rice and wheat. The rest portion of the study area is largely composed of two land covers which are residential and build-up land and water [25]. The study area located in Changjiang delta plain with elevation gradient ranging from 2.8 m to 3.5 m. The climate is subtropical monsoon with an average annual temperature of 16.8 degrees Celsius. Average annual precipitation for the study area is 112 cm and the average annual rainfall days is 137 [26]. Qingpu district is the western suburb of Shanghai and just 20 km away from

downtown. With the rapid urban sprawl of Shanghai, residential and build-land extended over 100 km^2 from 1990 to 2005, which caused the dominant land change of crops land and forest loss [27]. It is representative of Chinese land cover change pattern for the rapid economic development in the last few decades. For the purpose of paying equal attention to farmland protection and development maintenance, the region represents an ideal case study to evaluate semi-automatic land cover classification method applying in numbers of regions of China.

Data sources. Three types of data were used for evaluation, including the 1:100000 land cover maps from two different phases and multispectral images from Environment Satellites 1 (HJ-1). Land covers of 2005 and 2009 were visually interpreted and classified using Landsat TM images and field survey. The land cover maps were produced by the Chinese Academy of Sciences

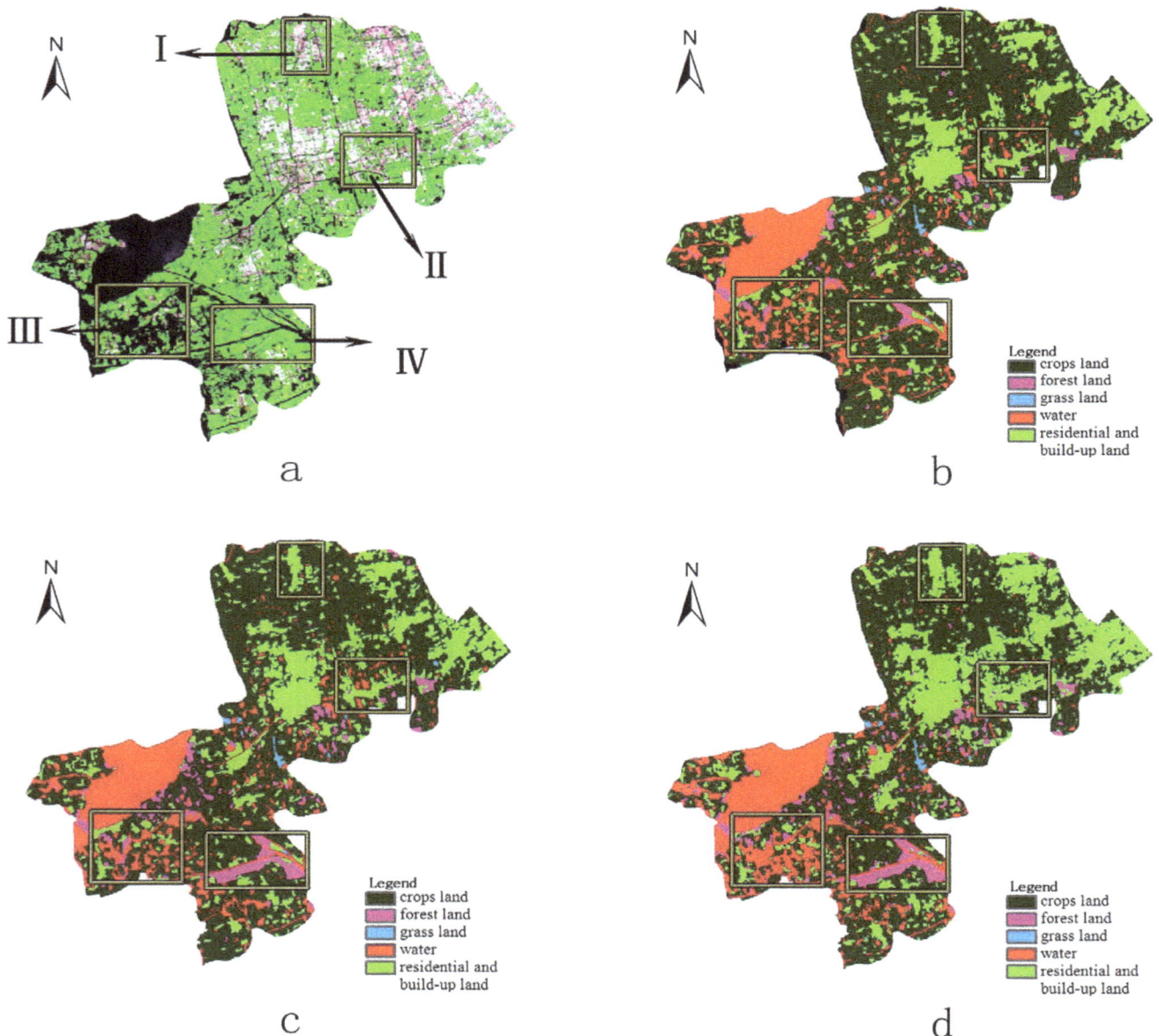

Figure 5. Comparison of land cover classification of Qingpu District. (a) HJ-1 image of 2009.9.22;(b) land cover of 2005;(c) land cover of 2009;(d) land cover classified by the proposed method.

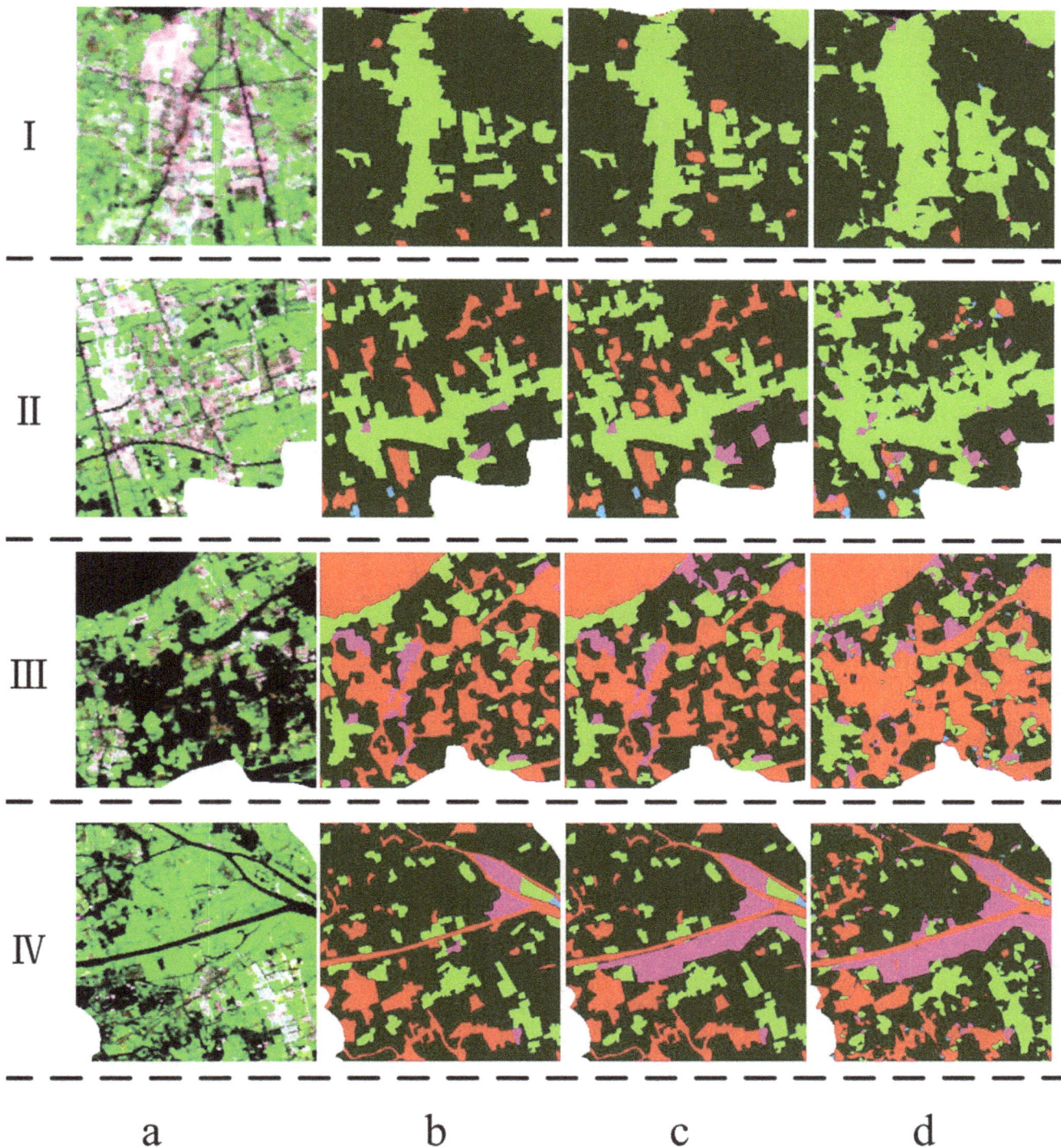

Figure 6. Comparison between the land cover types in the yellow rectangle in Fig. 5. (a) HJ-1 image of 2009.9.22;(b)land cover of 2005;(c)land cover of 2009;(d)land cover classified by the new method.

with consistent classification scheme, whose overall accuracy is 95% for all land use classes validated by intensive field surveys. The dataset includes five land cover types: farmland, grassland, forest, residential and construction land, and water. The map of land cover from 2005 was used as prior knowledge for hyperspace analysis and segmentation, while the land cover map of 2009 was used as a reference map for validation. The HJ-1 images of September 22, 2009 were geometrically and radiometrically corrected. Geometric correction was performed by use of 1:50,000 topographic map and 15 ground control points with a global Root Mean Square Error (RMSE) less than half of spatial resolution (15 m). The HJ-1 images consist of four spectral bands: three visible bands and a near infrared (NIR) band.

Table 1. Statistics of five land cover classes of the three classification results.

		Crops land	Grass land	Forest land	Water	residential and build-up land
visual interpretation land cover of 2005	area(km²)	425.7	2.1	14.3	110.3	117.9
	proportion(%)	63.5	0.3	2.1	16.5	17.6
visual interpretation land cover of 2009	area(km²)	391.2	2.1	26.7	108.6	141.0
	proportion(%)	58.4	0.3	4.0	16.2	21.1
Land cover of new method	area(km²)	360.9	3.0	29.9	113.8	163.0
	proportion(%)	53.8	0.4	4.5	17.0	24.3

Three-dimensional Feature Space Analysis

There are five land cover types in the study area: farmland, grassland, forest, water, and residential and construction land (urban space). The spectral Digital Number (DN) values of each type were retrieved with the method described in sections 2.1 and 2.2. Fig. 4 shows the three-dimensional scatter plots and feature space of five types of land cover of interest regions. For this purpose, land cover data derived by visual interpretation of 2005 and HJ-images of September 22, 2009 were adopted.

The results in Fig. 4 show that the spectral DN values of the five land cover types mostly cluster in three-dimensional ellipsoidal spaces. As the sampling regions were retrieved automatically, there are some outliers for each land cover type, which will affect the accuracy of the final classification. In the study area, most of the forest samples were outside of the ellipsoid (Fig. 4.2), accounting for up to 27.6% of the total samples. This may be because the change in forest cover did not follow the ecological rule, due to the high rate of urbanization in Shanghai, causing some large forest patches also to be less ecologically stable. Other land cover types present similar outliers but with fewer samples outside the ellipsoid, which may reduce the accuracy of the classification of forest land. Farmland is the dominant land cover type in the Qingpu District with sufficient samples for our classification method. As shown in Fig. 4.1, there are nearly 14.3% of samples are outside the automatically retrieved three-dimensional ellipsoid space. The area of grassland in the Qingpu District is quite small and spread among the forest and farmland; therefore, there are few grassland samples extracted from the images (Fig. 4.3), which also affects the final classification result. Water, such as rivers and lakes, is the third largest land cover type in the study area. Due to its low reflectivity, water areas are the easiest type to identify from remote sensing images by various other methods. As shown in Fig. 4.4, the DN values of water samples are quite low, and 91.7% pixels of samples in the regions of interest are inside the ellipsoid, ensuring the precision of its classification result. The second major land cover type of residential and construction area (urban land) of study area are of great three-dimensional spatial clustering (Fig. 4.5). Only 4.6% of sample pixels in the region of interest are outliers. In the Qingpu District, urban land is one of the most rapidly expanding land cover types, due to the rapid urbanization of Shanghai City. So within a period of time, the center of a large urban patch is unlikely to be transformed to a different land cover type, and the automatically retrieved samples in this area are well-represented spectrally.

Three-dimensional feature space analysis of the five types of land covers shows that the automatic retrieval of the spectral value of a region of interest can provide support to the successive land cover classification. Fig. 4 shows that samples of forest and grassland in the Qingpu District are less concentrated than the other three land cover types, which limits the accuracy of the final classification result of these two types of land covers. However, the method developed in this study is based on spectral pattern recognition, and the early visual interpretation of land cover data that is used as prior knowledge will offset the spectral bias, ensuring the accuracy of the final results.

Evaluation

Two types of land cover maps were compared to evaluate the performance of the final classification of the new method: (1) visual interpretation of land cover classes of 2009 is recognized as relatively exact data and (2) the automatically classified land cover of 2009 based on an HJ-1 image following our new method. Fig. 5 illustrates the data sources including the HJ-1 image (Fig. 5-a), and visual interpretation of land cover of 2005 (Fig. 5-b). From the HJ-1 image, the water (black pixels), vegetated land (green pixels) and urban land (other colors pixels) can be easily classified. This determination provides an intuitive basis for the accuracy assessment of classification. Fig. 5-c shows the results of 2009 from implementation of the visual interpretation classification, and Fig. 5-d shows the results using our new method. It should be noted that there are large numbers of pixels (depicted by yellow rectangles) where a significant change in land cover type is found by simply comparing Figs. 5-a and 5-b. Figs. 5-c and 5-d effectively reflect this change from 2005 to 2009.

To better express the performance of this new method in detail, Fig. 6 zooms in the images of the four yellow rectangles in Fig. 5. The results of visual interpretation and our method match the HJ-1 images well, which represents the land cover change during the time interval. In certain ways, Fig. 6-d matches Fig. 6-a better than Fig. 6-c. For Fig. 6-c is visual interpretation classified land cover data based on Landsat TM images of whole year of 2009. The borders of land cover patches in Fig. 6-c are sharper than those derived using the new method for the different implemented algorithms, which is consistent with the actual land cover of the Earth's surface. In this study, the 2009 visual interpretation data were considered to be the reference for assessing the accuracy of the results of the new method.

Assessment of Overall Spatial Accuracy

Fig. 5 and Fig. 6 show that the result of our method closely matches the HJ-1 image and visually interpreted land cover in intuitive view. To quantitatively assess the spatial accuracy of classification land cover from the simple method, Table 1 provides statistics of the five land cover classes of the Qingpu District.

The area of five land cover classes for 2009 derived from visual interpretation and our method were compared, and there was little bias between two sets of results. According to the visually interpreted data, the large areas of changed land cover in the Qingpu District from 2005 to 2009 are farmland and residential and construction land. The area proportion of farmland to whole region declined from 63.5% to 58.4%, whereas the proportion of residential and construction land grew from 17.6% to 21.1%. It was consistent with the rapid urbanization of Shanghai, which could be interpreted as the urban land overtaking farmland. The proportions of farmland and residential and construction land using our new method are 53.8% and 24.3%, respectively. Compared to visually interpreted results, the errors of the two land cover classifications are 5.4% and 3.2%. The other three land cover classes (grassland, forest and water) changed less from 2005 to 2009. The biases in proportion of the three land covers derived from our automatic classification method are 0.1, 0.5 and 0.8 compared to the visually interpreted land cover of 2009.

Pixel-by-pixel Analysis of the Agreement of Results

For the next phase of analysis, each pixel of the land cover classification maps using the new method was compared to its counterpart in the visually interpreted land cover map of 2009 on a pixel-by-pixel basis. Statistics on the accuracy of the findings are summarized by presenting Cohen's Kappa [28]. Although overall accuracy is often used as a standard indicator of map quality, many contend that Kappa provides a better overall measure because it also incorporates information on the errors of omission and commission [29,30].

Table 2 shows the land cover confusion matrix of results of our method and visual interpretation classification of 2009. The overall accuracy is 87.2%, which can be calculated from Table 2. According to analysis of Table 2, the misclassification rates of land cover classes of forest and grassland are 24.9% and 25.3%, respectively, which are higher than the misclassification rates of the other three types of land cover classes. The reason for this can be found in the automatic retrieval spectral values of the study region described in section 3.2. Three-dimensional scatter plots and the feature space of forest and grassland (Figs. 4.2 and 4.3) show that the two land cover classes are not as representative as the other three land cover types. However, forest and grassland occupied small proportions of the entire Qingpu District, which reduces the inaccurate effect on the map agreement. The other relatively map omission and commission error existed among crops land, water and residential and build-up land. In Table 2, some pixels of crops land were classified to water and residential and build-up land. They were mainly caused by the land cover classification system, especially crops land (including two sub-categories of paddy field and dry farming field) and residential and build-up(including three sub-categories of urban land, rural residential and other build-up land). Urban land includes landscape such as water for rest, may be classified to categories of residential and build-up in visual interpretation classification. It will caused the mis-classification between water and residential and build-up land. Rural residential land always scatters inside of large area of crops land in China, which will increase map omission and commission error of these two land covers. In addition, paddy field,normally flood irrigated in China, may present similar spectral signature to water, which will introduce mis-classification between crops land and water. Acknowledging these spectral limitations, the accuracy of classification result of Qingpu distribution is acceptable.

The Kappa value(K_{hat} = 0.79) of the two land cover classification maps of 2009 are calculated based on values in Table 2. Blackman and Landis [31,32] graded Kappa value between 0 and 1 for the analysis of map agreement, which has become the standard for the agreement of maps in practice. The Kappa value

Table 2. Confusion matrix of two classification algorithms of Qingpu District, 2009.

		New method					
		Crops land	Forest land	Grass land	Water	residential and build-up land	x_{+i}
visual interpretation classification	Crops land	381111	1648	1247	22685	24942	431633
	Forest land	2696	22121	14	2592	2024	29447
	Grass land	328	116	1751	73	77	2345
	Water	8912	4947	308	98505	7182	119854
	residential and build-up land	7883	4367	17	2484	141340	156091
x_{i+}		400930	33199	3337	126339	175565	739370

Kappa value (K_{hat} = 0.79).

for the results of our method and the validation visual interpretation classification of 2009 is 0.79, which means the two maps agree well (a Kappa value between 0.61 and 0.8 is considered to be in good agreement [31,32]). The proposed approach gives acceptable land cover classification in the Qingpu District. Furthermore, residential and construction land (urban land) has the best precision of the five land cover types, which indicates that this method is sensitive to changes in residential and construction land.

We also tested the accuracy of methodology for three different regions chosen based on their degrees of land cover change. These results show that the proposed method performs well in regions with normal-to-high rates of land cover change, especially droved by rapid urban development. While its accuracy is slightly lower in regions with little land cover change, such as natural conservation areas. There is mainly result from changed rules described in step 2.4. The land cover of a natural conservation area is relatively stable, and the small land cover change areas are natural affected without rules which will increase error of ultimate result of land cover classification. Some following work, such as study of the zoning in land cover change and a better-planned rules repository, may be useful in improving the performance of the method in the future.

Conclusion

Land cover data are important for research on global environmental change [32]. Remote sensing images are important data sources for land cover mapping [33]; however, the existing methods for land cover classification based on remote sensing images are not sufficiently flexible or effective [34,35]. Based on early land cover data used for change detection and the assumption that land cover changes are ecologically stable and in accordance with the changing rules, a simple but robust automatic land cover classification method was proposed. The satellite images can be automatically classified using only the prior land cover map and current images with less human interaction or interpretation. By applying the newly developed method to the Qingpu District of Shanghai and by comparing the results with a visually interpreted land cover map, it was found that the two land cover maps closely agree with each other, and the new method is appropriate for land cover classification. The new approach demonstrates the following three advantages over existing methods: (1) it automatically obtains training samples with Geography Information System (GIS) and statistical technology;

(2) it uses prior land cover maps as background knowledge to guarantee the accuracy of the final classification result; and (3) it revises the classification result based on the changed land cover rule in certain areas, which avoids errors originating from relying solely on the spectral pattern of remotely sensed images, and it reflects societal influence on land cover change.

The approach proposed in the paper is similar with state-of-the-art Support Vector Data Description (SVDD) method for they both can learn from contaminated training data (both inners and outliers) in the feature space. However, we provide different solutions to this problem. In our paper, the results are competitive for the feature space was established using the prior land cover knowledge and current images, while the SVDD using difference among images of different times.

Whereas, there are four points that need to be mentioned when implementing this method: (1) the classification results would be more accurate if the early land cover data were more precise; (2) a long time interval between the early land cover data and the classification images is not advisable, and results of our experiments suggest that a data gap of no more than five years is acceptable for China; (3) the changed land cover rule needs to be constructed based on the land cover policies of the study area and the expert's knowledge; and (4) images classification of area with irregular land changes, especially the area of changes mainly happens in the domain(e.g., commercially managed forests, type changed crops), will introduce challenges to the method. When training samples collected away from edges (where land cover is more likely ecologically stabile) are less pure or mixed pixels of specific land cover categories, error will be caused in final mapping.

Acknowledgments

We thank Dr. Jingying Fu and Dr. Yaxin Wang in Institute of Geographical Sciences and Natural Resources Research, Chinese Academy of Sciences for their help in writing this paper. The authors would also like to thank the editors and anonymous reviewers for their helpful remarks.

Author Contributions

Conceived and designed the experiments: DJ DFZ XLX. Performed the experiments: DJ YHH YQZ. Analyzed the data: DJ YHH DFZ YQZ. Contributed reagents/materials/analysis tools: YHH HYR. Wrote the paper: DJ YHH.

References

1. Liu J, Deng X (2010) Progress of the research methodologies on the temporal and spatial process of LUCC. Chinese Sci Bull 53: 1–9.
2. Dolman AJ, Verhagen A, Rovers CA (2003) Global environmental change and land use. Dordrecht: Kluwer Academic Publishers.
3. Friedle M, Brodley C (1997) Decision tree classification of land cover from remotely sensed data. Remote Sensing and Environment 61: 399–409.
4. Rogan J, Chen DM (2004) Remote sensing technology for mapping and monitoring land-cover and land-use change. Progress in Planning 61: 301–325.
5. Liu J, Liu M, Tian H, Zhuang D, Zhang Z, et al. (2005) Spatial and temporal patterns of China's cropland during 1990–2000: An analysis based on Landsat TM data. Remote Sensing of Environment 98: 442–456.
6. Loveland T, Merchant J, Ohlen D, Brown J (1991) Development of a land-cover characteristics database for the conterminous U.S. Photogrammetric Engineering and Remote Sensing 57: 1453–1463.
7. Zhao Y (2003) The Application Principle and Method of Remote Sensing. Beijing: Science Press.
8. Zhang B (2005) Application of Fuzzy Mathematics to Classification Processing of Remote Sensing Digital Images. Journal of Tianjin Normal University (Natural Science Edition) 25: 69–72.
9. Zhang Y, Feng X, Ruan R (2003) Application of Back-Propagation Neural Network Supported by GIS in the Classification of Remote Sensing Image. Journal of Nanjing University (Natural Sciences) 39: 806–813.

10. Schneider A, Friedl MA, Potere D (2010) Mapping global urban areas using MODIS 500-m data: New methods and datasets based on'urban ecoregions'. Remote Sensing of Environment 114: 1733–1746.
11. Hansen M, Dubayah R, Defries R (1996) Classification trees: an alternative to traditional land cover classifiers. International Journal of Remote Sensing 17: 1075–1081.
12. Friedl M, McIver D, Hodges J (2002) Global land cover mapping from MODIS: algorithms and early results. Remote Sensing of Environment 83: 287–302.
13. Gong P, Mahler S, Biging G, Newburn D (2003) Vineyard identification in an oak woodland landscape with airborne digital camera imagery. International Journal of Remote Sensing 24: 1303–1315.
14. Lambin EF, Geist H (2006) Land-use and land-cover change: local processes with global impacts. New York: Springer.
15. Fortier J, Rogan J, Woodcock C, Runfola DM (2011) Utilizing temporally invariant calibration sites to classify multiple dates of satellite imagery. Photogrammetric Engineering & Remote Sensing 77(2): 181–189.
16. Liu J, Liu M, Deng X, Zhuang D, Zhang Z, et al. (2002) The land use and land cover change database and its relative studies in China. Journal of Geographical Sciences 12: 275–282.
17. Liu J, Liu M, Zhuang D, Zhang Z, Deng X (2003) Study on spatial pattern of land-use change in China during 1995–2000. Science in China, Series D 46 :373–384.

18. Yan HM, Liu JY, Huang HQ, Tao B, Cao MK (2009) Assessing the consequence of land use change on agricultural productivity in China. Global and Planetary Change 67: 13–19.

19. Ran YH, Li X, Lu L (2010) Evaluation of four remote sensing based land cover products over China. International Journal of Remote Sensing 31: 391–401.

20. Eastment HT, Krzanowski WJ (1982) Cross-validatory choice of the number of components from a principal component analysis. Technometrics 24: 73–74.

21. Eklundh L, Singh A (1993) A comparative analysis of standardised and unstandardised principal components analysis in remote sensing. International Journal of Remote Sensing 14: 1359–1370.

22. Jia X, Richards JA (1999) Segmented Principal Components Transformation for Efficient Hyperspectral Remote-Sensing Image Display and Classification. IEEE Transactions on Geoscience and Remote Sensing 37: 538–542.

23. Bovolo F (2009) A multilevel Parcel-based approach to change detection in very high resolution multitemporal images. IEEE Geoscience and Remote Sensing Letters 6: 33–37.

24. Tax D, Duin R (2004) Support vector data description. Machine Learning 54: 45–66.

25. Shanghai Bureau of Statistics (2006) Compilation of Shanghai's Second Agricultural Census, 1st edn, China Statistics Press,Beijing.

26. Qingpu government website. Available: http://www.shqp.gov.cn/gb/special/node_9082.htm. Accessed 2012 Jul 18.

27. Guo Y (2007) Study on the Land Use and Landscape Patterns of Qingpu Area in Shanghai Base on GIS[D]. East China Normal University, Shanghai.

28. Cohen J (1960) A coefficient of agreement for nominal scales. Educational and Psychological Measurement 20: 37–46.

29. Allouche O, Tsoar A, Kadmon R (2006) Assessing the accuracy of species distribution models: Prevalence, kappa and the true skill statistic (TSS). Journal of Applied Ecology 43: 1223–1232.

30. Foody G (2007) Map comparison in GIS. Progress in Physical Geography 31: 439–445.

31. Blackman N, Koval J (2000) Interval estimation for Cohen's Kappa as a measure of agreement. Statistics in Medicine 19: 723–741.

32. Landis J, Koch G (1977) The measurement of observer agreement for categorical data. Biometrics 33: 159–174.

33. Foley J, DeFries R, Asner G (2005) Global consequences of land use. Science 309: 570–574.

34. Compton J, John T, Thomas E (1985) African Land-Cover Classification Using Satellite Data. Science 227: 369–375.

35. Chen S, Tong Q, Guo H (1998) Study of Information Mechanism of Remote Sensing, 1st edn., Science Press, Beijing.

36. Liu J (2005) Study on the Spatial-Temporal Dynamic Change of Land-use Change and Diving Forces Analyses in 1990s, 1st edn., Science Press, Beijing.

5

Digitise This! A Quick and Easy Remote Sensing Method to Monitor the Daily Extent of Dredge Plumes

Richard D. Evans[1,2]*, Kathy L. Murray[1], Stuart N. Field[1,2], James A. Y. Moore[1], George Shedrawi[1], Barton G. Huntley[1], Peter Fearns[3], Mark Broomhall[3], Lachlan I. W. McKinna[3], Daniel Marrable[3]

1 Department of Environment and Conservation, Kensington, Western Australia, Australia, 2 Oceans Institute, University of Western Australia, Crawley, Western Australia, Australia, 3 Remote Sensing and Satellite Research Group, Department of Imaging and Applied Physics, Curtin University, Bentley, Western Australia, Australia

Abstract

Technological advancements in remote sensing and GIS have improved natural resource managers' abilities to monitor large-scale disturbances. In a time where many processes are heading towards automation, this study has regressed to simple techniques to bridge a gap found in the advancement of technology. The near-daily monitoring of dredge plume extent is common practice using Moderate Resolution Imaging Spectroradiometer (MODIS) imagery and associated algorithms to predict the total suspended solids (TSS) concentration in the surface waters originating from floods and dredge plumes. Unfortunately, these methods cannot determine the difference between dredge plume and benthic features in shallow, clear water. This case study at Barrow Island, Western Australia, uses hand digitising to demonstrate the ability of human interpretation to determine this difference with a level of confidence and compares the method to contemporary TSS methods. Hand digitising was quick, cheap and required very little training of staff to complete. Results of ANOSIM R statistics show remote sensing derived TSS provided similar spatial results if they were thresholded to at least 3 mg L^{-1}. However, remote sensing derived TSS consistently provided false-positive readings of shallow benthic features as Plume with a threshold up to TSS of 6 mg L^{-1}, and began providing false-negatives (excluding actual plume) at a threshold as low as 4 mg L^{-1}. Semi-automated processes that estimate plume concentration and distinguish between plumes and shallow benthic features without the arbitrary nature of human interpretation would be preferred as a plume monitoring method. However, at this stage, the hand digitising method is very useful and is more accurate at determining plume boundaries over shallow benthic features and is accessible to all levels of management with basic training.

Editor: Guy J-P. Schumann, NASA Jet Propulsion Laboratory, United States of America

Funding: This project was funded as part of the Dredging Audit and Surveillance Program by the Gorgon Joint Venture as part of the environmental offsets. The Gorgon project is a joint venture of the Australian subsidiaries of Chevron, Exxonmobil, Shell, Osaka Gas, Tokyo Gas and Chubu Electric Power. The funders had no role in study design, data collection and analysis, decision to publish, or preparation of the manuscript.

Competing Interests: The authors have declared that no competing interests exist.

* E-mail: richard.evans@dec.wa.gov.au

Introduction

Industrialization, growing populations, and expansion of mining throughout the world, are impacting on natural coastal environments. Australia's coastal environments are under increased pressure from development at present with enormous investment in the mining industry and the subsequent construction of harbours to service these ventures throughout Australia. Construction of pipelines, marinas and ports within these coastal environments often requires land reclamation and dredging of the sea floor with subsequent spoil disposal. The suspension of sediments due to these activities potentially has impacts on surrounding ecosystems, such as physical smothering of benthic communities, and decreased light availability which limits growth and metabolic processes of sea grass [1], algae [2], coral [3,4] [5,6] and larval fish [7]. Scientists and management agencies are under growing pressure to be able to understand the spatial extent and potential influences, particularly cumulative, of these activities within short time frames (daily) to enable appropriate response measures.

Projects that monitor environmental change are restricted by the availability of consecutive images (repeat capture schedule) over the required monitoring time period, and the spatial resolution available that is required to monitor the area of interest. Projects utilising high resolution imagery may be further limited by the cost and the time required to interpret such imagery with greater detail. Advances in technology and product availability, over the past two decades, has enabled satellite remote sensing products to become an accepted method for monitoring the spatial extent of terrestrial and aquatic impacts [8–12]. Imagery from satellite sensors such as Landsat Thematic Mapper (TM) and Enhanced Thematic Mapper+(ETM+) can be downloaded gratis and have a pixel resolution of 30×30 m. Landsat satellites have a repeat capture of every 16 days (see http://glovis.usgs.gov/), limiting the power to detect rapid temporal change. Other higher spatial resolution sensors such as Advanced Land Observing System's Advanced Visible and Near Infrared Radiometer type 2 sensor (ALOS AVNIR-2), Quick Bird 2 (QB2) or World View 2 (WV2) (10×10 m, 2.5×2.5 m, and 2×2 m pixel resolution respectively) vary in regular repeat capture capability, ranging from four times per year to every couple of days, however the use of these sensors is cost prohibitive. When designing a monitoring project using remote sensing, an adaptive approach is required.

Imagery captured by the Moderate Resolution Imaging Spectroradiometer (MODIS) sensor is free to access and is well suited for monitoring daily events that occur on a wide spatial scale. The MODIS sensors, on board the Terra (EOS AM-1) and Aqua (EOS PM-1) satellites, achieve near-daily global coverage, capturing data across 36 spectral bands. MODIS data may be used in a number of ways ranging from observing atmospheric conditions to terrestrial and oceanic processes with pixel resolutions ranging from 1×1 km to 250×250 m pixels at nadir [13]. NASA provides MODIS imagery freely in raw format or as processed mosaiced georeferenced true colour enhanced imagery, making it very accessible for product generation or visual interpretation. Other studies have indicated MODIS band 1 (620–670 nm, centred on 645 nm) is a useful proxy for visualising elevated TSS loads in Australian marine waters [14,15]. MODIS band 1 reflectance can be used to calculate L2 TSS products from regionally tuned empirical models to provide quantitative estimates of sediment load [16–18]. Similar approaches have been used for dredge plume monitoring in Australian waters by Islam et al. (2007) [19] where a TSS algorithm was developed for MODIS data during 2006 dredge operations at Hay Point, Queensland. However there are difficulties in applying Level 2 products in turbid and shallow waters. This can result very limited data quality or limited quantities of data where L2 processing fails. Hand digitising of MODIS imagery has also recently been used as a management tool for monitoring the spatial extent and distribution of sediment laden riverine flood plumes within the Great Barrier Reef World Heritage Area, Australia [15,20].

Visual interpretation and hand digitization of either high resolution aerial photography or high, medium, or low resolution satellite imagery is a common method used for assessing environmental impacts in near-real time, as well as accessing historical imagery from archives to compare to targeted captures in the present. The hand digitising approach is a simple standard method adopted by some government agencies for monitoring, management and compliance in terrestrial environments: For example, in emergency fire digitising [21–23], deforestation [24], and mangrove growth and reduction [25], urban development [26], compliance prosecution cases for illegal land clearing [27] and over-flooding on Alaskan ice [28]. The visual interpretation of daily MODIS satellite imagery with manual digitization was considered to be a quick, relatively accurate and easy method to complement in-situ biological monitoring of coral reef communities and physical monitoring (sediment traps) of a dredge plume.

The objective of the digitisation monitoring method was to gain an understanding of the daily spatial extent and the temporal frequency dynamics of the plume (the sum of the plume's daily presence). Analysis of the spatial/temporal frequency of the plume has the potential to highlight areas most affected, and create a better understanding of the influence this environmental change has on monitored sites of high biological significance. The results of the digitization method were compared to recognized semi-automated methods, using calibrated MODIS imagery to produce a total suspended solids (TSS) product. This paper highlights the strengths and weaknesses of each technique.

Methods

Location and Background of Study Area

The Montebello and Barrow Islands are situated in the Pilbara Offshore marine bioregions [29], approximately 1,600 km north of Perth, Western Australia, 120 km WNW of Dampier and 80 km NW of Cape Preston [30]. The waters and reefs surrounding these remote islands are characterised by geomor-

phological and oceanographic conditions which provide a high diversity of mainly tropical fauna including both widely distributed and endemic species [30,31]. The marine environment is generally considered to be in a relatively undisturbed condition as a result of low human usage and strict management controls on industry activities in the area [31]. The Montebello/Barrow Islands Marine Protected Areas (MBIMPAs), incorporating the Montebello Marine Park, Barrow Island Marine Park and the Barrow Island Marine Management Area, were gazetted in 2004 and the conservation and management objectives for these reserves are expressed in the management plan for the MBIMPA [31] (Fig. 1). The Department of Environment and Conservation (DEC) monitors coral, fish, macroalgae and macro invertebrates at selected sites of significance throughout the MPAs (Fig. 2). The Gorgon Project (GP), which is based on Barrow Island (20.80°S, 115.40°E), is one of the world's largest natural gas projects and the largest single resource natural gas project in Australia's history. The GP included a dredging program that involved the removal and dumping of ~ 7.6 million m^3 of marine sediment over a period of approximately 18 months and started in May 2010 [32]. Modelling of the likely plume dispersal and likely impacts completed by Chevron Australia Pty Ltd prior to the commencement of dredging operations predicted the dispersal both north and south from the dredging activities dependent on the prevalent winds [32].

Satellite Sensors

Near-daily MODIS true colour mosaic imagery (available free on the internet) provided a simple and rapid way to observe the dredge plume extent adjacent to Barrow Island. In addition, two captures of ALOS AVNIR-2 before (18 November 2006 and 23 November 2008) and one during (29 August 2010) the dredging project assisted the interpretation of the plume extent with the MODIS imagery. When available, cloud free Landsat 5 TM and 7 ETM+ images were also acquired and visually enhanced specifically for the water around Barrow Island, again to assist the interpretation of MODIS images.

Once acquired, MODIS imagery were reprojected into GDA94 MGA zone 50 and displayed in a spatial viewer with limited functionality maintained by the DEC. The dredging plume boundary was interpreted by manually digitising (drawing) a vector (digital polygon) around the plume at a scale of 1:450000. This was found to be the optimal scale for digitising the area with 250 m by 250 m pixels resolution of MODIS. A new polygon was created for each visible plume, and then attributed with the date (Julian day) of the MODIS image being interpreted and comments. A strict file structure and naming convention was employed to aid the quick generation of "clean" datasets that were easy to display when required for quality assurance. Some MODIS images had no plumes and others had more than one plume. A year of MODIS imagery pre-dredging was also acquired and interpreted for naturally occurring plumes.

Interpretation

The visual interpretation of daily MODIS imagery required a set of guidelines to be developed to ensure a conservative interpretation of plume boundaries. Potential guidelines include: a) comparison of imagery captured on plume-free days under similar tidal and meteorological conditions; b) only digitising plume areas that the observer had complete confidence i.e. not reef or bottom features.

Officers with little or no Geographical Information Systems (GIS) experience were trained in basic GIS skills to conduct the visual interpretation, digitising and editing of the daily plume

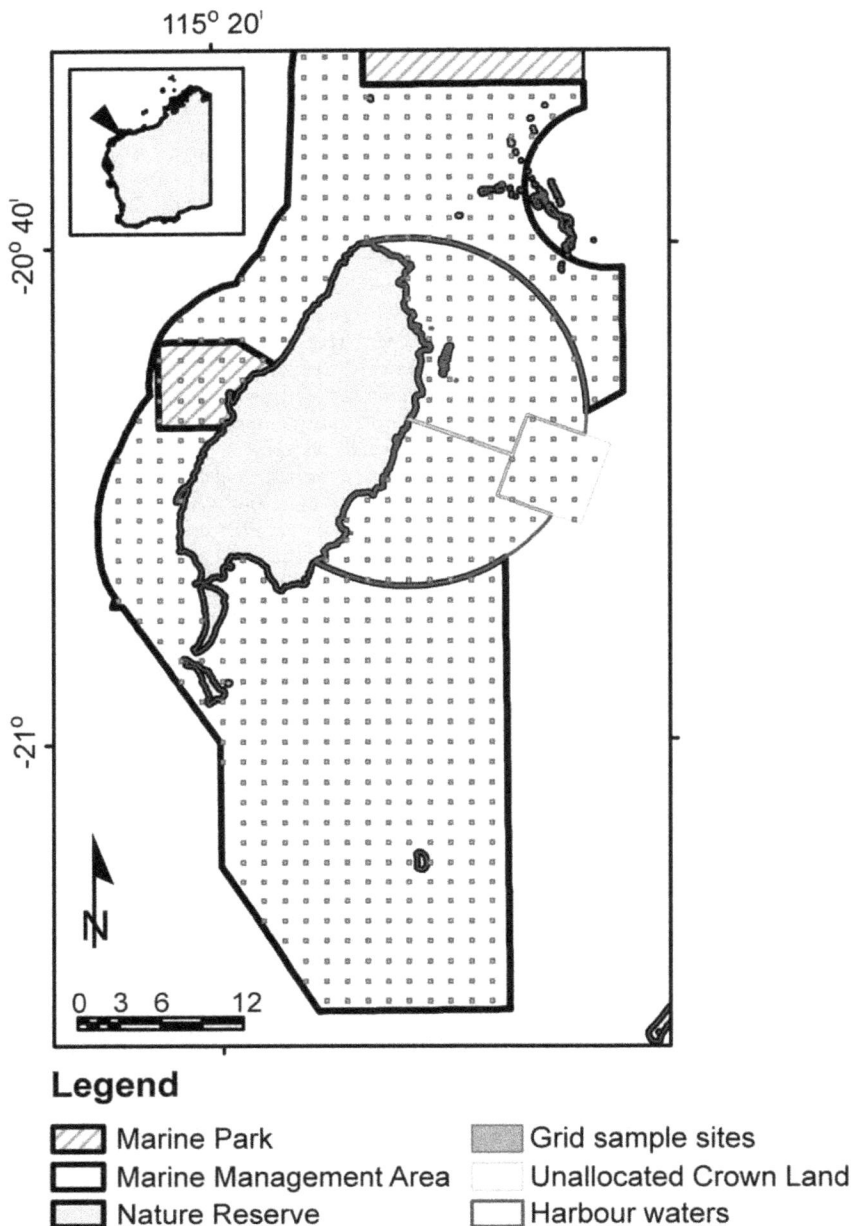

Figure 1. Map of the marine management areas surrounding Barrow Island and the sampling area for the grid point comparison.

boundary. To assist with training and quality control, the month (31 days) of October 2010 was digitised by three officers independently. The resulting boundaries were compared before the whole dataset was interpreted by one of the three observers applying strict rules developed during the one month trial.

Plume Digitising Rules

- Three potential sources of plume were identified – Marine Offloading Facility (both extractive and land reclamation activities), turning circle (extractive activities) and the spoil ground (deposition)
- If not joined, each plume was digitised separately for each location

- Plume sources were combined in one digitised vector if the plume connects them
- All visible plumes, including both primary and subsequent plumes, were included when the observer was confident it was not substrate,
- When in doubt, areas were excluded from the polygon. Multiple images (MODIS, Landsat and ALOS) of the location without plume were used to identify natural substrate features.

Hotspot Analysis

The definition of hotspot analysis in a GIS context was defined as 'The highest frequency occurrence of plume coverage at a geographic position within the time period assessed [34]. A

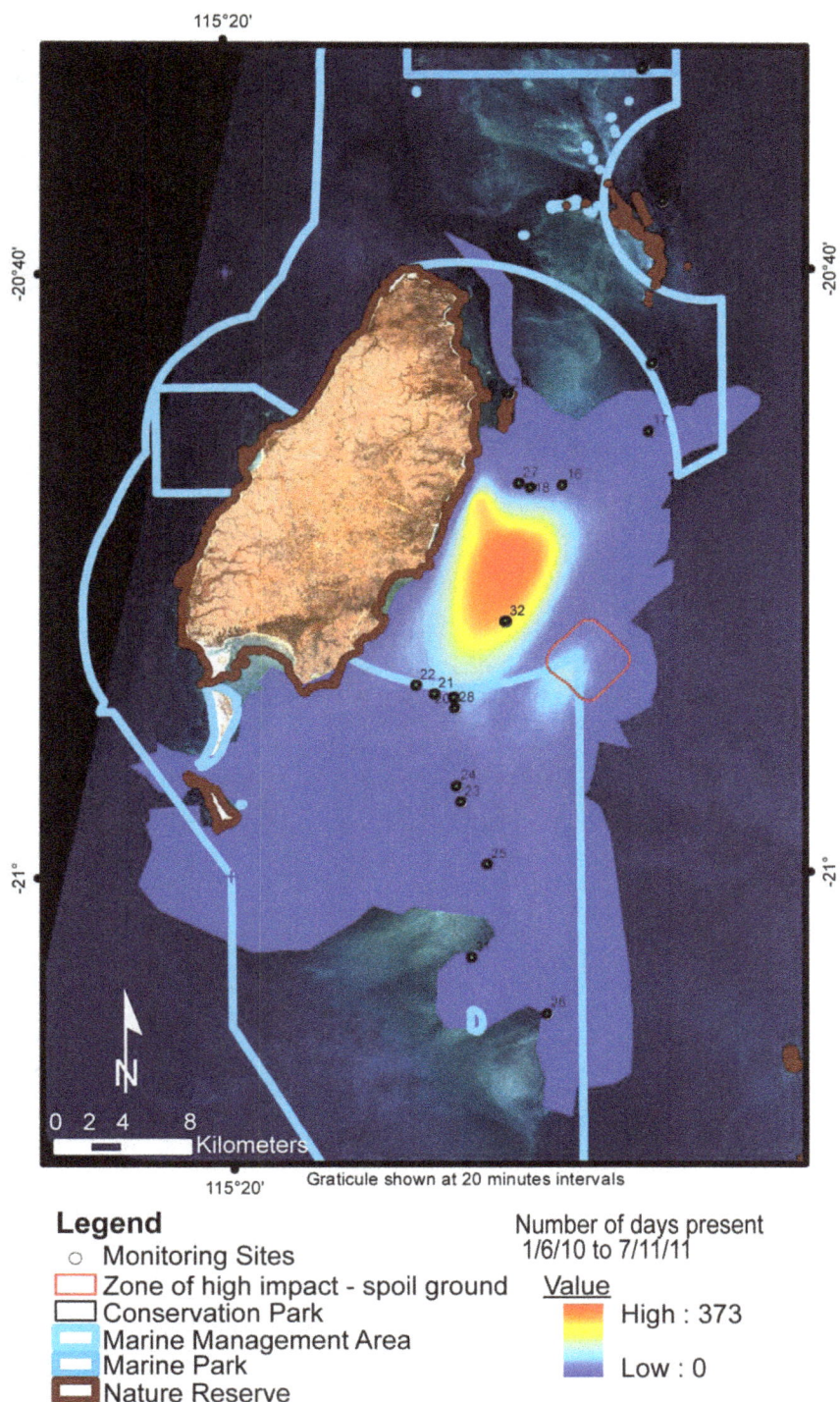

Figure 2. Hotspot analysis of hand digitised vectors representing the presence of dredge plume originating from the Barrow Island dredge site. Colours in the centre of the dispersal kernel represent higher numbers of days covered, decreasing to the light purple colour with low coverage. Points represent DEC monitoring sites.

hotspot analysis was run on the cumulative daily digitised plume boundaries to provide a dataset describing the number of days the plume was present at any position within the Barrow Island Marine Park and surrounds. This involved appending the datasets in ArcGIS [35] and determining the frequency of plume presence (days) in IDRISI [36] for the entire period of the dredge operations (525 days). DEC monitoring sites were buffered by 100 m and intersected with the resulting frequency dataset for the entire dredge period to extract the frequency of days each site was under the influence of the plume. This method was also used to determine the amount of time the plume covered the Montebello and Barrow Islands Marine Protected Areas. This was calculated by determining the area of plume that overlapped the boundary of the Barrow Island Marine Management Area to the north and south of the Barrow Island Port boundary.

Total Suspended Solids

Seawater samples were collected for 13 sites, each in clean 1 L bottles. The GPS coordinates of each sample site were also recorded. On return to shore, TSS samples were filtered under low vacuum onto dry, pre-weighed Whatman GF/F filters (47 mm diameter) with a nominal pore size of 0.7 μm. Filters were dried for 24 hours and subsequently weighed to determine the TSS load in units of mg L^{-1}.

Above-water Radiometry

Above-water hyperspectral radiometric data were collected using a DALEC three-channel radiometer developed at Curtin University and available from In-situ Marine Optics (http://insitumarineoptics.com/). The DALEC instrument measures reflectance properties in a similar fashion to satellite-borne sensors. However, the DALEC is under-atmosphere and thus, atmospheric corrections do not have to be considered when processing the data. The DALEC sensor collects radiometric data over the spectral range 400–900 nm with a nominal resolution of 3 nm. The DALEC continually logs radiometric data and GPS coordinates so that locations of interest can be identified and processed. DALEC hyperspectral data were convolved with MODIS spectral response functions to give under-atmosphere remote sensing reflectance, R_{rs}, data with the same spectral resolution and band response as the MODIS sensor.

TSS Model Development

Using the measured TSS from the filtered water and the coincident DALEC R_{rs} data, an empirical algorithm was developed which was then applied to MODIS imagery. A non-linear least squares fitting algorithm was used to develop a model which relates TSS concentration to at-surface DALEC-synthesised MODIS band 1 reflectances (See Figure S1). The model fitted through the observed data is shown in Figure 3 (r-squared = 0.959).

MODIS TSS Algorithm

On the 9th August, 2011 a field campaign (according to DEC permit numbers SC001263 & SW013765) was undertaken at Barrow Island during which a total of 13 sites were sampled. For each site, at-surface radiometric data were collected with co-incident TSS data. These in-situ data were used to develop a regional algorithm for determining TSS concentration from MODIS data for the Barrow Island dredge plume using the processes describe herein. MODIS data were accessed from the West Australian Satellite Technology and Applications Consortium (WASTAC) archive (http://www.wastac.wa.gov.au/) over a geographic domain containing Barrow Island. MODIS data were processed using the SeaWiFS Data Analysis System (SeaDAS) [37] to produce atmospherically corrected band 1 (645 nm) remote sensing reflectance. The algorithm in equation (1.) with the regionally fitted coefficients was applied to the MODIS R_{rs} data on a pixel-by-pixel basis to derive the L2 MODIS TSS product. During processing, nuisance/bad pixels containing land and/or clouds were masked out. No attempt to account for shallow water reflectance was incorporated into the processing. These data were then georeferenced and regridded before the region surrounding Barrow Island was extracted for this comparison.

Comparison of Total Suspended Sediment Data with the Results of the Digitised Method

TSS images were only compared with hand digitised MODIS true colour images if they had no distortion due to low view angle, were cloud free and provided full images covering the area of interest. The comparison used a grid sampling method of 623 sites covering the Barrow Island Marine Management Area and the Barrow Port Area (Fig. 1). Both datasets were intersected with the grid points to return a binary value for each point to represent presence (1) or absence (0) of the plume in the digitised and TSS dataset. The TSS data were then converted to binary data for 6 thresholds of TSS, >1 mg L^{-1}, >2 mg L^{-1}, >3 mg L^{-1}, >4 mg L^{-1}, >5 mg L^{-1}, >6 mg L^{-1} so both the digitised data and the TSS data samples could be directly compared as presence/absence data. Comparisons of digitised data to the range of TSS values were analysed using a one-way analysis of similarities (ANOSIM) in PRIMER [38] to determine the R statistic and any

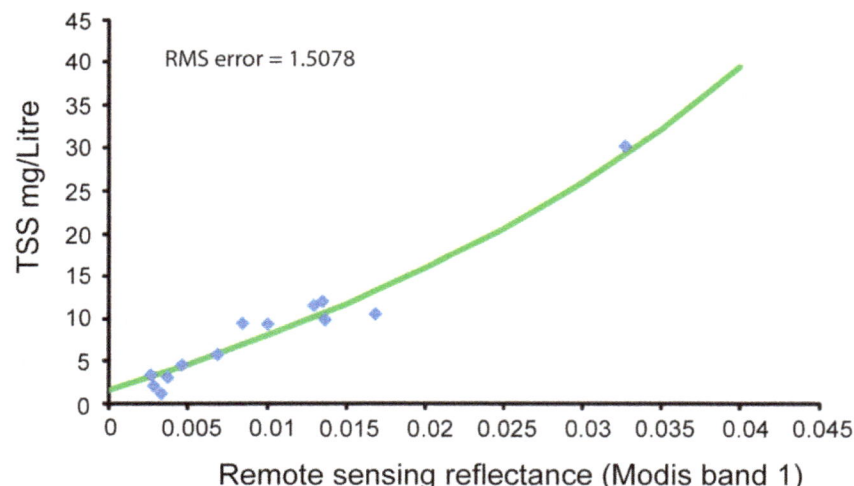

Figure 3. Plot of observed TSS relative to DALEC-synthesised band 1 reflectance (blue diamonds) from waters adjacent to Barrow Island. The TSS model is shown fitted through the observed data (green line).

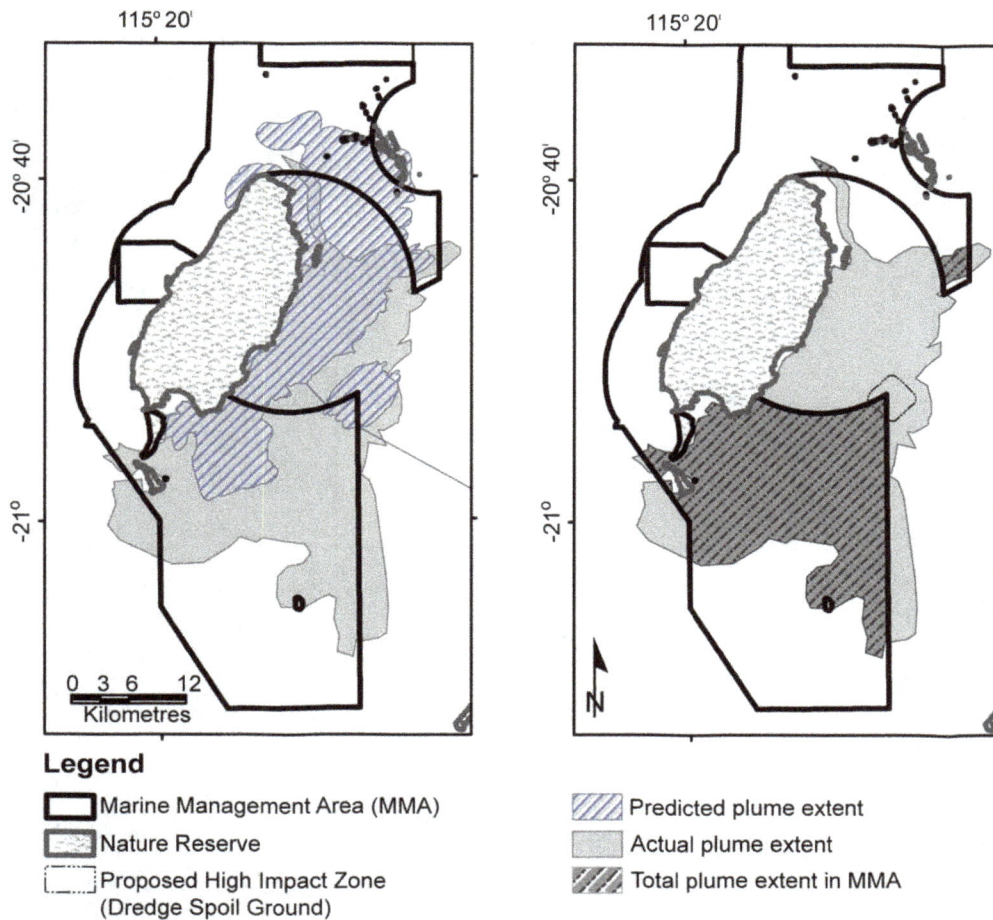

Figure 4. Map showing Montebello and Barrow Island Protected Areas, and a) the extent of the modelled plume and the actual dredge plume extent as defined by hand digitising MODIS images; and b) The overlap of the dredge plume into the Barrow Island Marine management area, over the duration of dredge period as defined by hand digitising MODIS images.

significant differences between the data sets. The R statistic is a useful tool to compare the degree of separation of two data sets and is as important (if not more so) as the statistical significance [39]. The R statistic in ANOSIM provides a result between 1 and −1, results closer to zero suggest more similarity between two data sets [38,39]. The data sets were also represented in a non-metric Multidimensional Scaling (MDS) ordination plot to visualize the differences between the digitising and each of the TSS data sets. The non-metric MDS is a visual representation of the similarity of more than one data set based on the chosen factors. There is no metric to the axes and hence no labeling. In this study, the factors are the presence/absence of plume recorded using hand digitizing and several levels of TSS measurement. Points become closer the more similar they are. The more dispersed the points become the more dissimilar are the shapes of the plumes over the grid points in Figure 1 (See [39] for more detail).

Results

Image Interpretation

Three observers digitised one month of daily true colour MODIS imagery captured during the dredging operation. Inter-observer variability was not significantly different (p = 0.647) and resulted in an average of 53 km^2 area of plume digitised with an average standard deviation (SD) of 10 km^2 (18%), a minimum SD of 0.4 km^2 (0.7%) and a maximum SD of 36 km^2 (68%). The main difference between interpreters was the level of conservatism surrounding what was regarded as either concentrated or marginal plume (areas where the digitiser was unsure of the plumes presence). Observers attempted to be conservative in their interpretation, but some were more so than others. For example, often conservative observers digitised the concentrated plume and not the areas of marginal plume (areas that were questionably plume or reef). A set of rules were generated based on the inter-observer comparison to improve consistency and confidence of the sole observer who digitised the entire dredge period (approximately 18 months).

Plume Monitoring

The plume was monitored using satellite MODIS imagery from 19 May 2010 to 7 November 2011 for the entirety of the dredging operations resulting in 538 possible days where the plume could be observed and digitised. It should be noted that no plume was observed until 1 June 2010. Both MODIS Aqua and Terra satellite imagery were employed for interpretation, with the choice of each dependent on image quality. Vectors were not digitised for 24% (127 days) of the total dredge period due to cloud cover and poor satellite image coverage (i.e. the sensor did not capture the

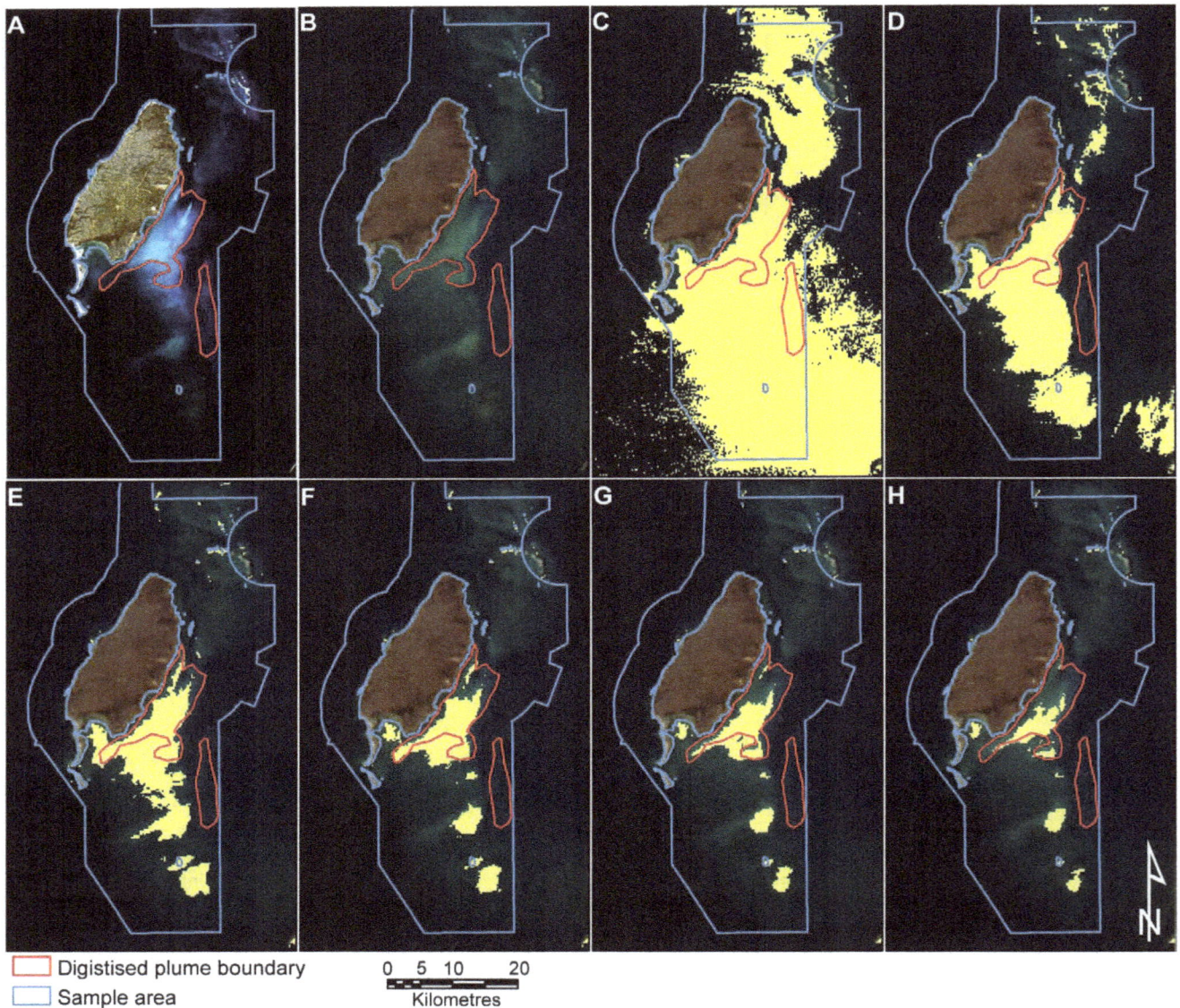

Digistised plume boundary
Sample area

0   5 10   20
Kilometres

Figure 5. Series of images showing how the digitised plume of a MODIS image relates to a clearer high resolution image and the range of TSS thresholds analysed in this study. A) Landsat image B); MODIS digitised interpretation; C) TSS threshold at >1 mg L^{-1}; D) TSS threshold at >2 mg L^{-1}; E) TSS thresholded at >3 mg L^{-1}; F) TSS threshold at >4 mg L^{-1}; G) TSS threshold at >5 mg L^{-1}; H) TSS threshold at >6 mg L^{-1}. Red line is the outline of the hand digitised plume from the MODIS imagery. Purple Line is the Barrow Island Marine Management Area and includes the Barrow Port area.

Barrow Island region of interest). The number of days individual DEC in-situ monitoring sites were covered by the plume ranged from <1% to 70% of the days digitised (Fig. 2).

Plume dispersal modelling completed by Chevron Australia Pty Ltd, prior to the commencement of dredging operations predicted coverage of 465 km^2 moving both north and south from the dredging activities dependent on the prevalent winds [33]. Our study found that the plume moved predominantly southward with minimal days of northward movement. Plume digitising showed that 279 km^2 overlapped with the modelling and 455 km^2 did not (Fig. 4A). A hotspot analysis was undertaken to highlight how often sites of high biological significance were covered by the plume. Individual DEC coral reef monitoring sites were covered from 1 to 296 of the 411 days observed (Fig. 2). During the period of the dredge operation, the cumulative plume coverage over the marine

management area (MMA) reached 395 km^2, and the total MMA within the expected high impact area of the spoil ground was 2.3 km^2 (see Fig. 4B).

Hand Digitising vs Remote Sensing TSS Comparison

Hand digitising (MODIS ONLY) enabled interpretation of 304 days of useful imagery compared to 116 days derived from processing MODIS data to a TSS product that was suitable to interpret clearly with full coverage, free of cloud and distortion due to low viewing angles. The analysis to compare the results of both methods was only carried out for days that both methods were able to interpret the MODIS data. The approach adopted was to compare the spatial extent of the plume derived by each method. The delineation of plume extent based on the remote sensing TSS

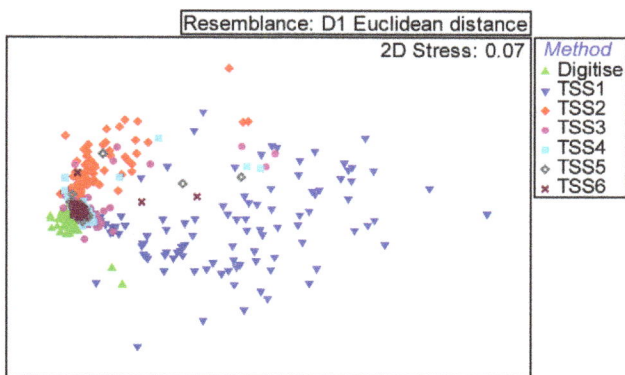

Figure 6. Non-metric Multidimensional scaling plot showing the similarity of points representing the spatial distribution of the dredge plume per day for hand digitised plume vectors and six levels of TSS measured from MODIS imagery.

approach depends upon selection of the TSS threshold. For this work we compared images based on hand digitizing and remote sensing TSS with TSS thresholds of 1, 2, 3, 4, 5 and 6 mg L^{-1} (See Fig. 5). Each point in the MDS plot represents the spatial distribution of grid points covered by the plume on a day (Fig. 6). The MDS plot represents the similarity of plume presence detected by the methods for each day. The very low stress (0.07) indicates that the data were well represented in 2 dimensions. The MDS illustrates a cluster of points in the left of the panel which represents the days when each of the methods had a similar result. This was likely to represent the primary main dredge plume area. However, consistent differences were found as there were no complete overlaps between the digitising of the plume and any of the different TSS concentration threshold extents. The widespread TSS points throughout the rest of the panel suggest highly variable plume spatial extents at different TSS thresholds. The high correlation (by visual inspection) of the location of TSS features and the locations of shallow bathymetric features suggests that the high variability in plume extent is probably due to days where the shallow benthic features were incorrectly interpreted as plume. This is particularly the case on strong wind days and after large rain periods with increased river outflow and natural turbidity moving from inshore areas to the Southern Barrow Shoals. It is important to note that image quality plays a big part in the amount of shallow water benthic habitat observed using TSS. Image quality can be affected by satellite angle, sun angle, tides and wind energy.

The results of the ANOSIM show that TSS 3 mg L^{-1} is most similar to the hand digitised method (Table 1). However, the differences between TSS 2 to 4 mg L^{-1} are so slight that either of these results may be equally representative depending on the day analysed. For example, TSS 4 mg L^{-1} is probably the most similar based on the visual representation of Julian day 205 (24 July) 2010

(Fig. 5). There is, however, no exact visual matchup, which arises from the lack of any TSS reading of the plume from the spoil ground (except TSS1) and the inclusion of benthic habitats (particularly areas of sand) in shallow areas of the Southern Barrow Shoals at all TSS levels. TSS1 consistently overestimated the dredge plume and TSS 5 and 6 underestimated the plume near the dredge location. This highlights the issues of determining spatial extent of dredge plumes in clear water shallow environments with TSS.

Discussion

Shallow water benthic communities suffer from both loss of light and smothering from sediments derived from dredging activities. Quantifying the extent to which an area is impacted by a dredge plume is critical to attributing changes detected to the benthic community from these events. In this paper we have shown that the hand digitising approach described is a useful tool for monitoring dredge plume movements in shallow coral reef environments (Table 2). Using visual inspection of MODIS images to digitise the plume one can discern, with a reasonable level of certainty, the differences between sediment plumes generated by dredging activities from natural features such as shallow water habitats and/or wave and wind generated resuspension of sediments. Through the use of reference images, including high resolution satellite imagery (i.e. Landsat and ALOS imagery), topographic maps and aerial photography [26,32] the image digitiser can distinguish the boundaries of dredge generated plume event from these natural features. Further confirmation of the plume can always be achieved through the incorporation of in-situ light and turbidity loggers and sediment traps where available.

MODIS images are captured at a relatively low resolution of 250 m×250 m (compared to ALOS at 10 m) sometimes providing images of poorer quality. However, twice daily passes of the MODIS satellite over the study area increased the opportunity to capture at least one good quality daily image. While poorer images were often un-useable with the TSS method, the hand digitising method had a greater chance of using these images through the comparison with other pre-dredge MODIS and higher resolution images that assisted the interpreter. This resulted in a greater number of days interpreted with the hand digitising methods compared with the semi-automated TSS method (304 hand digitized versus 116 TSS of the total 411 days captured). Furthermore, the plume area was limited to a relatively small area around Barrow Island requiring only 1 to 3 polygons digitised per day. Focusing on a small area of potential plume extent allowed the interpreter to quickly become familiar with the appearance of natural benthic features in the imagery at times with or without dredge plume and under different weather conditions. This is in stark contrast to the extent and number of units processed during fire digitising which is done within Australia on a regular basis [21]. Therefore, a digitised visual interpretation of the plume in MODIS imagery was a time efficient option.

Table 1. The R statistic shows the similarity of the spatial distribution of plume estimated by a range of Total Suspended Sediment values compared to hand digitised vectors of dredge plume due to the Barrow Island dredge operations.

TSS	1	2	3	4	5	6	7	8	9
R Statistic	0.539	0.484	<u>0.439</u>	0.484	0.527	0.539	0.545	0.554	0.555

The lowest R statistic (most similar) is underlined.

Table 2. Summary of the two methods used to characterize the extent of a dredge plume at Barrow Island, NW Australia.

	Digitising	TSS
Information Provided	Temporal coverage in a spatial context	Estimate of concentration temporally and spatially
Staff training	Basic	Professional expertise: University or technical remote sensing
Ability to differentiate shallow benthos from plume	Relatively accurate with some site familiarity. Can be conservative.	Without Mask/threshold: overestimate dredge plume; With Mask/threshold: Potential to underestimate dredge plume.
Time	1–2 weeks to digitize a whole year	Developing method takes time. Once process is working and running, results could be near real time
Fieldwork requirement	Not essential but helpful to validate observations.	Ground truthing results derived from an algorithm is essential. Can require extensive costly fieldwork.
Arbitrariness	High	Nil
Ability to interpret	High	Moderate
Reprocessing	Slow	Rapid

This study has shown that once people are trained to interpret MODIS imagery within an ecological and environmental context, within the limitations of the image quality, the confidence in the human interpretation of a boundary was greater than the semi-automated remote sensing approach [40]. The dredge plume hand digitisation methodology developed for this project was found to be time efficient for an inexperienced GIS user recently trained in the method. In comparison, remote sensing TSS algorithm approaches use techniques that require specialised skills to develop the regionally tuned models and to produce code to process satellite data with the models. The development of regionally tuned models requires fieldwork, which can be costly and time consuming. The models are tuned to the constituents of suspended sediment and if these change due to the stratification of the sea floor then this would affect the accuracy of the TSS estimates from MODIS data. Application of these procedures in an automated computer processing environment provides the advantage of rapid processing time, and enables faster delivery of products. Developing such a system however requires significant technical expertise and may be outside the financial capacity of many government and non-government organisations.

At present, the semi-automated TSS approach has difficulty discerning between shallow reef habitat and dredge plume (Fig. 5). This can be improved by either masking out (removing or ignoring) any shallow water environments or thresholding (reducing the sensitivity) the level of TSS detected by the algorithm. However, both these methods limit the ability to detect plume coverage over the shallow areas. Statistically, there is an apparent similarity between the hand digitistation and semi-automated TSS methods at TSS threshold values from 2 to 4 mg L^{-1} using ANOSIM, however, overlaying hand digitised vectors with TSS thresholded images (Fig. 5) showed that separating the dredge plume from shallow benthic features was something that the eye could achieve more consistently than a semi-automated algorithm. This is highlighted in the example provided (Fig. 5). Removing shallow areas from the analysis would provide false negatives, while hand digitsing allows the interpretation of the plume over the shallow benthic environments. Improvements to differentiate shallow water habitats are in development to counter this issue (unpublished: McKinna), however the rapid semi-automated method will remain limited in relation to this inaccuracy and would become time consuming to correct with visual inspection of daily MODIS imagery to determine if there is plume crossing the reef. This is similar for hand digitizing, but organizations do not require highly trained and specialized staff to complete such an approach.

The benefit of the remote sensing TSS algorithm approach is that it offers a consistent quantitative estimate of sediment concentration and therefore gives an indication of potential impact a dredge plume may have on benthic organisms such as coral communities in the Barrow Island region. Hand digitising does not provide concentration estimates and can be used only as a spatial recognition tool for "snapshot monitoring". However, temporal snapshot monitoring provides invaluable information for managers or researchers to understand the extent of plume coverage over significant sites, such as coral monitoring sites, fish aggregation spawning sites or marine park boundaries. Another advantage of the TSS approach is the ability to perform reprocessing of the entire MODIS imagery archive if improvements to the technique are made in the future. This processing can be run on a high performance computing system allowing an entire 10-year archive to be processed in a little over a week.

Conclusion

Ultimately the improvement of the semi-automated method for the process of extracting the plume boundaries would be ideal, however, with the tools presently available, the hand digitizing technique provides some important benefits to the monitoring of dredge plume events particularly in relation to monitoring shallow benthic communities. Not only has this project shown that hand digitising of dredge plumes, particularly in shallow water environments, is a useful tool, it has also provided a dataset that can be used to improve future dredge plume modelling in the region and assist as a reference dataset to improve automated remote sensing methods in the future. This research has demonstrated that sometimes the simplest methods can provide time and cost effective information to help manage impacts to our natural environment. Furthermore, it highlights the need to improve semi-automated processes in this field.

Supporting Information

Figure S1   The model used to relate TSS to MODIS band 1 reflectance, $R_{rs}(b1)$, where, a_w and b_{bw} are the spectral absorption and scattering properties of pure water

respectively. The coefficients c_0, c_1, and c_2 are constants with values of 0.1172490, 0.00479719, and -0.00629920 respectively.

Acknowledgments

We would like to thank Katherine Zdunic for remote sensing assistance; and Chris Simpson, Matt Williams and four anonymous reviewers for helpful comments on this manuscript. In situ research was conducted under permit No. SC001263 & SW013765. This project was completed as part of the Dredging Audit and Surveillance Program by the Gorgon Joint Venture as part of the environmental offsets. The Gorgon project is a joint venture of the Australian subsidiaries of Chevron, Exxonmobil, Shell, Osaka Gas, Tokyo Gas and Chubu Electric Power.

Author Contributions

Conceived and designed the experiments: RDE KLM SNF GS PF MB. Performed the experiments: RDE KLM DM GS MB. Analyzed the data: KM RDE PF LIWM MB JAYM BGH. Contributed reagents/materials/analysis tools: KM RDE PF LIWM MB JAYM GS BGH. Wrote the paper: RDE KLM SNF JAYM. Contributed to the first draft and to the corrections of the manuscript: KM RDE SNF PF LIWM MB JAYM DM GS BGH.

References

1. Erftemeijer PLA, Lewis RR III (2006) Environmental impacts of dredging on seagrasses: A review. Mar Pollut Bull 52: 1553–1572.
2. Lyngby JE, Mortensen SM (1996) Effects of Dredging Activities on Growth of *Laminaria saccharina*. Mar Ecol 17: 345–354.
3. Vargas-Ángel B, Peters EC, Kramarsky-Winter E, Gilliam DS, Dodge RE (2007) Cellular reactions to sedimentation and temperature stress in the Caribbean coral *Montastraea cavernosa*. J Invertebr Pathol 95: 140–145.
4. Bak RPM (1978) Lethal and sublethal effects of dredging on reef corals. Mar Pollut Bull 9: 14–16.
5. Babcock R, Davies P (1991) Effects of sedimentation on settlement of *Acropora millepora*. Coral Reefs 9: 205–208.
6. Gilmour J (1999) Experimental investigation into the effects of suspended sediment on fertilisation, larval survival and settlement in a scleractinian coral. Mar Biol 135: 451–462.
7. Partridge GJ, Michael RJ (2010) Direct and indirect effects of simulated calcareous dredge material on eggs and larvae of pink snapper *Pagrus auratus*. J Fish Biol 77: 227–240.
8. Gutman G, Masek JG (2012) Long-term time series of the Earth's land-surface observations from space. Int J Remote Sens 33: 4700–4719.
9. Ni W, Jupp DLB (2000) Spatial variance in directional remote sensing imagery - recent developments and future perspectives. Remote Sens Rev 18: 441–479.
10. Satapathy DR, Katpatal YB, Wate SR (2008) Application of geospatial technologies for environmental impact assessment: an Indian Scenario. Int J Remote Sens 29: 355–386.
11. Scott LP, Dirk P, Alan AK, Yuni K, Warren BC (2007) Moderate resolution remote sensing alternatives: a review of Landsat-like sensors and their applications. J Appl Remote Sens 1: 012506. Available: http://remotesensing.spiedigitallibrary.org/article.aspx?articleid=707762. Accessed 2012 Nov 12.
12. William BG (2007) Remote sensing in the coming decade: the vision and the reality. J Appl Remote Sens 1: 012505. Available: http://remotesensing.spiedigitallibrary.org/article.aspx?articleid=707761. Accessed 2012 Nov 12.
13. MODIS-Website. Available: http://modis.gsfc.nasa.gov/about/. Accessed 2012 Nov 12.
14. Lambrechts J, Humphrey C, McKinna L, Gourge O, Fabricius KE, et al. (2010) Importance of wave-induced bed liquefaction in the fine sediment budget of Cleveland Bay, Great Barrier Reef. Estuar Coast Shelf Sci 89: 154–162.
15. Devlin MJ, McKinna LW, Álvarez-Romero JG, Petus C, Abott B, et al. (2012) Mapping the pollutants in surface riverine flood plume waters in the Great Barrier Reef, Australia. Mar Pollut Bull 65: 224–235.
16. Rodriguez-Guzman V, Gilbes-Santaella F (2009) Using MODIS 250 m Imagery to Estimate Total Suspended Sediment in a Tropical Open Bay. Int J Systems Applic, Eng & Develop 3: 36–44. Available: http://www.universitypress.org.uk/journals/saed/saed-58.pdf. Accessed 2012 Nov 12.
17. Miller RL, McKee BA (2004) Using MODIS Terra 250 m imagery to map concentrations of total suspended matter in coastal waters. Remote Sens Environ 93: 259–266.
18. Kutser T, Metsamaa L, Vahtmae E, Aps R (2007) Operative Monitoring of the Extent of Dredging Plumes in Coastal Ecosystems Using MODIS Satellite Imagery. J Coast Res SI50: 180–184.
19. Islam A, Wang L, Smith C, Reddy S, Lewis A, et al. (2007) Evaluation of satellite remote sensing for operational monitoring of sediment plumes produced by dredging at Hay Point, Queensland, Australia. J Appl Remote Sens 1: 011506. Available: http://remotesensing.spiedigitallibrary.org/article.aspx?articleid=707754. Accessed 2012 Nov 12.
20. Bainbridge ZT, Wolanski E, Álvarez-Romero JG, Lewis SE, Brodie JE (2012) Fine sediment and nutrient dynamics related to particle size and floc formation in a Burdekin River flood plume, Australia. Mar Poll Bull 65: 236–248.
21. Heath B, Craig R, Marsden J, Steber M, Smith R, et al. (2002) Continental Fire monitoring. Proceedings of the 11th Australasian Remote Sensing and Photogrammetry Conference (ARSPC), Brisbane. 476–485.
22. Smith R, Craig R, Steber M, Marsden J, Carolyn M, et al. (1998) Use of NOAA-AVHRR for fire management in Australia's tropical savannas; Proceedings of the 9th Australasian Remote Sensing and Photogrammetry Conference (ARSPC), Sydney. 2101.
23. Qian YG, Kong XS (2012) A method to retrieve subpixel fire temperature and fire area using MODIS data. Int J Remote Sens 33: 5009–5025.
24. Mas JF, Puig H, Palacio JL, Sosa-Lopez A (2004) Modelling deforestation using GIS and artificial neural networks. Env Modelling & Software 19: 461–471.
25. Alonso-Perez F, Ruiz-Luna A, Turner J, Berlanga-Robles CA, Mitchelson-Jacob G (2003) Land cover changes and impact of shrimp aquaculture on the landscape in the Ceuta coastal lagoon system, Sinaloa, Mexico. Ocean Coast Manag 46: 583–600.
26. Wilson JS, Lindsey GH (2005) Socioeconomic Correlates and Environmental Impacts of Urban Development in a Central Indiana Landscape. J Urban Plan and Develop 131: 159–169.
27. Purdy R (2010) Using Earth Observation Technologies for Better Regulatory Compliance and Enforcement of Environmental Laws. J Environ Law 22: 59–87.
28. Dickins D, Hearon G, Morris K, Ambrosius K, Horowitz W (2011) Mapping sea ice overflood using remote sensing: Alaskan Beaufort Sea. Cold Regions Sci and Tech 65: 275–285.
29. IMCRA (2006) A Guide to the Integrated Marine and Coastal Regionalisation of Australia Version 4.0. Department of the Environment and Heritage, Canberra, Australia. Available: http://www.environment.gov.au/coasts/mbp/publications/imcra/imcra-4.html. Accessed 2012 Nov 12.
30. Wells FE, Berry PF (2000) The physical environments, marine habitats, and characteristics of the marine fauna. In: Survey of the Marine Fauna of the Montebello Islands, Western Australia and Christmas Island, Indian Ocean. Records of the Western Australian Museum 9–13.
31. DEC (2007) Management Plan for the Montebello/Barrow Islands Marine Conservation Reserves 2007–2017. Prepared for Marine Parks and Reserves Authority by the Department of Environment and Conservation, Perth, Western Australia, Management Plan 55. 125 pp. Available: http://www2.dec.wa.gov.au/component/option,com_hotproperty/task,view/id,117/Itemid,1584/. Accessed 2012 Nov 12.
32. Chevron-Australia (2009) Gorgon Gas Development and Jansz Feed Gas Pipeline Dredging and Spoil Disposal Management and Monitoring Plan. Chevron Australia, Perth, Western Australia. (G1-NT-PLNX0000373). 255.
33. Modarres R, Patil GP (2007) Hotspot detection with bivariate data. J Stat Plan Inference 137: 3643–3654.
34. ESRI (1999–2006) ARCGIS 9.2 Build 1500, Patent No. 5 710 835, ESRI Inc. United States of America.
35. Eastman JR (2009) IDRISI Taiga Worcester, MA: Clark University.
36. Baith K, Lindsay R, Fu G, McClain R (2001) Data analysis system developed for ocean color satellite sensors. Eos Trans AGU 82: 202, doi:210.1029/1001EO00109.
37. Clarke KR, Gorley RN (2006) PRIMER v6: User manual/Tutorial. Primer-E, Plymouth.
38. Clarke KR, Warwick RM (2001) Change in marine communities: an approach to statistical analysis and interpretation, 2nd edition. PRIMER-E: Plymouth.
39. Johnston CA (1998) Geographic Information Systems in Ecology. Oxford: Blackwell Science Ltd. Ch7,128–130 p.
40. Pope RM, Fry ES (1997) Absorption spectrum (380–700 nm) of pure water. II. Integrating cavity measurements. Appl Opt 36: 8710–8723.

6

Artificial Regulation of Water Level and Its Effect on Aquatic Macrophyte Distribution in Taihu Lake

Dehua Zhao*, Hao Jiang, Ying Cai, Shuqing An

Department of Biological Science and Technology, Nanjing University, Nanjing, China

Abstract

Management of water levels for flood control, water quality, and water safety purposes has become a priority for many lakes worldwide. However, the effects of water level management on the distribution and composition of aquatic vegetation has received little attention. Relevant studies have used either limited short-term or discrete long-term data and thus are either narrowly applicable or easily confounded by the effects of other environmental factors. We developed classification tree models using ground surveys combined with 52 remotely sensed images (15–30 m resolution) to map the distributions of two groups of aquatic vegetation in Taihu Lake, China from 1989–2010. Type 1 vegetation included emergent, floating, and floating-leaf plants, whereas Type 2 consisted of submerged vegetation. We sought to identify both inter- and intra-annual dynamics of water level and corresponding dynamics in the aquatic vegetation. Water levels in the ten-year period from 2000–2010 were 0.06–0.21 m lower from July to September (wet season) and 0.22–0.27 m higher from December to March (dry season) than in the 1989–1999 period. Average intra-annual variation (CV_a) decreased from 10.21% in 1989–1999 to 5.41% in 2000–2010. The areas of both Type 1 and Type 2 vegetation increased substantially in 2000–2010 relative to 1989–1999. Neither annual average water level nor CV_a influenced aquatic vegetation area, but water level from January to March had significant positive and negative correlations, respectively, with areas of Type 1 and Type 2 vegetation. Our findings revealed problems with the current management of water levels in Taihu Lake. To restore Taihu Lake to its original state of submerged vegetation dominance, water levels in the dry season should be lowered to better approximate natural conditions and reinstate the high variability (i.e., greater extremes) that was present historically.

Editor: Christopher Fulton, The Australian National University, Australia

Funding: This work was financially supported by the National Natural Science Foundation of China (No. 31000226)(http://www.nsfc.gov.cn) and the State Key Development Program for Basic Research of China (2008CB418004). The funders had no role in study design, data collection and analysis, decision to publish, or preparation of the manuscript.

Competing Interests: The authors have declared that no competing interests exist.

* E-mail: dhzhao@nju.edu.cn

Introduction

Because of the important ecological and socioeconomic functions of aquatic macrophytes, such as stabilization of sediments, regulation of the nutrient cycle, slowing of water currents and fishery maintenance, numerous studies over the past three decades have focused on the dynamics of aquatic macrophytes in freshwater ecosystems and identification of the forces driving their abundances and distributions [1–4]. Water quality degradation of the world's freshwater ecosystems over the past decades has led to extensive decreases in the area occupied by aquatic macrophytes as well as species losses [5,6]. Promoting the recovery of aquatic macrophytes has become a critical step in the restoration and rehabilitation of these degraded aquatic ecosystems [7–9].

Water levels, which are controlled by both natural conditions (e.g., meteorological and catchment characteristics) and local human activities (e.g., flood-control projects and artificial water transfer) [10], have been thought to be responsible for the variability in biomass and species composition of aquatic macrophytes in many freshwater ecosystems of the world [10–16]. Although artificial management and manipulation of water levels have been practiced widely, the effect of managed water levels on aquatic macrophytes has not been fully understood in most cases because of the complex relationship between macrophytes and water level [10,14,17,18].

Taihu Lake is the third-largest freshwater lake in China, occupying a surface area of 2,425 km^2 [19–21]. Due to rapid industrialization and urbanization, nutrient concentrations have increased continually during the past decades, and eutrophication has become a dominant water quality problem [22]. In an effort to recover the degraded aquatic ecosystem of Taihu Lake, numerous costly water conservation projects have been implemented in recent years. Planting and restoration of aquatic macrophytes for the purpose of removing excess nutrients are key facets of most of these projects [8,23,24].

Meanwhile, large amounts of water have been flushed into the lake from the Yangtze River since 2001 under the premise of "conquering the unmoving with the moving, diluting the polluted with the clean, supplementing low flow with ample flow" to improve water quality and control algal blooms [25,26]. Following the notorious blue-green algal bloom that occurred in the summer of 2007 and which resulted in serious drinking water shortages in Wuxi City [27,28], one of the most economically developed cities in Jiangsu Province, even more water was pumped into the lake [19]. Concurrently, more than 28,000 km of sea walls, river banks, embankments and polder dikes were built to control flooding [19].

As a result, water levels and their dynamics, especially intra-annual dynamics, have changed substantially in Taihu Lake.

Despite the considerable changes in the water levels in Taihu Lake, little attention has been focused on the effects on aquatic macrophytes, even though inter- and intra-annual water levels have been identified as one of the most important forces driving variability in aquatic macrophyte distribution [10,14]. Because aquatic macrophytes are distributed over such a large area (i.e. hundreds of square kilometers) [22], slight variations in water levels are likely to have precipitated a decrease or increase in the distribution of the macrophytes on the scale of tens of kilometers, greatly affecting their combined ability to remove nutrients and act as a source of food for fisheries [29,30]. Many water conservation projects focusing on planting aquatic macrophytes have been conducted [23,24,31], but it is likely more economical to protect and restore the existing communities of aquatic macrophytes. Protection and restoration, however, requires that increased attention be focused on understanding the effects of inter- and intra-annual water levels on aquatic macrophytes in the lake.

Although some authors have found correlations between the variation in aquatic vegetation and water levels in regard to aquatic systems at large temporal scales, most of those studies were based on either limited short-term or discrete long-term data [12,13,32], and thus the results are either narrowly applicable or easily confounded by the cumulative effect of other environmental factors with gradual temporal variation, such as trophic status. In this project, we mapped aquatic vegetation distribution between 1989 and 2010 based on remote sensing images with spatial resolutions from 15 to 30 m. Our objective was to determine the effect of managed water level on the distribution and composition of aquatic macrophytes in Taihu Lake.

Materials and Methods

2.1 Ethics Statement

No specific permits were required for the described field studies. The location studied is not privately-owned or protected in any way, and the field studies did not involve endangered or protected species.

2.2 Study Area

The Taihu Lake catchment plays an important role in China's political economy, containing 3.7% of the country's population, creating 11.6% of Chinese gross domestic product (GDP) and contributing 19% of total revenue while comprising only 0.4% of the land area of China [19]. Since the 1950s and especially since the 1980s, human activities have placed increased pressure on the lake's ecological components [21]. Our study area was limited to the areas identified in the remote sensing images as being covered by water in winter, the season when water levels were lowest and the topsoil of most patches of emergent vegetation was dry. As a result, most emergent vegetation was excluded from this work. We chose to exclude most emergent vegetation in order to reduce the effects of human activities on the relationship between aquatic vegetation and water levels since human activities have drastically altered the emergent vegetation of Taihu Lake through large-scale construction of embankments and buildings, as well as vegetation restoration or destruction [31].

Pen-fishing has also had a large influence on the distribution of aquatic vegetation due to farmers' activities such as planting and harvest. The area subject to pen-fishing has varied dramatically over the past two decades, i.e. a gradual increase between 1990 and 2005 followed by a sudden decrease after 2007, when an extensive blue-green algal bloom occurred and resulted in serious drinking water shortages in Wuxi City [27,28]. However, pen-fishing activities have been limited primarily to the East Bay of Taihu Lake [31,33]. Therefore, the East Bay was also excluded from this study to minimize the confounding influence of pen-fishing on aquatic macrophyte distribution.

From 1960 to 2000, human activities resulted in a worsening of the water quality of Taihu Lake at an approximate rate of one grade every 10–15 years [23]. To improve water quality, much effort has been expended on lake restoration, especially after 2000. Artificial management of water levels through pumping of water as well as construction of embankments and dams was a common strategy for improving water quality and controlling blue-green algal bloom. In particular, increased amounts of water were pumped into Taihu Lake from the Yangtze River after 2000 [19,25]. Therefore, our study period ranged from 1989 to 2010 to encompass the ten years before and after 2000, i.e., Period 1 (1989–1999) and Period 2 (2000–2010). However, high-quality remote sensing images were available for only sixteen of the years between 1989 and 2010.

2.3 Field Surveys

The aquatic vegetation of Taihu Lake was grouped into four types: emergent, floating-leaf, floating and submerged [22,31]. Dominant species in the lake included *Phragmites communis*, *Zizania latifolia*, *Nymphoides peltatum*, *Trapa natans*, *Potamogeton malaianus* and *Vallisneria spiralis*, as identified by field observations as well as previously published studies [30,31]. We divided the aquatic vegetation into two types according to their spectral characteristics. Type 1 represented the typical green vegetation identified in remote sensing images by lower red band reflectance paired with higher near-infrared band reflectance than other ground cover types and included emergent, floating-leaf and floating vegetation having some green leaves above the water surface. As previously noted, most emergent vegetation was excluded from Type 1. Type 2 consisted of the submerged vegetation, which had all green leaves submerged beneath the water surface, thus distinguishing it from typical green vegetation in remote sensing images. Because Type 1 vegetation had a higher signal intensity than Type 2, areas containing both Type 1 and Type 2 vegetation were classified as emergent vegetation.

To obtain data for developing and validating models to identify aquatic vegetation, we conducted field surveys on 14–15 September 2009 and 27 September 2010. A total of 783 samples were collected in open water or aquatic vegetation (mostly floating-leaf or submerged) of Taihu Lake, including the East Bay. An additional 182 samples of reed (emergent) vegetation or terrestrial areas (e.g., shoreline roads and buildings such as docks, businesses and factories) were obtained from a 1:50,000 land use and land cover map due to logistical difficulties in maneuvering a boat in the dense reed vegetation. A total of 426 and 539 ground truth samples were collected in 2009 and 2010, respectively (Fig. 1). At each field sampling plot, photographs were taken using a digital camera (IXUS 950, Canon) held at about 1.2 m above the water surface, with the camera axis angled about 30 degrees down from the horizon. The position of each photograph was geo-located using a portable GPS receiver with an accuracy of 3 m. In the laboratory, all the photographs were interpreted visually and classified as Type 1, Type 2 or open water sediment.

2.4 Image Processing

Multispectral TM, ETM+, SPOT-4, HJ, and CBERS remote sensing images were used in this study, with spatial resolutions ranging from 15 to 30 meters. Following the recommended standard for aquatic remote sensing [34], we selected images

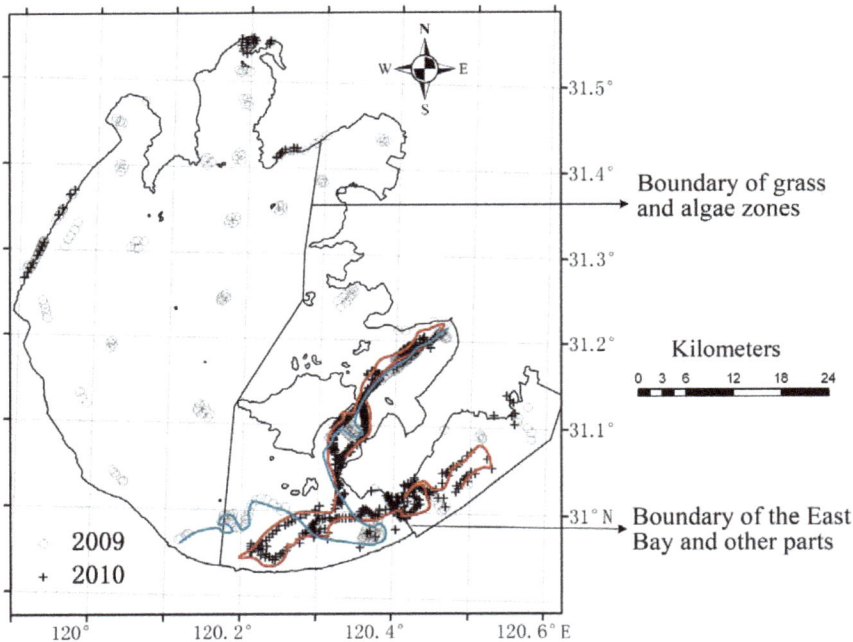

Figure 1. The study area showing the distribution of 965 ground truth samples (426 in 2009 and 539 in 2010) in Taihu Lake, China.

containing no more than 10% cloud cover at the study area. The cosine approximation model (COST; Chavez, 1996), which has been implemented successfully in other aquatic remote sensing studies [35,36], was used to apply atmospheric corrections to all the images used. Prior to atmospheric correction, cloud-contaminated pixels were removed from all images using interactive interpretation. Geometric correction was applied using second-order polynomials with accuracy higher than 0.5 pixel.

Table 1. Remote sensing images used in this study with associated dates.

Years	Image-1	Image-2	Image-3	Image-4
1989	TM, 1/14	TM, 7/17	TM, 10/21	
1991	TM, 1/12	TM, 7/23	TM, 8/24	
1992	TM, 2/16	TM, 7/25	TM, 8/10	
1995	TM, 2/24	TM, 8/3	TM, 8/19	
1996	TM, 1/10	TM, 7/20	TM, 9/6	
1998	TM, 1/31	TM, 8/11	TM, 7/10	
2000	ETM+, 3/18	ETM+, 8/8	CBERS, 9/16	ETM+, 10/12
2001	ETM+, 1/15	ETM+, 7/26	ETM+, 9/28	
2002	ETM+, 2/3	ETM+, 7/13	TM, 8/22	
2003	ETM+, 2/6	ETM+, 8/1	SPOT, 8/23	
2004	ETM+, 2/9	ETM+, 8/3	CBERS, 8/8	ETM+, 8/19
2005	ETM+, 3/31	ETM+, 6/19	ETM+, 9/7	
2006	ETM+, 3/2	CBERS, 8/6	TM, 9/18	ETM+, 9/28
2008	ETM+, 2/20	CBERS, 7/24	ETM+, 8/14	HJ, 9/23
2009	ETM+, 1/13	ETM+, 8/25	HJ, 9/10	
2010	ETM+, 3/13	ETM+, 8/20	ETM+, 9/21	

Because they contain information on seasonal dynamics of both aquatic vegetation and related environmental factors, multiple intra-annual remote sensing images can provide higher accuracy for the identification of aquatic vegetation than a single image [37,38]. Therefore, we used combinations of winter (between January and March when the biomass of aquatic vegetation was lowest) and summer (between June and October when the biomass of aquatic vegetation was highest) images from each study year between 1989 and 2010. For each study year, at least three clear Landsat images were selected (one from winter and the others from summer) and formed into at least two pairs in which the winter image was paired with each summer image. A total of 36 pairs were used in this study (Table 1).

Aquatic vegetation was mapped using each image combination, so at least two vegetation maps were obtained for each year. Maps for the same year were superimposed and combined according to the following rules: (1) In the grass type zone of Taihu Lake (i.e. the eastern portion in Fig. 1), if a pixel was classified as aquatic vegetation in either map within a single study year, it was classified as aquatic vegetation in the final map. This rule was established primarily because human activities such as harvesting of aquatic vegetation might decrease the distribution of aquatic vegetation, and because particulate matter that is suspended very high in the water column might obscure the submerged vegetation, resulting in underestimation of submerged vegetation [22] at some time during the growing season. (2) In the algae type zone (i.e. the remaining portions of the lake not in the grass zone in Fig. 1), a pixel was regarded as aquatic vegetation only when it was classified as aquatic vegetation in all maps for a single study year. This rule aimed to reduce classification interference from algal blooms, which occur frequently between May and October [27,28,39] and is based on the probability being much lower that algal blooms will appear twice in the same location than that aquatic macrophytes will. The ground truth samples from 2009 and 2010 were used to evaluate the accuracy of the final aquatic vegetation classifications derived from the 2009 and 2010 image pairs.

2.5 Analytical Methods

2.5.1 Identification of aquatic vegetation in remote sensing images.
Classification tree (CT) analysis, which uses recursive dichotomous partitioning of the data according to calculated thresholds, has been used successfully for the identification of aquatic vegetation because of its flexibility with regard to the inclusion of data from multiple sources and of multiple types, such as spectral signals, environmental variables and other variables related to aquatic vegetation growth [38,40–44]. Zhao et al. (2012) developed an improved CT modeling algorithm for identifying emergent, floating-leaf and submerged vegetation from remote sensing images both from different times [45] and from different sensor (Zhao et al. A method for application of classification tree models to map aquatic vegetation using remotely sensed images from different sensors and dates. Sensors, Re-submitted after revision). Because we divided the aquatic vegetation into only two types in this study (i.e. Type 1 and Type 2), minor modifications were made to the CT model structure (Fig. 2).

We obtained the quantitative thresholds for the CT base model structure and thus the final CT models by applying CT analysis to the 2009 image pairs (Fig. 2), attaining an overall classification accuracy of 94.0%, with classification accuracies of 95.6% and 88.8% for Type 1 and Type 2 vegetation, respectively. When the CT models were applied to the image pairs of 2010, overall accuracy was 93.3%, with classification accuracies of 94.2% and 87.9% for Type 1 and Type 2 vegetation, respectively (Table 2). These results suggested that our CT model could be used to

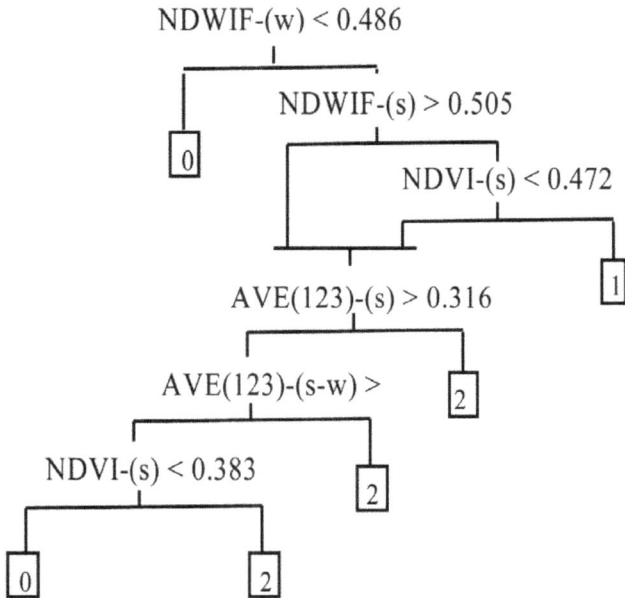

Figure 2. Classification tree models established for Type 1 and Type 2 aquatic vegetation. The numbers 1 and 2 in the end nodes of the classification trees represent Type 1 and Type 2 vegetation, respectively, whereas 0 represents other types. Variables used are the Modified Normalized Difference Water Index (MNDWI), the Normalized Difference Vegetation Index (NDVI) and the average reflectance of the blue, green and red image bands (AVE123). Variables were calculated by season (s = summer, w = winter) or differences among seasonal values (e.g., s-w).

effectively identify the aquatic vegetation in Taihu Lake. Therefore, we used the models to map the distribution of aquatic vegetation in Taihu Lake from 1989 to 2010.

2.5.2 Evaluation of water level effects on aquatic vegetation.
The annual Coefficient of Variation (CV_a) was calculated to describe the inter-annual fluctuation of water levels:

$$CV_a = \frac{\sqrt{\frac{\sum (x - \overline{x})^2}{n-1}}}{\overline{x}} \times 100\% \qquad (1)$$

where x and $\overline{x}$ are average monthly water level and average annual water level, respectively.

We used regression analysis to investigate the effects of water level variation and CV_a fluctuation on the distribution of aquatic vegetation through time. However, using un-transformed values for water level and area of aquatic vegetation is unlikely to reflect the true relationship because of the inevitable temporal autocorrelation of aquatic vegetation as well as the confounding influence of gradual changes in environmental factors such as water nutrient content, chemical oxygen demand (COD) and water clarity. Therefore, we transformed the water level and aquatic vegetation area variables using the variability from one year to the next before performing the regression analysis. This was accomplished by subtracting the previous year values from the focus year values using the same time period. Thus, water level variability was calculated as:

$$V_{wl} = WL_i - WL_{i-1} \qquad (2)$$

Where WL_i and WL_{i-1} are the average water levels or CV_a in the focus year (i) and the year previous to the focus year (i−1), respectively. However, if data for the previous year were missing, data from two years prior to the focus year (i−2), or three years prior (i−3) if data were also missing for i−2, were used instead. Variability in aquatic vegetation area was calculated as:

$$VA_{ava} = \frac{(AVA_i - AVA_{i-1})}{AVA_{i-1}} \qquad (3)$$

Where AVA_i and AVA_{i-1} are aquatic vegetation areas in the focus year (i) and year previous to the focus year (i−1), respectively. Similar to Equation (2), if aquatic vegetation data were not available for the previous year, it was replaced by the data for the closest year for which data were available.

Because most of the images used to map aquatic vegetation distribution were dated prior to October with only two exceptions (Table 1), water levels between October and December in a certain year did not influence aquatic vegetation area of the same year. However, October-December water levels probably influenced the aquatic vegetation of the following year. Therefore, October through December water levels of the year previous to the focus year were used to analyze relationships between monthly water levels and aquatic vegetation.

Results

3.1 Temporal Dynamics of Water Level

Between 1989 and 2010, annual average water levels fluctuated between 2.86 m and 3.33 m, with no significant inter-annual trend (Fig. 3A). The average water level in Period 2

Table 2. Confusion matrix of the CT models developed in this paper as applied to 2009 and 2010 data, respectively (in number of field samples).

			Prediction				
			Type 1	Type 2	Other types	Classification accuracy (%)	Overall accuracy (%)
2009	Truth	**Type 1**	130	4	2	95.6	94.0
		Type 2	5	103	8	88.8	
		Other types	1	4	145	96.7	
2010	Truth	**Type 1**	175	7	4	94.1	93.3
		Type 2	6	102	8	87.9	
		Other types	0	8	186	95.9	

(2000–2010) was 3.18 m, slightly higher than that in Period 1 (i.e. 1989–1999, average = 3.10 m). However, substantially different intra-annual dynamics were observed between the two temporal periods (Fig. 3B), with more stable water levels in Period 2 than in Period 1. In Period 1, monthly water levels ranged from 2.57 m to 4.61 m with the annual Coefficient of Variation (CV_a) ranging from 3.06% (1994) to 18.41% (1999) and averaging 10.21%. However, in Period 2, monthly water levels ranged from 2.76 m (2006) to 3.98 m (2009), with CV_a ranging from 2.65% to 7.94% and averaging 5.41% (Fig. 3C). CV_a in Period 1 was significantly higher than that in Period 2 ($p = 0.01$). For July, August and September, monthly water levels in Period 2 were 0.064 to 0.21 m lower than those in Period 1, whereas monthly water levels in Period 2 for the remaining months were 0.042 to 0.27 m higher than those in Period 1. Thus, our results indicate differences in intra-annual dynamics of water level between Period 1 and Period 2.

Monthly precipitation in Period 1 was slightly higher (1.49 to 30.48 mm) than that in Period 2 for January, March, June, July, August and September; for the other six months, monthly precipitation in Period 1 was slightly lower (0.03 to 9.80 mm) than that in Period 2 (Fig. 3D). Annual precipitation in Period 1 (1175.4 mm) was slightly higher than that in Period 2 (1132.1 mm). Thus, precipitation and water level showed different intra-annual variation patterns between the two periods. These results suggested that climatic conditions were not responsible for the greater stability of intra-annual water level in Period 2 than in Period 1.

3.2 Temporal Dynamics of Aquatic Vegetation Distribution Area

From 1989 to 2010, aquatic vegetation was distributed primarily in the eastern part of the lake (Fig. 4). The spatial pattern of distribution experienced some changes, with aquatic vegetation shifting gradually from the northeast to the southeast.

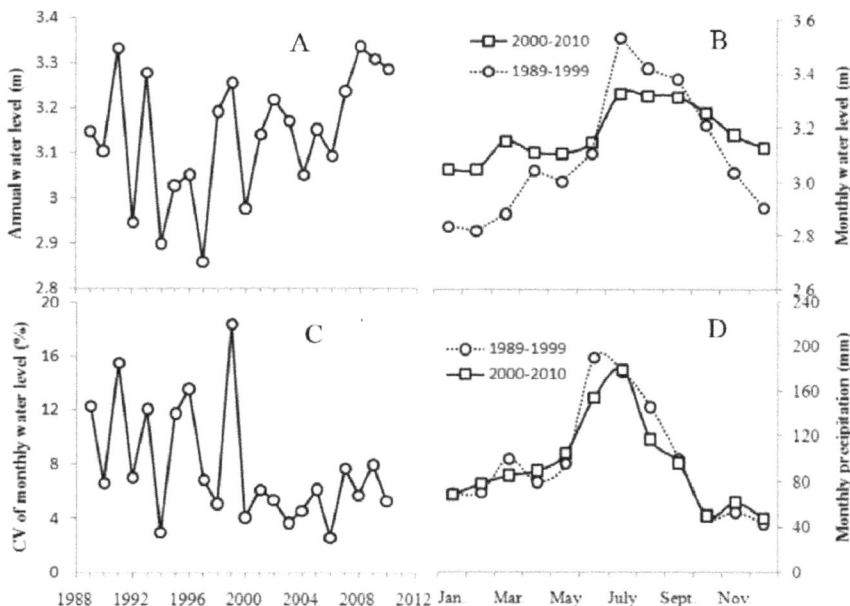

Figure 3. Inter- and intra- annual dynamics of water level and precipitation in Taihu Lake. (A) annual water level dynamics between 1989 and 2010; (B) average monthly water levels for the two 10-year time periods examined in this study; (C) intra-annual fluctuation of water levels (CV_a) from 1989–2010; and (D) average monthly precipitation for the two 10-year time periods examined in this study.

Substantial changes were observed in both distribution area and composition of aquatic vegetation during the study period (Fig. 5). The area covered by Type 1 vegetation ranged from 5.80 km^2 (1996) to 142.5 km^2 (2009), with an average of 57.3 km^2, whereas Type 2 vegetation covered an area ranging from 68.3 km^2 (1991) to 190.8 km^2 (2001), with an average of 137.4 km^2. The total aquatic vegetation area (i.e., the sum of Type 1 and Type 2 vegetation) ranged from 77.9 km^2 (1991) to 282.0 km^2 (2005), with an average of 194.7 km^2, and the ratio of Type 2 to Type 1 vegetation ranged from 0.94 (2010) to 23.4 (1996), with an average of 6.07. Significant temporal dynamics were observed for each variable from 1989 to 2010 ($p = 0.01$). Both total aquatic vegetation area and area of Type 1 vegetation increased significantly over the study period ($p = 0.01$). Area of Type 2 vegetation increased before 2001 and then decreased ($p = 0.01$), and the ratio of Type 2 to Type 1 vegetation decreased steadily over the study period ($p = 0.01$).

The area covered by aquatic vegetation also differed between the two inter-annual temporal periods examined. Average Type 2 vegetation area increased from 99.2 km^2 in Period 1 to 160.4 km^2 in Period 2, an increase of 61.6%, and average Type 1 vegetation area increased 7.14-fold, from 10.5 km^2 in Period 1 to 85.4 km^2 in Period 2. Total vegetation area increased 124.0%, from 109.7 km^2 in Period 1 to 245.8 km^2 in Period 2. Finally, the ratio of Type 2 to Type 1 vegetation decreased from 11.8 in Period 1 to 2.62 in Period 2.

3.3 Relationship between Water Level and Aquatic Vegetation Area

Firstly, we investigated whether the annual average water levels and intra-annual fluctuation of water levels (CV$_a$) influenced inter-annual aquatic vegetation area and its derivations. No significant correlations were found between V$_{wl}$ and VA$_{ava}$ using either annual averages or CV$_a$ for either vegetation type, total vegetation area, or the ratio of Type 2 to Type 1 vegetation (Table 3).

Secondly, we tested the correlations between monthly average water levels and aquatic vegetation (Fig. 6). Significant positive correlations were found between Type 1 vegetation area and monthly water levels from December to March ($p = 0.05$), whereas significant negative correlations were found between Type 2 vegetation area and monthly water levels from January to April ($p = 0.05$). Total vegetation area was negatively correlated with monthly water levels from March to April ($p = 0.05$), and the ratio of Type 2 to Type 1 vegetation was negatively correlated with monthly water levels from November to March ($p = 0.05$). These results suggested that water levels in late winter and early spring (traditional dry season) significantly influenced aquatic vegetation area.

Discussion

4.1 Temporal Changes

We found significant relationships between water level from December to March and the area of Type 1 vegetation, with

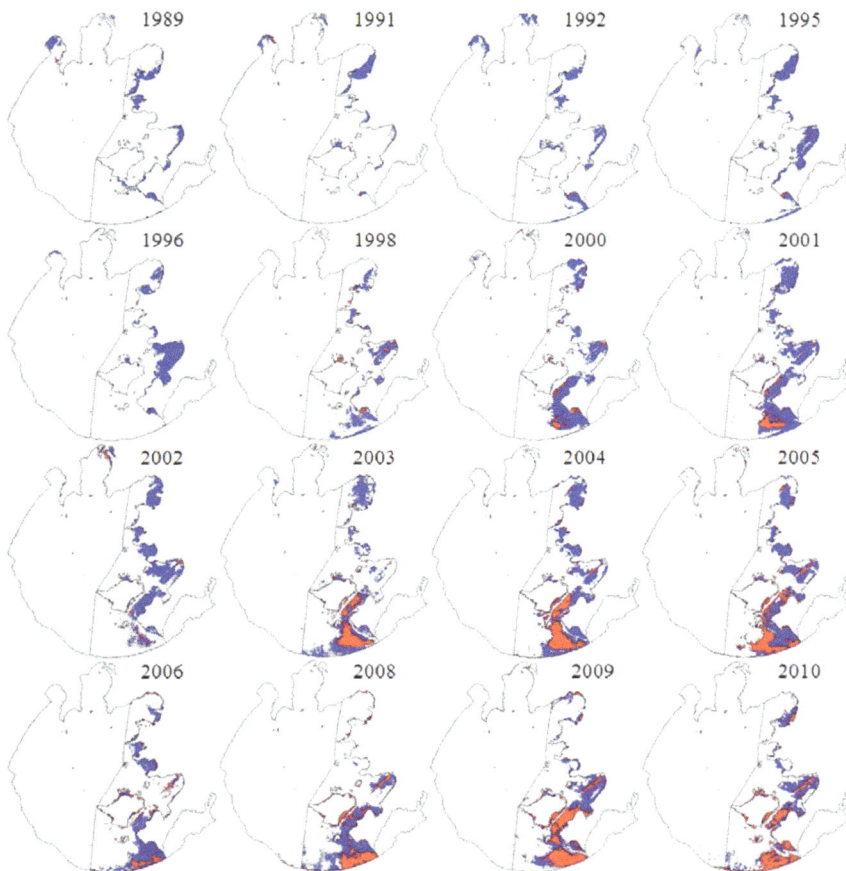

Figure 4. Distribution of aquatic vegetation of Type 1 (red) and Type 2 (blue) between 1989 and 2010. Type 1 vegetation consisted of emergent, floating-leaf and floating vegetation, whereas Type 2 consisted of submerged vegetation.

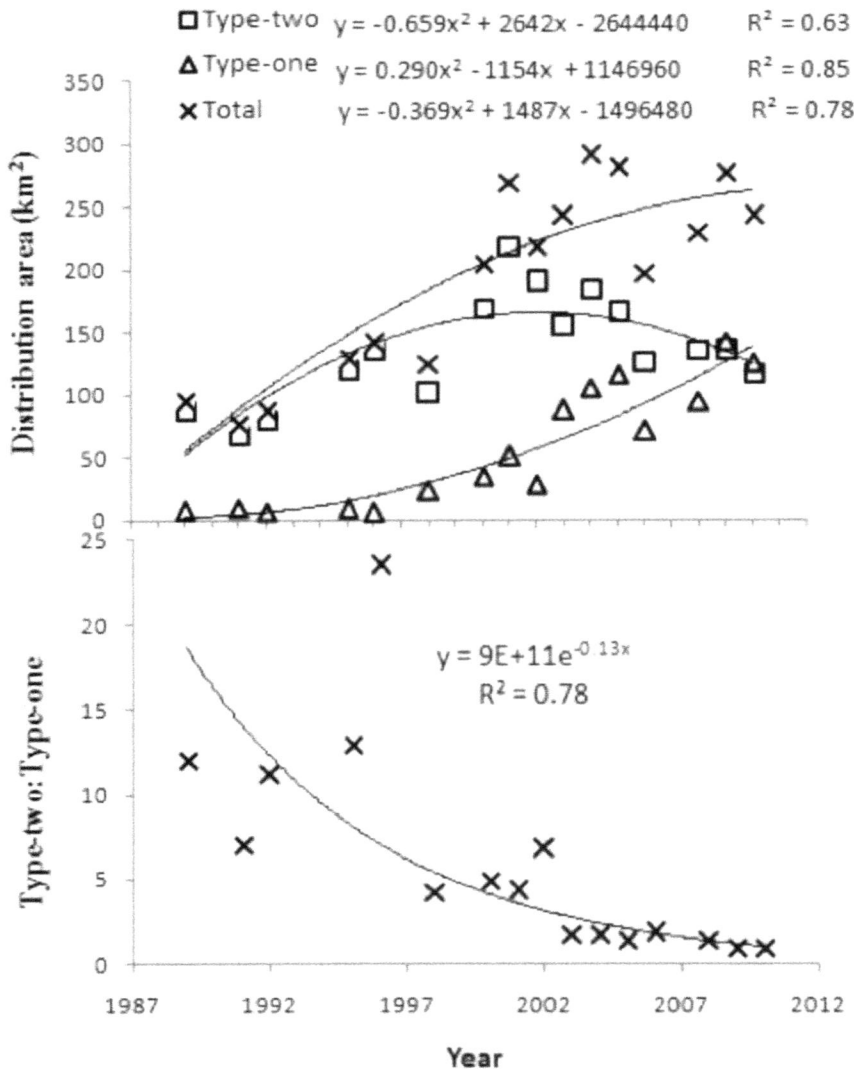

Figure 5. Temporal trends (1989–2010) of the distribution area of aquatic vegetation components and derivatives. Type 1 vegetation consisted of emergent, floating-leaf and floating vegetation, whereas Type 2 consisted of submerged vegetation.

increases in Type 1 vegetation occurring steadily from 1989 to 2010 (and consequently for the 2000–2010 period relative to the 1989–1999 period) and coinciding with increased water levels during the dry season. The aquatic macrophyte communities in

Table 3. Linear correlation coefficients between variation of aquatic vegetation (VA$_{ava}$) and variation of water level (V$_{wl}$) (n = 15).

Aquatic Vegetation Parameter	Annual average	CV$_a$
Type 2	−0.33	0.29
Type 1	0.21	−0.34
Total	−0.25	0.33
Type 2:Type 1	−0.40	0.02

Type 1 vegetation consisted of emergent, floating-leaf and floating vegetation, whereas Type 2 consisted of submerged vegetation.

very shallow lakes such as Taihu Lake [19] may be especially sensitive to variation in water levels. In addition, according to the stable state theory proposed by Scheffer et al. (2003) [46], nutrient enrichment should shift the vegetation in freshwater systems toward dominance by floating plants; in fact, nutrient concentrations, as well as area of Type 1 vegetation, have shown a gradual increase in Taihu Lake over the past 20 years [23,47]. Therefore, nutrient enrichment and higher dry-season water levels may represent two of the most important factors responsible for the temporal shifts in Type 1 vegetation distribution observed in this study.

We found significant negative relationships between Type 2 vegetation and dry-season monthly water level as well as slightly higher dry-season water levels in 2000–2010 than in 1989–1999. Therefore, water level couldn't explain the 61.6% increase of Type 2 in 2000–2010 over that in 1989–1999. According to an alternative equilibrium theory [48,49], submerged vegetation responds in a non-linear way to eutrophication, first increasing then decreasing. Our findings support this theory, with nutrient enrichment increasing over the past 20 years in Taihu Lake and

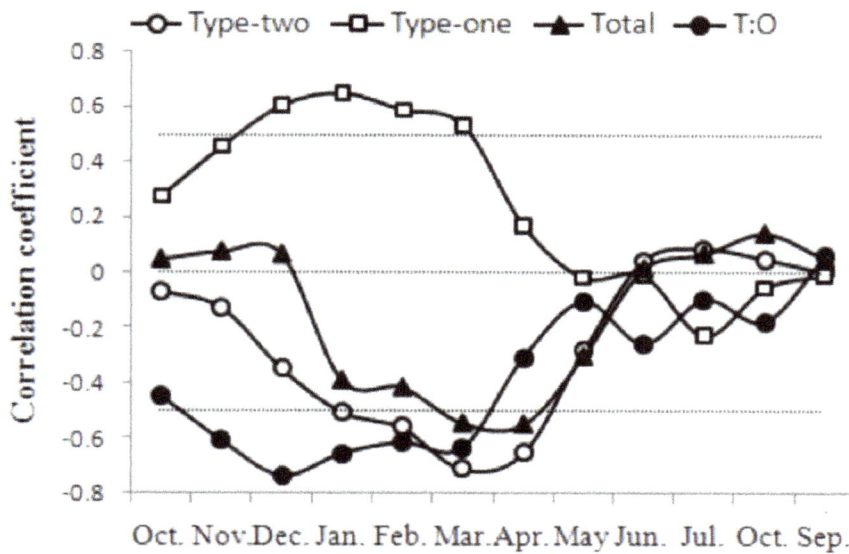

Figure 6. Linear correlation coefficients between monthly water level (in V_{wl}) and the distribution area of aquatic vegetation components (in VA_{ava}) during the time period 1989–2010. Type 1 vegetation consisted of emergent, floating-leaf and floating vegetation, whereas Type 2 consisted of submerged vegetation. T:O is the ratio of Type 2 to Type 1 vegetation area.

Type 2 vegetation increasing in area until about 2002 and decreasing thereafter. Nutrient enrichment is thought to be responsible for the expansion of aquatic vegetation in numerous lakes throughout the world and probably acted as one of most important driving factors of vegetation dynamics in our study [50,51]. We speculate that water level probably acted as the dominant factor determining Type 2 vegetation area on a one- to two-year temporal scale, whereas gradual changes in eutrophication probably acted as the dominant driving factor on a longer (~10 years) temporal scale.

The methodology used for identifying the effect of water level on aquatic vegetation can have a large influence on the results obtained. Despite a few successful reports [10,11], using non-transformed values of water level and aquatic vegetation area in regression models is unlikely to reveal the actual temporal relationship between the two variables in periods when confounding factors such as nutrient concentration vary considerably. Using transformed parameters such as the V_{wl} and VA_{ava} variables we used in this study instead of non-transformed water level and aquatic vegetation area values has two advantages: (1) it can alleviate the problem of temporal autocorrelation in aquatic vegetation area and its derivatives; and (2) it can help distinguish influences of water level from gradually and continuously increasing or decreasing factors such as water eutrophication [23].

4.2 Mechanisms for the Influence of Water Level on Aquatic Vegetation

Underwater light availability, which is affected strongly by water level, is an important mechanism influencing aquatic vegetation. Decreases in available light have been found to be of the most important factors resulting in species disappearance and biodiversity loss both for aquatic and terrestrial vegetation [52,53]. Many studies have found underwater light to be closely correlated with water level [4,10,13,16], especially for turbid waters such as Taihu Lake [25,54]. Generally, increases in water level will reduce underwater light availability, especially at the bottom. If under-

water light decreases below the threshold of a species' minimum light requirements, the species will disappear from the community.

Changes in light availability resulting from water level variability can easily explain the negative correlation found between area of Type 2 vegetation, which is submerged, and water level in traditional annual dry seasons. However, the positive correlation between Type 1 vegetation area and water level during the dry season cannot be explained as a direct result of light availability because of the positive effect of light availability on aquatic vegetation growth [55]. A more plausible explanation of the positive correlation would be due to competitive interactions between Type 1 and Type 2 vegetation [56]. According to observations from our field campaign, Type 1 species such as *Nymphoides peltatum* and Type 2 species such as *Potamogeton malaianus* grew widely in mixed communities in Taihu Lake [30], promoting strong competition for space and light. Because Type 2 vegetation is more sensitive to underwater light restrictions, which can be exacerbated by increases in water level, than Type 1 vegetation [55,57,58], water level increases can strengthen competitive ability of Type 1 vegetation by inhibiting the growth of Type 2 vegetation, ultimately resulting in the correlation patterns observed.

Our results indicated that water level influenced aquatic vegetation in dry seasons more so than in rainy seasons. This may be a consequence of the phenology of aquatic plants, which are most sensitive to light conditions in the germination and initial growth stages [59]. Most species in Taihu Lake survive winter with tuber-like buds in silt and germinate in the early spring [30,31]. Water level can directly influence the germination of buds by altering light availability at the lake bottom. Upon entering the rapid growth period, *Nymphoides peltatum* and *Potamogeton malaianus*, two of the most widely distributed species, become less sensitive to water level variability because of their strong morphological plasticity [60]. Our results were consistent with Blindow et al. (1993) and Paillisson and Marion (2006) [11,49], who found that spring water level influenced the growth of aquatic vegetation.

4.3 Management Implications

Our results suggest that regulation of the distribution of aquatic vegetation is feasible through management of water levels in Taihu Lake. Since 2000, and especially since 2007 when a severe blue-green algal bloom resulted in serious drinking water shortages in Wuxi City [27,28], multiple costly water conservation projects have been conducted in Taihu Lake. Restoration of aquatic vegetation through artificial planting or other means has been one of most common approaches [31] because of the purification function of aquatic vegetation. The significantly better water quality in the eastern coastal area of Taihu Lake relative to other parts of the lake is a solid example of this purification function [60]. Our results indicate that decreasing water levels in the dry season could increase the area occupied by aquatic vegetation in tens of square kilometers, which is a difficult goal to achieve using other restoration strategies such as direct planting.

In addition to regulation of the area occupied by aquatic vegetation generally, our results suggest that artificial control of distinct aquatic vegetation components is possible by regulating water levels. This finding is potentially very useful for lake management because of the different ecological and socioeconomic functions performed by the different aquatic vegetation types such as submerged vs. floating-leaf vegetation [29,30]. For example, submerged vegetation is one of the most important food resources for breeding crabs, which represent an annual harvest value from Taihu Lake of more than two hundred million dollars [29]. Farmers usually must plant submerged vegetation in the lake to act as a food source for the crabs, but water level regulation may represent a more economical and effective alternative. Additionally, decreasing water levels in the dry season will bring intra-annual water level fluctuations closer to natural conditions (i.e. large intra-annual fluctuations [14]) while restoring Taihu Lake to its original state of dominance by submerged vegetation.

Finally, our results suggest that regulation of water levels could be used to better control algal blooms in the lake. One of the main objectives of the current water level regulation strategies, such as the flushing of water into the lake from the Yangtze River, is to control algal blooms [19,25,26]. Because algal blooms usually occur between May and October in Taihu Lake, decreases in water level between late winter and early spring will not reduce the effectiveness of the current strategy for controlling algal blooms. On the contrary, careful reductions in water level between late winter and early spring are beneficial to the control of algal blooms in the lake because of their ability to increase the distribution of aquatic vegetation, which will in turn reduce nutrient levels.

Author Contributions

Conceived and designed the experiments: DZ SA. Performed the experiments: DZ HJ. Analyzed the data: DZ YC. Wrote the paper: DZ HJ.

References

1. Orth RJ, Moore KA (1983) Chesapeake Bay: an unprecedented decline in submerged aquatic vegetation. Science 222: 51–53.
2. Jackson JBC, Kirby MX, Berger WH, Bjorndal KA, Botsford LW, et al. (2001) Historical overfishing and the recent collapse of coastal ecosystems. Science 293: 629.
3. Franklin P, Dunbar M, Whitehead P (2008) Flow controls on lowland river macrophytes: A review. Sci Tot Environ 400: 369–378.
4. van der Heide T, van Nes EH, van Katwijk MM, Olff H, Smolders AJP (2011) Positive feedbacks in seagrass ecosystems–evidence from large-scale empirical data. PloS one 6: e16504.
5. Gullström M, Lundén B, Bodin M, Kangwe J, Ohman MC, et al. (2006) Assessment of changes in the seagrass-dominated submerged vegetation of tropical Chwaka Bay (Zanzibar) using satellite remote sensing. Estuar Coast Shelf Sci 67: 399–408.
6. Chambers P, Lacoul P, Murphy K, Thomaz S (2008) Global diversity of aquatic macrophytes in freshwater. Hydrobiologia 595: 9–26.
7. Qin B (2009) Lake eutrophication: Control countermeasures and recycling exploitation. Ecol Eng 35: 1569–1573.
8. Li EH, Li W, Wang XL, Xue HP, Xiao F (2010) Experiment of emergent macrophytes growing in contaminated sludge: Implication for sediment purification and lake restoration. Ecol Eng 36: 427–434.
9. Lorenz AW, Korte T, Sundermann A, Januschke K, Haase P (2012) Macrophytes respond to reach-scale river restorations. J Appl Ecol 49: 202–212.
10. Paillisson JM, Marion L (2011) Water level fluctuations for managing excessive plant biomass in shallow lakes. Ecol Eng 57: 241–247.
11. Paillisson JM, Marion L (2006) Can small water level fluctuations affect the biomass of Nymphaea alba in large lakes? Aquat Bot 84: 259–266.
12. Liira J, Feldmann T, Mäemets H, Peterson U (2010) Two decades of macrophyte expansion on the shores of a large shallow northern temperate lake-a retrospective series of satellite images. Aquat Bot 93: 207–215.
13. O'Farrell I, Izaguirre I, Chaparro G, Unrein F, Sinistro R, et al. (2011) Water level as the main driver of the alternation between a free-floating plant and a phytoplankton dominated state: a long-term study in a floodplain lake. Aquat Sci 73: 275–287.
14. Van Geest G, Coops H, Roijackers R, Buijse A, Scheffer M (2005) Succession of aquatic vegetation driven by reduced water-level fluctuations in floodplain lakes. J Appl Ecol 42: 251–260.
15. Geest GJV, Wolters H, Roozen F, Coops H, Roijackers R, et al. (2005) Water-level fluctuations affect macrophyte richness in floodplain lakes. Hydrobiologia 539: 239–248.
16. Bain MB, Singkran N, Mills KE (2008) Integrated ecosystem assessment: Lake Ontario water management. PloS one 3: e3806.
17. Coops H, Vulink JT, Van Nes EH (2004) Managed water levels and the expansion of emergent vegetation along a lakeshore. Limnologica 34: 57–64.
18. Wantzen KM, Rothhaupt KO, Mörtl M, Cantonati M, Tóth LG, et al. (2008) Ecological effects of water-level fluctuations in lakes: an urgent issue. Hydrobiologia 595: 1–4.
19. An S, Wang RR (2009) The human-induced driver on the development of Lake Taihu: In Lee, Xuhui ed. Lectures on China's Environment, Yale School of Forestry and Environmental Studies.
20. Zhang YL, Dijk MA, Liu ML, Zhu GW, Qin BQ (2009) The contribution of phytoplankton degradation to chromophoric dissolved organic matter (CDOM) in eutrophic shallow lakes: Field and experimental evidence. Water Res 43: 4685–4697.
21. Zhao DH, Cai Y, Jiang H, Xu DL, Zhang WG, et al. (2011) Estimation of water clarity in Taihu Lake and surrounding rivers using Landsat imagery. Adv Water Resour 34: 165–173.
22. Ma RH, Duan HT, Gu XH, Zhang SX (2008) Detecting aquatic vegetation changes in Taihu Lake, China using multi-temporal satellite imagery. Sensors 8: 3988–4005.
23. Qing BQ (2009) Progress and prospect on the eco–environmental research of Lake Taihu. J Lake Sci 21: 445–455.
24. Pan G, Yang B, Wang D, Chen H, Tian B, et al. (2011) In-lake algal bloom removal and submerged vegetation restoration using modified local soils. Ecol Eng: 302–308.
25. Hu W, Zhai S, Zhu Z, Han H (2008) Impacts of the Yangtze River water transfer on the restoration of Lake Taihu. Ecol Eng 34: 30–49.
26. Li Y, Acharya K, Yu Z (2011) Modeling impacts of Yangtze River water transfer on water ages in Lake Taihu, China. Ecol Eng 37: 325–334.
27. Guo L (2007) Doing battle with the green monster of Taihu Lake. Science 317: 1166.
28. Yang M, Yu J, Li Z, Guo Z, Burch M, et al. (2008) Taihu Lake not to blame for Wuxi's woes. Science 319: 158.
29. Gu X, Zhang S, Bai X, Hu W, Hu Y, et al. (2005) Evolution of community structure of aquatic macrophytes in East Taihu Lake and its wetlands. Acta Ecol Sinica 25: 1541–1548.
30. He J, Gu XH, Liu GF (2008) Aquatic macrophytes in East Lake Taihu and its interaction with water environment. J Lake Sci 20: 790–795.
31. Liu WL, Hu WP, Chen YG, Gu XH, Hu ZX, et al. (2007) Temporal and spatial variation of aquatic macrophytes in west Taihu Lake. Acta Ecol Sinica 27: 159–170.
32. Otahelová H, Otahel J, Pazúr R, Hrivnák R, Valachovic M (2011) Spatio-temporal changes in land cover and aquatic macrophytes of the Danube floodplain lake. Limnologica 41: 316–324.
33. Yang Y, Jiang N, Yin L, Hu B (2005) RS-based dynamic monitoring of lake area and enclosure culture in East Taihu Lake. J Lake Sci 17: 133–138.
34. Kloiber SM, Brezonik PL, Bauer ME (2002) Application of Landsat imagery to regional-scale assessments of lake clarity. Water Res 36: 4330–4340.
35. Chavez PS (1996) Image-based atmospheric corrections-revisited and improved. Photogramm Eng Rem Sens 62: 1025–1035.
36. Wu G, de Leeuw J, Skidmore AK, Prins HHT, Liu Y (2007) Concurrent monitoring of vessels and water turbidity enhances the strength of evidence in remotely sensed dredging impact assessment. Water Res 41: 3271–3280.
37. Ozesmi SL, Bauer ME (2002) Satellite remote sensing of wetlands. Wetlands Ecol Manage 10: 381–402.

38. Davranche A, Lefebvre G, Poulin B (2010) Wetland monitoring using classification trees and SPOT-5 seasonal time series. Remote Sens Environ 114: 552–562.

39. Lu N, Hu WP, Deng JC, Zhai SH, Chen XM, et al. (2009) Spatial distribution characteristics and ecological significance of alkaline phosphatase in water column of Tahihu Lake. Environ Sci 30: 2898–2903.

40. Brown EC, Story MH, Thompson C, Commisso K, Smith TG, et al. (2003) National Park vegetation mapping using multitemporal Landsat 7 data and a decision tree classifier. Remote Sens Environ 85: 316–327.

41. Baker C, Lawrence R, Montagne C, Patten D (2006) Mapping wetlands and riparian areas using Landsat ETM+ imagery and decision-tree-based models. Wetlands 26: 465–474.

42. Wright C, Gallant A (2007) Improved wetland remote sensing in Yellowstone National Park using classification trees to combine TM imagery and ancillary environmental data. Remote Sens Environ 107: 582–605.

43. Wei A, Chow-Fraser P (2007) Use of IKONOS imagery to map coastal wetlands of Georgian Bay. Fisheries 32: 167–173.

44. Midwood JD, Chow-Fraser P (2010) Mapping floating and emergent aquatic vegetation in coastal wetlands of Eastern Georgian Bay, Lake Huron, Canada. Wetlands 30: 1–12.

45. Zhao DH, Jiang H, Yang TW, Cai Y, Xu DL, et al. (2012) Remote sensing of aquatic vegetation distribution in Taihu Lake using an improved classification tree with modified thresholds. J Environ Manage 95: 98–107.

46. Scheffer M, Szabó S, Gragnani A, Van Nes EH, Rinaldi S, et al. (2003) Floating plant dominance as a stable state. Proceedings of the national academy of sciences 100: 4040–4045.

47. Zhu GW (2008) Eutrophic status and causing factors for a large, shallow and subtropical Lake Taihu, China. J Lake Sci 20: 21–26.

48. Scheffer M, Hosper S, Meijer M, Moss B, Jeppesen E (1993) Alternative equilibria in shallow lakes. Trends Ecol Evol 8: 275–279.

49. Blindow I, Andersson G, Hargeby A, Johansson S (1993) Long-term pattern of alternative stable states in two shallow eutrophic lakes. Freshwater Biol 30: 159–167.

50. Scheffer M, Carpenter S, Foley JA, Folke C, Walker B (2001) Catastrophic shifts in ecosystems. Nature 413: 591–596.

51. Schindler DW (2006) Recent advances in the understanding and management of eutrophication. Limnol Oceanogr 51: 356–363.

52. Hautier Y, Niklaus PA, Hector A (2009) Competition for light causes plant biodiversity loss after eutrophication. Science 324: 636–638.

53. Michelan TS, Thomaz S, Mormul RP, Carvalho P (2010) Effects of an exotic invasive macrophyte (tropical signalgrass) on native plant community composition, species richness and functional diversity. Freshwater Biol 55: 1315–1326.

54. Zhang S, Liu A, Ma J, Zhou Q, Xu D, et al. (2010) Changes in physicochemical and biological factors during regime shifts in a restoration demonstration of macrophytes in a small hypereutrophic Chinese lake. Ecol Eng 36: 1611–1619.

55. Squires M, Lesack L, Huebert D (2002) The influence of water transparency on the distribution and abundance of macrophytes among lakes of the Mackenzie Delta, Western Canadian Arctic. Freshwater Biol 47: 2123–2135.

56. Szabo S, Scheffer M, Roijackers R, Waluto B, Braun M, et al. (2010) Strong growth limitation of a floating plant (*Lemna gibba*) by the submerged macrophyte (*Elodea nuttallii*) under laboratory conditions. Freshwater Biol 55: 681–690.

57. Jin X, Yan C, Xu Q (2007) The community features of aquatic plants and its influence factors of lakeside zone in the north of Lake Taihu. J Lake Sci 19: 151–157.

58. Havens KE (2003) Submerged aquatic vegetation correlations with depth and light attenuating materials in a shallow subtropical lake. Hydrobiologia 493: 173–186.

59. Tuckett R, Merritt D, Hay F, Hopper S, Dixon K (2010) Dormancy, germination and seed bank storage: a study in support of ex situ conservation of macrophytes of southwest Australian temporary pools. Freshwater Biol 55: 1118–1129.

60. Liu WL, Hu WP, Chen Q (2007) The phenotypic plasticity of *Potamogeton malaianus* Miq. on the effect of sediment shift and Secchi depth variation in Taihu Lake. Ecol Environ 16: 363–368.

7

Efficient Video Panoramic Image Stitching Based on an Improved Selection of Harris Corners and a Multiple-Constraint Corner Matching

Minchen Zhu[1], Weizhi Wang[2], Binghan Liu[1]*, Jingshan Huang[3]

1 College of Mathematics and Computer Science, Fuzhou University, Fuzhou, Fujian, China, **2** College of Civil Engineering, Fuzhou University, Fuzhou, Fujian, China, **3** School of Computing, University of South Alabama, Mobile, Alabama, United States of America

Abstract

Video panoramic image stitching is extremely time-consuming among other challenges. We present a new algorithm: (i) Improved, self-adaptive selection of Harris corners. The successful stitching relies heavily on the accuracy of corner selection. We fragment each image into numerous regions and select corners within each region according to the normalized variance of region grayscales. Such a selection is self-adaptive and guarantees that corners are distributed proportional to region texture information. The possible clustering of corners is also avoided. (ii) Multiple-constraint corner matching. The traditional Random Sample Consensus (RANSAC) algorithm is inefficient, especially when handling a large number of images with similar features. We filter out many inappropriate corners according to their position information, and then generate candidate matching pairs based on grayscales of adjacent regions around corners. Finally we apply multiple constraints on every two pairs to remove incorrectly matched pairs. By a significantly reduced number of iterations needed in RANSAC, the stitching can be performed in a much more efficient manner. Experiments demonstrate that (i) our corner matching is four times faster than normalized cross-correlation function (NCC) rough match in RANSAC and (ii) generated panoramas feature a smooth transition in overlapping image areas and satisfy real-time human visual requirements.

Editor: Derek Abbott, University of Adelaide, Australia

Funding: This research was supported by Natural Science Foundation of the Fujian Province, China (under Grant Numbers 2012J01263 and the Fujian Province, China (under Grant Number 2009J1007). The funders had no role in study design, data collection and analysis, decision to publish, or preparation of the manuscript.

Competing Interests: The authors have declared that no competing interests exist.

* E-mail: lbh@fzu.edu.cn

Introduction

Video panoramic image stitching is based on the similarity of overlapping regions between adjacent images. State-of-the-art image registration (a.k.a. image alignment) algorithms can be classified into three different categories: intensity-based (for example, [1,2]), frequency-domain-based (for example, [3]), and feature-based (for example, [4–11]). Intensity-based algorithms usually involve a large amount of computation and do not handle well the image alignment after rotation and scaling. Algorithms based on the frequency domain are in general faster and more appropriate for small translation, rotation, and scaling, but their performance is degraded when dealing with smaller overlapping regions. Therefore, more research has been focused on feature-based algorithms, which make use of a small number of invariant points, lines, or edges to align images. The computational complexity is reduced due to less information needing to be processed. In addition, they are robust to changes in image intensity. However, two issues have been identified for many feature-based algorithms. (i) Alignment outcomes are vulnerable to numerous factors (e.g., noise, information distribution pattern in images, and so forth) and result in a relatively low accuracy. In the area of video panorama stitching, more often than not, the overlapping between adjacent images is relatively small. Therefore, the successful stitching relies heavily on the accuracy and robustness of selected features. (ii) Many existing algorithms make use of an exhaustive search based on template matching. The computation, although already decreased to some extent, is still intensive, which does not meet the real-time requirement usually found in video panorama stitching.

Based on the above analysis on existing methodologies, we herein present a new algorithm. Our algorithm is classified as a feature-based algorithm and is motivated by the need to stitch adjacent images in both cases of (i) when there exist small overlapping areas and (ii) when their difference in translation, rotation, and scaling is small. Our algorithm has two preferable features, i.e., a self-adaptive selection of Harris corners and a multiple-constraint corner matching. (i) We utilize Harris corners, obtained with low computational complexity and high robustness, as features to align images. We fragment each original image into several regions and select corners within each region according to the normalized variance of region grayscales. Such a self-adaptive selection guarantees that corners are distributed such that more will be selected from regions with richer texture information. In addition, our algorithm avoids the possible clustering of corners. (ii) Upon filtering out a large number of inappropriate corners according to their position information, we generate an initial set of matching-corner pairs based on grayscales of adjacent regions around each corner. Finally we apply multiple constraints on every two candidate pairs, e.g., their midpoints, distances, and slopes, to remove incorrectly matched pairs. As such, we are able to significantly reduce the number of iterations needed in traditional Random Sample Consensus (RANSAC) algorithm [12]. The video panoramic image stitching can then be performed in a more efficient manner. Moreover, our algorithm is robust to errors caused by the camera shake because optimal parameter values will be obtained in a self-adaptive manner during the image stitching.

The rest of this paper is organized as: Section "Related Work" briefly discusses related work; Section "Materials and Methods" introduces in detail our methodology; Section "Results and Discussion" describes experimental results; Section "Limitations of the Study, Open Questions, and Future Work" discusses some open questions along with future research directions; and finally Section "Conclusions" concludes.

Related Work

Reddy and Chatterji [3] discussed an extension of the well-known phase correlation technique to cover translation, rotation, and scaling during the image matching. Fourier scaling properties and Fourier rotational properties were used to find scale and rotational movement, and the phase correlation technique determined the translational movement. The algorithm presented in this work is characterized by its insensitivity to translation, rotation, scale, and noise as well as by its low computational cost. The authors believed that their matching algorithm was exact except for scaling where it was found to deviate in the third place of decimal probably due to the nonlinear processing.

Since to solve the problem of modeling and analyzing pushbroom sensors commonly used in satellite imagery is difficult and computationally intensive, the authors presented a simplified model of a pushbroom sensor (the linear pushbroom model) in [1]. Their model has the advantage of computational simplicity while at the same time giving very accurate results compared with the full orbiting pushbroom model. Besides remote sensing, the linear pushbroom model is also useful in many other imaging applications, and this model leads to theoretical insights that are approximately valid for the full model as well.

Consistency of image edge filtering is of prime importance for 3D interpretation of image sequences using feature tracking algorithms. To cater for image regions containing texture and isolated features, the authors in [4] presented a combined corner and edge detector based on the local auto-correlation function. They successfully demonstrated that their proposed detector is able to perform with good consistency on natural imagery.

Lacey, Pinitkarn, and Thacker [12] compared the use of RANSAC for the determination of epipolar geometry for calibrated stereo reconstruction of 3D data with more conventional optimisation schemes. The authors illustrated the poor convergence efficiency of RANSAC and highlighted the need for an a-priori estimate of outlier contamination proportion. An algorithm was suggested to make better use of the solutions found during the RANSAC search while giving a convergence criteria that is independent of outlier proportion.

A review of recent and classic image registration methods was presented in [13]. Approaches reviewed were classified according to their nature (area-based and feature-based) and four basic steps of the image registration procedure: feature detection, feature matching, mapping function design, and image transformation and resampling. The authors also discussed problematic issues of image registration and provided an outlook for the future research.

Sivic and Zisserman [5] described an approach to object and scene retrieval that searches for and localizes all occurrences of a user outlined object in a video. The object was represented by a set of viewpoint invariant region descriptors so that recognition can proceed successfully despite changes in viewpoint, illumination, and partial occlusion. The authors used the temporal continuity of the video to track regions to reject unstable regions and reduce the noise effect in the descriptors.

The performance of descriptors computed for local interest regions was compared in [6]. Mikolajczyk and Schmid believed that descriptors should be distinctive and at the same time robust to changes in viewing conditions as well as to errors of the detector. Their evaluation used as criterion recall with respect to precision and was carried out for different image transformations. In addition, they also proposed an extension of

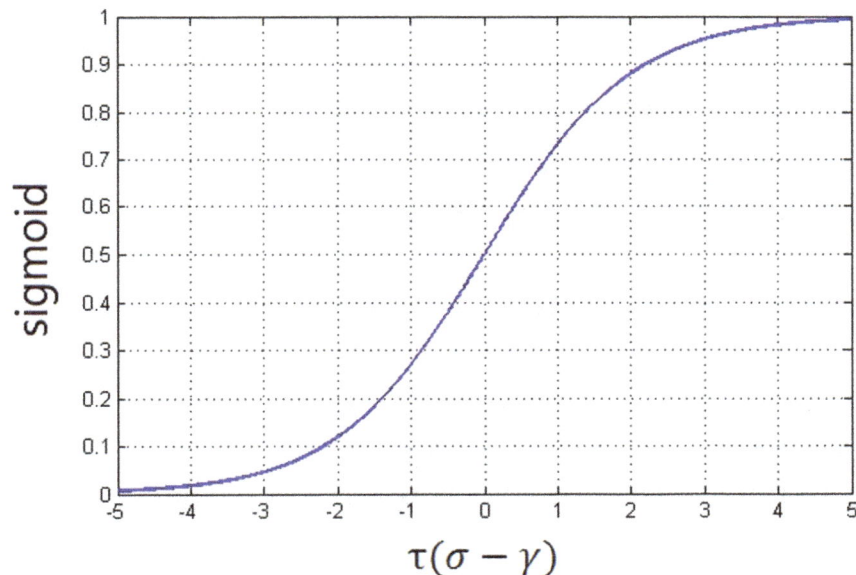

Figure 1. The Sigmoid function, where it is clearly demonstrated that the critical value range of τ is [−5, 5].

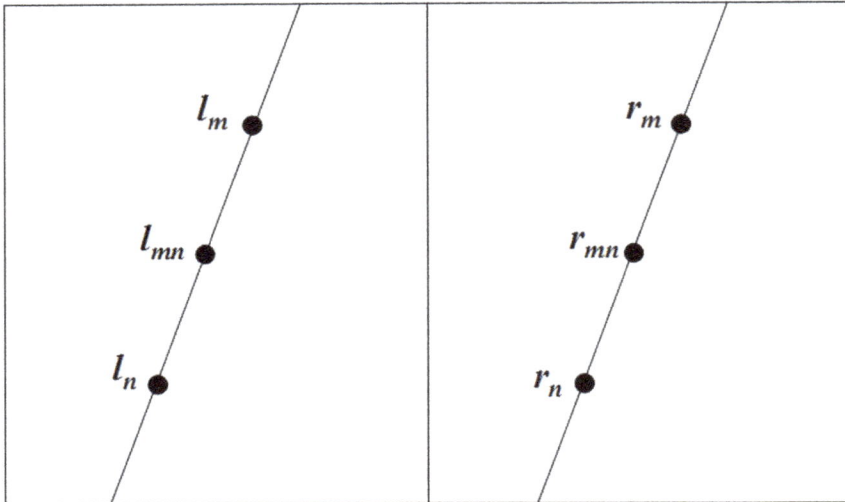

Figure 2. Two initially matching-corner pairs, (l_m, r_m) **and** (l_n, r_n), **along with their respective midpoints:** l_{mn} **between** l_m **and** l_n **and** r_{mn} **between** r_m **and** r_n.

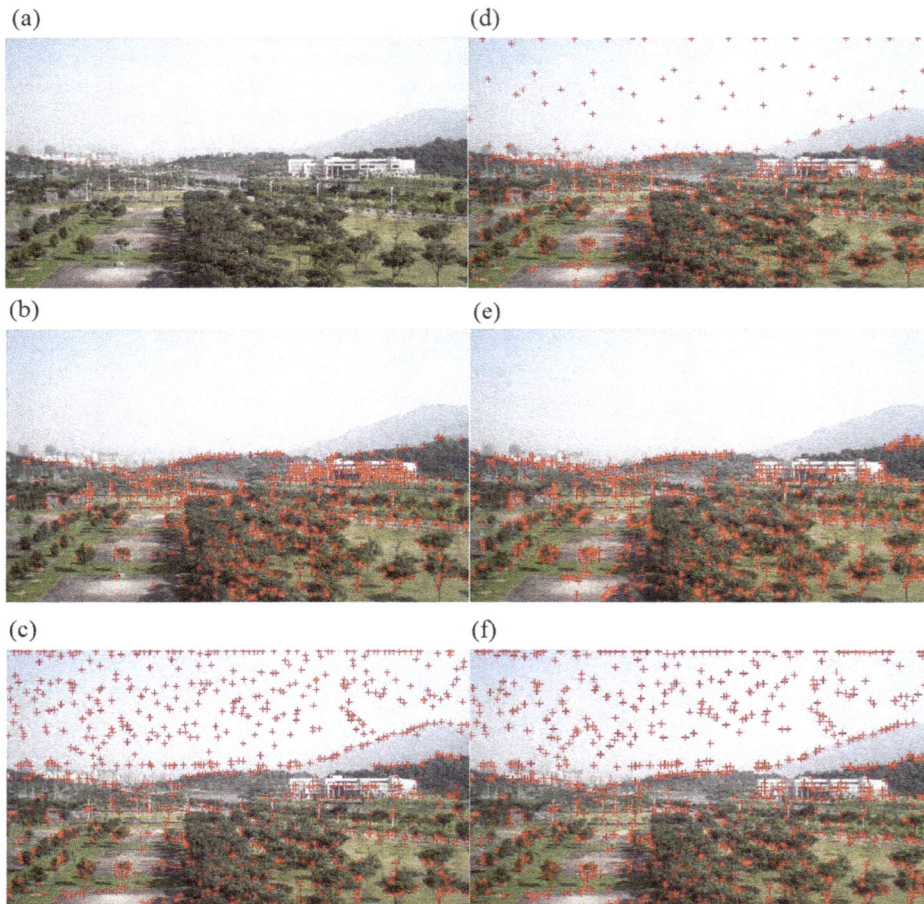

Figure 3. Experiments on corner selection: (a) Original image; (b) Harris detector; (c) The regional corner-selection algorithm; (d) Our algorithm with $\tau = 10$; **(e) Our algorithm with** $\tau = 30$; **(f) Our algorithm with** $\tau = 1$.

(a)

(b)

(c)

(d)

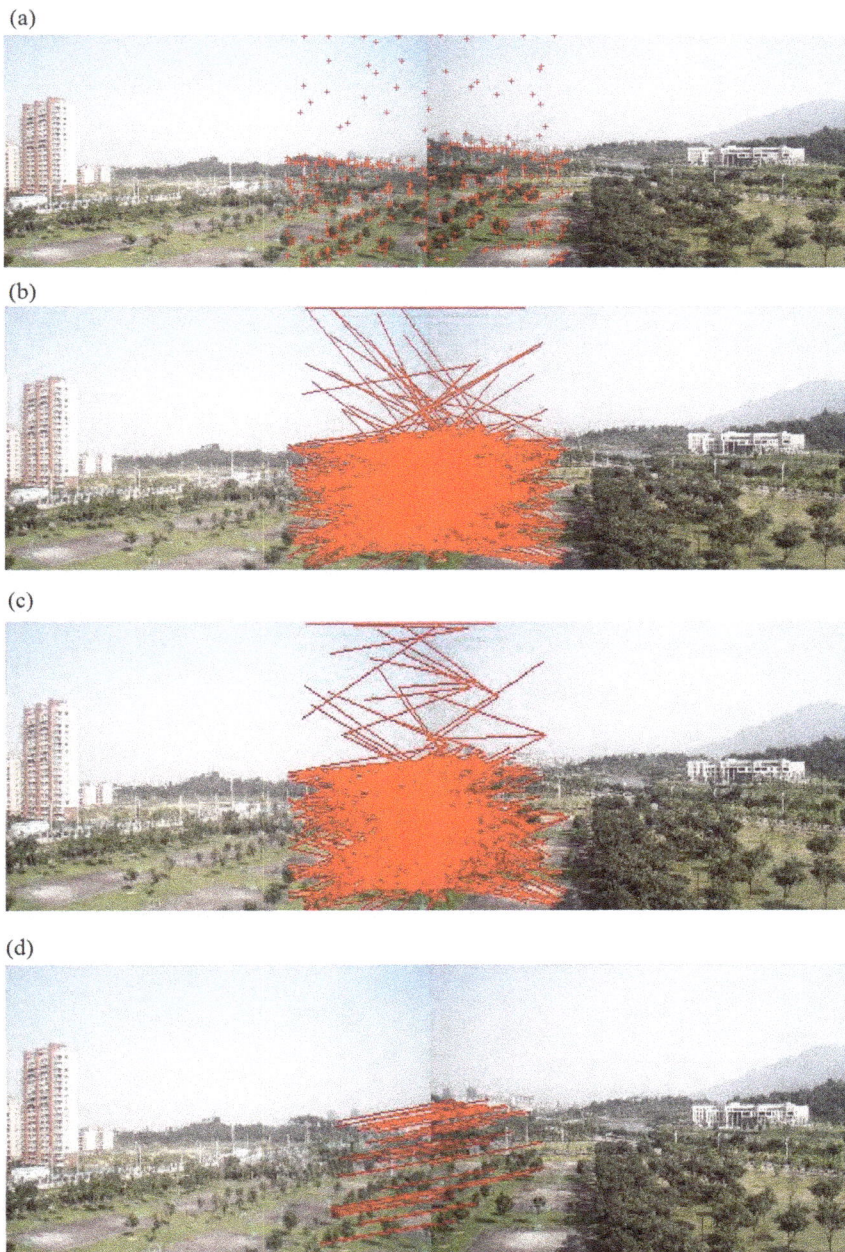

Figure 4. The first experiment on the corner matching efficiency: (a) Selected corners (using our algorithm) from two original images; (b) Corner matching by traditional RANSAC algorithm (NCC rough match); (c) Initial set of matching-corner pairs in our algorithm; (d) Final set of matching-corner pairs in our algorithm.

the SIFT descriptor and showed that it outperforms the original method.

A method for image similarity measure was presented in [2], where a hand-drawn rough black-and-white sketch was compared with an existing database of full color images (art works and photographs). The system created the evaluation of nonprecise, easy-to-input sketched information and can then provide the user with options of either retrieving similar images in the database or ranking the quality of the sketch against a given standard. Alternatively, the inherent pattern-matching capability of the

system can be utilized to allow detection of distortion in any given real time-image sequences in vision-driven ambient intelligence applications.

Nister, Naroditsky, and Bergen [7] presented a system to estimate the motion of a stereo head or a single moving camera based on video input. The system operates in real-time with low delay and the motion estimates were used for navigational purposes. Point features were matched between pairs of frames and linked into image trajectories at video rate. Robust estimates of the camera motion were then produced from the feature tracks

(a)

(b)

(c)

(d)

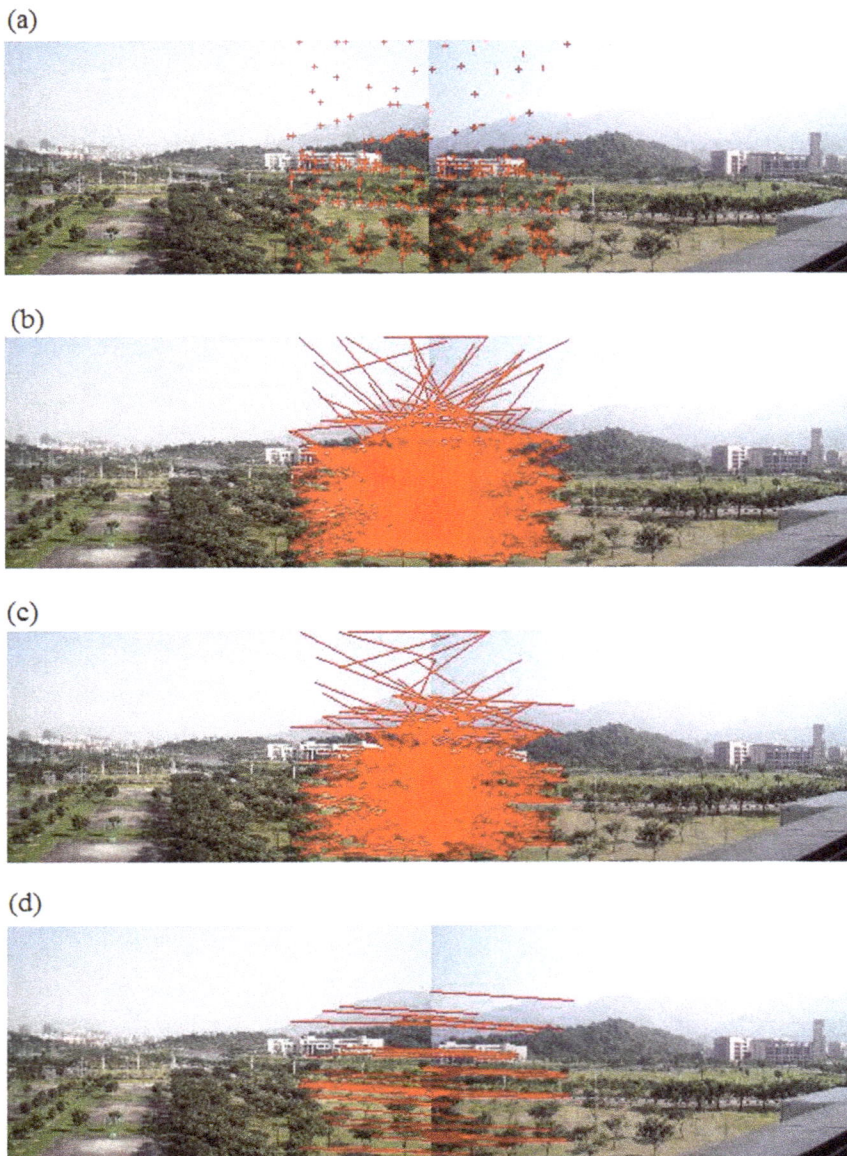

Figure 5. The second experiment on the corner matching efficiency: (a) Selected corners (using our algorithm) from two original images; (b) Corner matching by traditional RANSAC algorithm (NCC rough match); (c) Initial set of matching-corner pairs in our algorithm; (d) Final set of matching-corner pairs in our algorithm.

using a geometric hypothesize-and-test architecture. The authors successfully applied the pose estimation method to video from aerial, automotive, and handheld platforms.

Zhu et al. [8] introduced the concept of distribution quality to quantify the distribution of seed points for triangle constrained image-matching propagation. An intensive experimental analysis

Table 1. Efficiency comparison between traditional RANSAC and our algorithm (I).

	Total number of corner pairs to be processed	Total number of generated corner pairs	Time spent (second)
NCC rough match in traditional RANSAC	72,900(= 270×270)	528	7.20
Our algorithm (set T)	17,993	456	1.81
Our algorithm (set T^*)	**456**	**41**	**0.05**

Table 2. Efficiency comparison between traditional RANSAC and our algorithm (II).

	Total number of corner pairs to be processed	Total number of generated corner pairs	Time spent (second)
NCC rough match in traditional RANSAC	72,900(= 270×270)	534	6.80
Our algorithm (set T)	16,176	453	1.63
Our algorithm (set T^*)	**453**	**45**	**0.08**

was illustrated using two different stereo aerial images and, based on the experimental results, a seed point selection strategy was proposed. An automatic selection method was then introduced to provide good distribution quality for a defined number of seed points. Seed points with proper distribution are able to provide better matching results and the distribution quality is a useful descriptor to the distribution of seed points.

Zhao et al. [9] presented a regional corner-selection algorithm to overcome the disadvantage found in traditional Harris detector, i.e., corners tend to cluster around regions with richer texture information. They first fragmented the original image into several regions, then selected a fix number of corners in each region and ultimately applied a scaling parameter to finalize the corner selection. Their algorithm is able to select corners with relatively high quality but there exist some issues. We will further discuss this algorithm in later sections because it is closely related to our proposed algorithm in this paper.

In [14], the authors defined digital image mosaics as image registration and fusion of many overlapping images or multiple video frames into a single wide field of view image or dynamic panoramic image. They introduced four basic types of image mosaics based on the characters of images to be mosaic. The state-of-the-art image mosaics research was summarized and new research directions were discussed.

The authors in [10] presented a fast image mosaic algorithm based on feature points matching. They proposed two algorithms. The first one is to import a new filtering method for matching points by choosing pairs of correlated feature points with a clustering algorithm aiming at the disadvantage of the RANSAC algorithm. The second algorithm presents a new method of blending images by using optimum path best-matched-line combined with pixel brightness weighting function in HSI color space. Their methodologies were able to remove the gusting phenomenon and result in a brightness blending image, and, in a more efficient manner.

Cao et al. [11] proposed to extract features from edges of an image so that the multi-scale image fusion and mosaicing can be

performed. An edge-smoothing pyramid was built and stable features of image registration were extracted. They then reused the multi-scale representation to fuse registered images and thus eliminated the cost of mosaicing. Results indicated that their algorithm can eliminate false feature matches, enhance the image transformation precision, and reduce the computation cost of registration and following mosaic.

Materials and Methods

Improved Harris Corner Selection

The Harris algorithm [4] detects corners by calculating the intensity change of a pixel after it is shifted in some direction. According to [4], for a given pixel (x,y), an autocorrelation function of the surrounding area of (x,y), $E_{x,y}$, is defined in Eq. (1), where u and v are small shifts in x and y coordinates, respectively; and $w(x,y)$ is the Gaussian function to filter out noise.

$$E_{x,y} = \sum_{u,v} w(u,v)[I_{x+u,y+v} - I_{u,v}]^2 \qquad (1)$$

The Taylor expansion of $E_{x,y}$ is shown in Eq. (2), and $\hat{M}(x,y)$ is the autocorrelation matrix (2×2 and symmetric), calculated as in Eq. (3), where $\frac{\partial I}{\partial x}$ and $\frac{\partial I}{\partial y}$ are partial derivatives of the pixel in x and y coordinates, respectively; $w(x,y)$ is the Gaussian function to filter out noise; and $\otimes$ is the convolution operation.

$$E_{x,y} \cong [uv]\hat{M}(x,y)\begin{bmatrix} u \\ v \end{bmatrix} \qquad (2)$$

Table 3. Efficiency comparison between traditional RANSAC and our algorithm (Overall).

	Total number of corner pairs to be processed	Total number of generated corner pairs	Time spent (second)
	Average/Variation	**Average/Variation**	**Average/Variation**
NCC rough match in traditional RANSAC	72,900/0	532/16	7/0.07
Our algorithm (set T)	17,087/37	455/4.5	1.73/0.02
Our algorithm (set T^*)	**455/4.5**	**43/6**	**0.07/0.003**

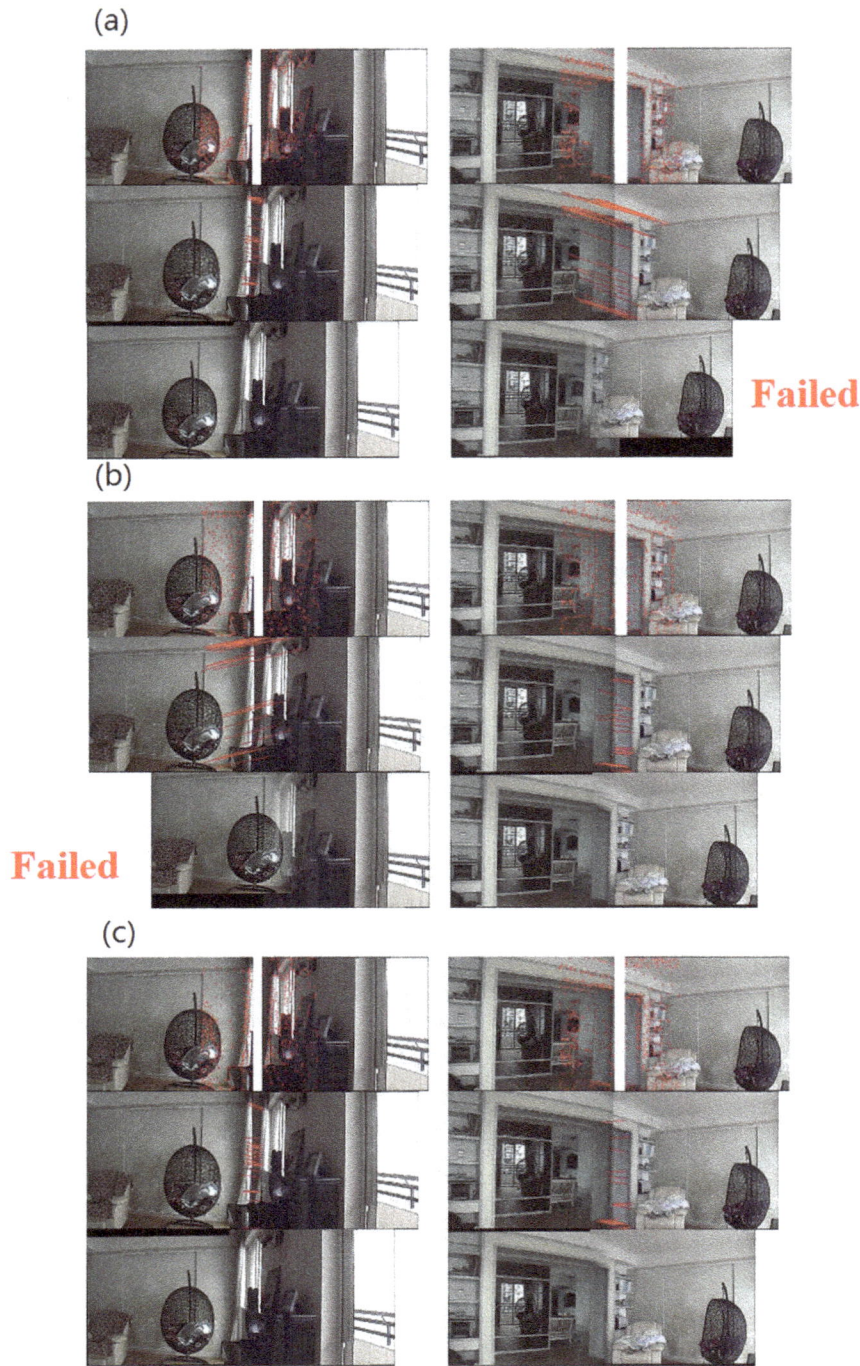

Figure 6. Comparison of stitching results from three different algorithms (I): (a) Traditional Harris detector; (b) The regional corner-selection algorithm in [9]; (c) Our presented algorithm.

$$\hat{M}(x,y) = \sum_{x,y} w(x,y) \otimes \begin{bmatrix} \left(\dfrac{\partial I}{\partial x}\right)^2 & \dfrac{\partial I}{\partial x} \cdot \dfrac{\partial I}{\partial y} \\ \dfrac{\partial I}{\partial x} \cdot \dfrac{\partial I}{\partial y} & \left(\dfrac{\partial I}{\partial y}\right)^2 \end{bmatrix} \quad (3)$$

The autocorrelation matrix, $\hat{M}$, can be used to detect corners. To be more specific, Eq. (4) defines a response function for the matrix, where $\det(\hat{M})$ and $\mathrm{tr}(\hat{M})$ are the determinant and trace of $\hat{M}$, respectively; and η is a parameter in the range of $[0.04, 0.06]$. If the R value of a pixel is greater than a threshold, such a pixel will be selected as a Harris corner. Note that R depends on various characteristics of actual images, size and texture for example. As a

(a)

(b)

(c)

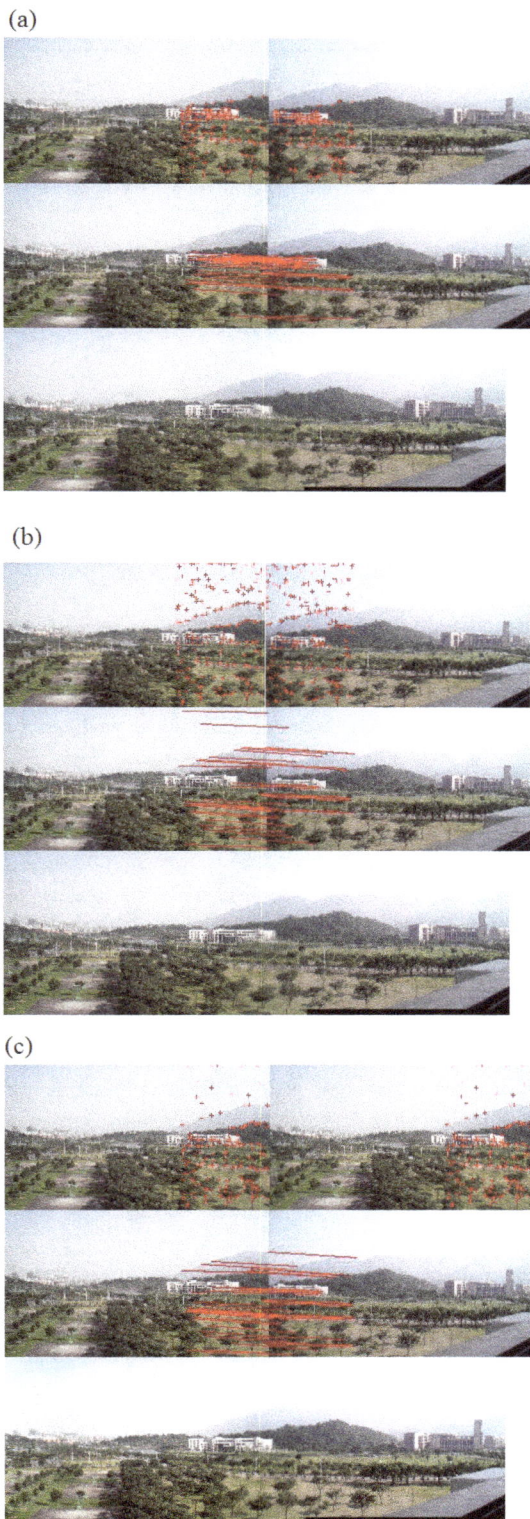

Figure 7. Comparison of stitching results from three different algorithms (II): (a) Traditional Harris detector; (b) The regional corner-selection algorithm in [9]; (c) Our presented algorithm.

result, R is usually determined indirectly: pixels are sorted in a descending order of their R values; then the first Sum (a predefined total number, see greater details in Eq. (8)) pixels are selected as

corners. Notice that the discovery of meaningful corners does not depend on region boundaries because the calculation of R value is not tied with fragmented regions.

$$R = \det(\hat{M}) - \eta \cdot \mathrm{tr}^2(\hat{M}) \qquad (4)$$

Harris detector only involves the first-order difference and filtering operations of pixel grayscale, with low computational complexity. A large number of corners will be detected in regions with rich texture, whereas fewer corners will be selected in regions with less texture information. Therefore, selected corners are not evenly distributed. In other words, corners tend to cluster around regions with richer texture. Zhao et al. [12] proposed a regional corner-selection algorithm, where they fragmented the original image into several regions. A fix number of corners with top R values were selected in each region as candidate corners, and all such candidate corners (a total of Sum) were then sorted in their R values. Finally a scaling parameter, η, whose range is (0, 1), was applied to finalize the corner selection, i.e., a total of $\eta \times$ Sum corners will be generated. To assure that each region contains some finalized corners, they iteratively applied different η values in an ascending order and the iteration was terminated as soon as there was at least one finalized corner for each region. This algorithm is able to select corners with relatively high quality but some issues exist. (i) When corners are evenly distributed the iteration will be terminated early with a small η value, which results in fewer corners than necessary and will affect the corner matching thereafter. (ii) Corners selected in regions with rich texture contain redundant, neighboring corners. (iii) The value of η is biased to outlier regions, i.e., regions with extremely rich or poor texture.

We believe that the corner selection should be proportional to region texture information and, at the same time, should avoid the possible clustering as much as we can. To handle such a tradeoff, we propose to select corners in each region according to the normalized variance of region grayscales. Our corner selection is self-adaptive and decomposed into several steps.

- We fragment the original image into $M \times N$ regions. For each region, a texture weight, ω_i, is calculated according to the grayscale variance of that region. To be more specific, such a weight should be able to offset the region texture feature. In other words, the larger the grayscale variance in a region, the more influence on the weight of that region. Due to its characteristics of smoothness and progressiveness (Figure 1), we utilize the Sigmoid function (demonstrated in Eq. (5), where τ is a ratio parameter; σ is the variance; and γ is the offset amount) to normalize the grayscale variance in a region (demonstrated in Eq. (6), where i is an integer in the range of $(1, M \times N)$; σ_i is the grayscale variance of the ith region; and σ_{min}, σ_{max}, and $\bar{\sigma}$ are the minimum, maximum, and expected values of the grayscale variance, respectively). After we calculate the normalized variance of each region, $\hat{\sigma}_i$, it is in turn used to determine the weight for that region (Eq. (7)). Finally we use ω_i to determine the number of corners, B_i, in each region (Eq. (8)), where Sum is the predefined total number of corners for all regions.

- R value is calculated for each pixel in each region according to Eq. (4). Corners with maximum or minimum R values are eliminated.

Table 4. Qualitative and quantitative analysis on three different algorithms (I).

Algorithm	Left Part		Right Part	
	Successful stitching or not? (qualitative)	Similarity between neighboring regions to be stitched (quantitative)	Successful stitching or not? (qualitative)	Similarity between neighboring regions to be stitched (quantitative)
A (Traditional Harris detector)	Yes	0.9479	No	0.8613
B (Regional corner-selection algorithm)	No	0.8088	Yes	0.9827
C (Our presented algorithm)	Yes	0.9442	Yes	0.9788

• All pixels in each region are sorted in a descending order of their R values. The first B_i pixels in each region are then selected as corners.

$$\text{sigmoid} = \frac{1}{1 + \exp(-\tau(\sigma - \gamma))} \quad (5)$$

$$\hat{\sigma}_i = \frac{1}{1 + \exp\left(-\tau \frac{\sigma_i - \bar{\sigma}}{\sigma_{\max} - \sigma_{\min}}\right)} \quad (6)$$

$$\omega_i = \frac{\hat{\sigma}_i}{\sum\limits_{j=1}^{M \times N} \hat{\sigma}_j} \quad (7)$$

$$B_i = \omega_i \times \text{Sum} \quad (8)$$

Note that in Eq. (6) we set the value of τ to 10 because it is clearly demonstrated in Figure 1 that the critical variable range is the grayscale variance (i.e., the mean variance) as the offset amount because compared with the median, the mean can better offset outliner regions.

Multiple-Constraint Corner Matching

The traditional RANSAC algorithm is inefficient, especially when stitching a large number of images and when these images have similar features. It thus does not meet the real-time requirement commonly found in video panorama stitching. Note that in the field of video panorama stitching, more often than not, adjacent images have highly similar features with each other. Based on this insight, we propose to apply multiple constraints on candidate matching-corner pairs to remove incorrectly matched pairs. As such, we can significantly reduce the number of iterations needed in traditional RANSAC algorithm.

• Step 1: Corner similarity matrix between adjacent images. Suppose that there are two images to be aligned, I^l on the left hand side and I^r on the right hand side, both of which have the same resolution of $W \times H$. One corner from the left image is denoted I_i^l (with coordinates x_i^l and y_i^l, and the intensity of $G^l(x_i, y_i)$) and another corner from the right image is denoted I_j^r (with coordinates x_j^r and y_j^r, and the intensity of $G^r(x_j, y_j)$). These two corners, I_i^l and I_j^r, can be aligned with each other if the following constraints are satisfied: (i) the y coordinate difference of these two corners is less than a threshold; (ii) the x coordinate of the left corner is no less than that of the right corner; (iii) the difference of R values of these two corners is less than a threshold; and (iv) there is a high intensity correlation between these two corners. The first three constraints are formalized in Eq. (9), where λ_H and λ_R are thresholds in constraints 1 and 3, respectively. Note that constraint 2 requires knowing in advance which image is the left one and which is the right one. In other words, our algorithm is only working on camera panoramas where the panning direction (left or right) is known in advance and does not change. Fortunately, such a restriction is not difficult to be satisfied in real-world applications. As for the last constraint, we need to make use of the normalized cross-correlation (NCC) function described in [14]. Suppose that the similarity window size is $(2w+1) \times (2w+1)$, NCC is calculated as in Eq. (10), where $\begin{cases} D_l(i,u,v) = G^l(x_i+u, y_i+v) - \overline{G^l(x_i, y_i)} \\ D_r(j,u,v) = G^r(x_j+u, y_j+v) - \overline{G^r(x_j, y_j)} \end{cases}$, and $\overline{G^l(x_i, y_i)}$ and $\overline{G^r(x_j, y_j)}$ are the mean intensity of windows around corners I_i^l and I_j^r, respectively. Finally, we utilize Eq. (11) to calculate pairwise corner similarity and create a similarity matrix between adjacent images I^l and I^r.

Table 5. Qualitative and quantitative amalysis on three different algorithms (II).

Algorithm	Successful stitching or not?	Similarity between neighboring regions to be stitched
A (Traditional Harris detector)	Yes	0.9737
B (Regional corner-selection algorithm)	Yes	0.9735
C (Our presented algorithm)	Yes	**0.9842**

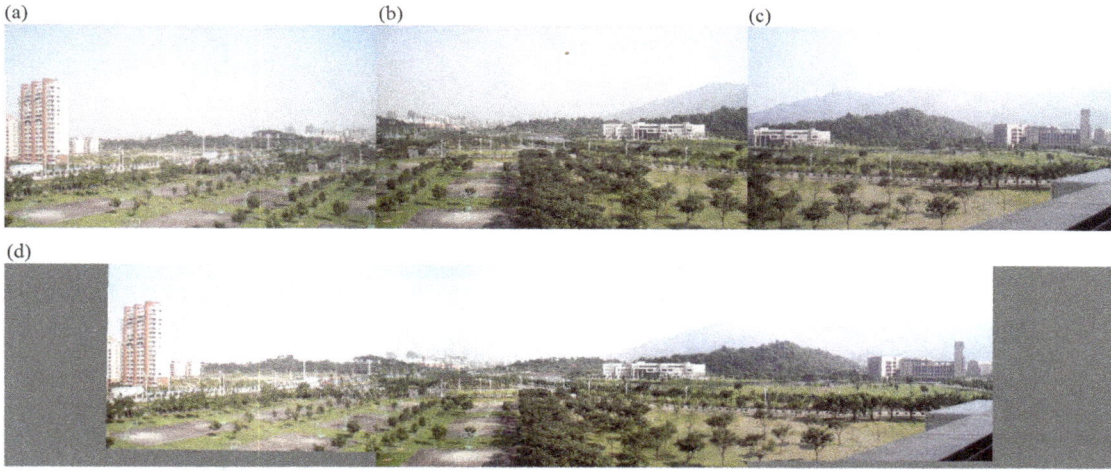

Figure 8. Experiments on the entire image-stitching algorithm (I): (a) Video image 1; (b) Video image 2 (reference image); (c) Video image 3; (d) Generated video panoramic image.

$$\begin{cases} \text{constraint1}: & |y_i^l - y_j^r| < \lambda_H \\ \text{constraint2}: & x_i^l \geq x_j^r \\ \text{constraint3}: & |R(I_i^l) - R(I_j^r)| < \lambda_R \end{cases} \quad (9)$$

$$NCC(I_i^l, I_j^r) =$$

$$\frac{\sum_{u=-w}^{w} \sum_{v=-w}^{w} [D_l(i,u,v) \cdot D_r(j,u,v)]}{\sqrt{[\sum_{u=-w}^{w} \sum_{v=-w}^{w} D_l(i,u,v)^2] \cdot [\sum_{u=-w}^{w} \sum_{v=-w}^{w} D_r(j,u,v)^2]}} \quad (10)$$

$$s(i,j) =$$

$$\begin{cases} |NCC(I_i^l, I_j^r)| & \text{if satifying three constrains in Eq.(9)} \\ & \text{and } |NCC(I_i^l, I_j^r)| > \lambda_n \\ 0 & \text{else} \end{cases} \quad (11)$$

- Step 2: Initial set of matching-corner pairs. A set of indexes of matching-corner pairs is generated by the following procedure: in each row of the similarity matrix obtained in Step 1, we find the column index such that the corresponding cell in the matrix has the maximum value for that row, and the pair of (row index, column index) is added into the set. After we process all rows in the matrix, we will obtain a set T^l. Such a procedure is formally described in Eq. (12), where Sum_l is the predefined total number of corners in the left image I^l.

$$T^l = \{(i,j) | \forall i \in [1, Sum_l], s(i,j) = \max(s(i,:)), s(i,j) \neq 0\} \quad (12)$$

Similarly, we can obtain another set T^r by searching the maximum row index for each column. Eq. (13) is the formal description of this procedure where Sum_r is the predefined total number of corners in the right image I^r.

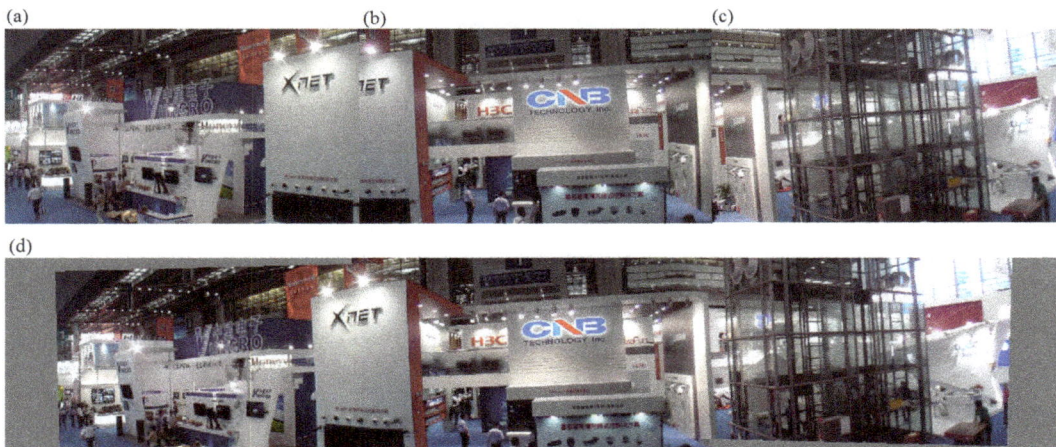

Figure 9. Experiments on the entire image-stitching algorithm (II): (a) Video image 1; (b) Video image 2 (reference image); (c) Video image 3; (d) Generated video panoramic image.

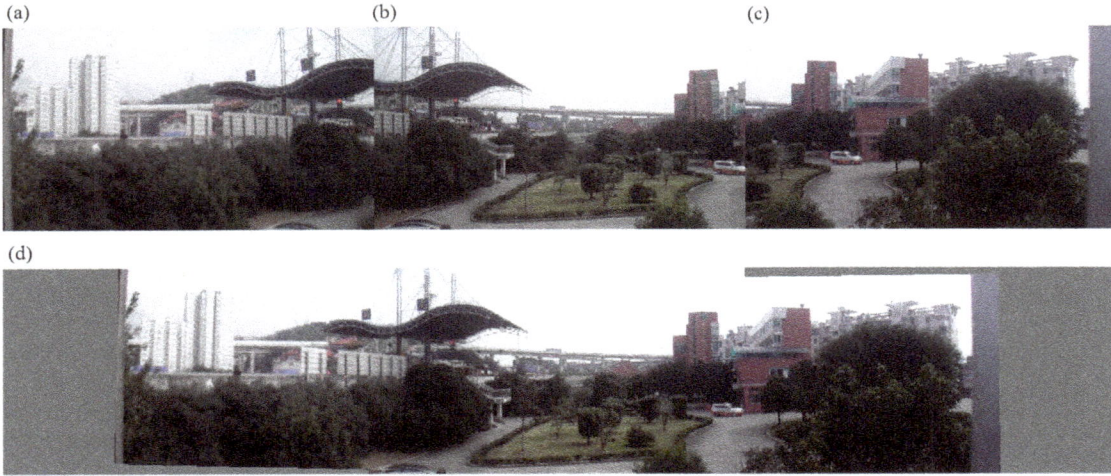

Figure 10. Experiments on the entire image-stitching algorithm (III): (a) Video image 1; (b) Video image 2 (reference image); (c) Video image 3; (d) Generated video panoramic image.

$$T^r = \{(i,j)|\forall j \in [1, \text{Sum}_r], s(i,j) = \max(s(:,j)), s(i,j) \neq 0\} \quad (13)$$

In general, Sum_l in Eq. (12) and Sum_r in Eq. (13) can take different values. In our algorithm we use the same value for these two parameters. Now we compare two sets, T^l and T^r. If a row index and a column index happen to have each other as the other component in a pair, their similarity should be adjusted. In other words, if two corners mutually find their "best" match as each other, such a pair will have an adjusted similarity value of 1. Eq. (14) formalizes this procedure of similarity adjustment. Note that this idea is also known as the "stable marriage" criteria, and was proposed in literature, [7] for example. Our methodology is different from existing methodologies in that (i) we do not discard

pairs not satisfying the stable marriage criteria and (ii) the calculation of the similarity measurement is only one of the steps in our algorithm.

$$s^*(i,j) = \begin{cases} 1 & \text{if } (i,j) \in T^l \cap T^r \\ s(i,j) & \text{else} \end{cases} \quad (14)$$

Finally we generate an initial set of matching-corner pairs T by a union of T^l and T^r (Eq. (15)). Note that this initial set of pairs is already reduced in size compared with NCC rough match in traditional RANSAC algorithm because as exhibited in Eq. (9) we have already filtered out some inappropriate corners according to their coordinate values in respective regions.

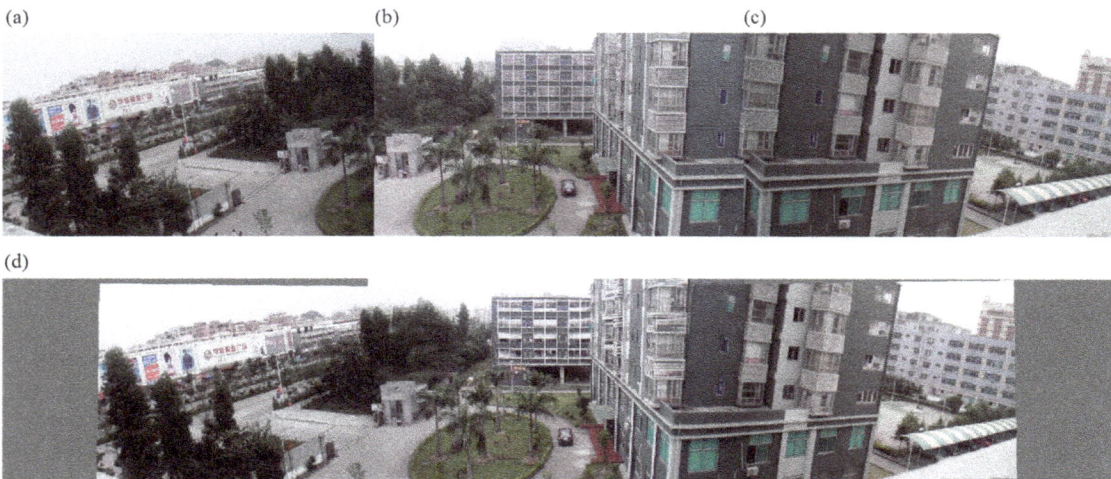

Figure 11. Experiments on the entire image-stitching algorithm (IV): (a) Video image 1; (b) Video image 2 (reference image); (c) Video image 3; (d) Generated video panoramic image.

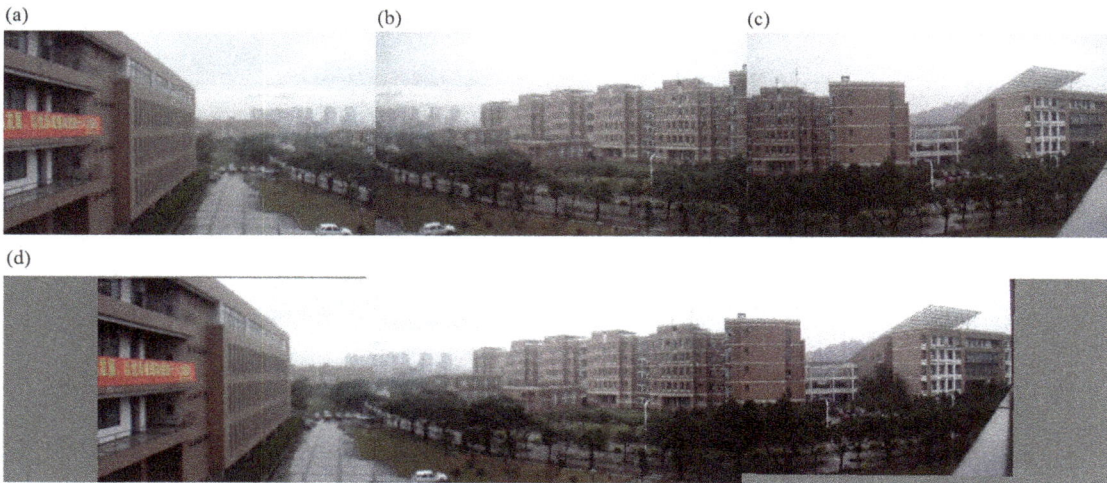

Figure 12. Experiments on the entire image-stitching algorithm (V): (a) Video image 1; (b) Video image 2 (reference image); (c) Video image 3; (d) Generated video panoramic image.

$$T = T^l \cup T^r = \{(i,j)|(i,j) \in T^l \text{ or} (i,j) \in T^r\} \qquad (15)$$

- Step 3: Multiple constraints on matching pairs. Considering two initial matching-corner pairs in Figure 2, (l_m, r_m) and (l_n, r_n), along with their respective midpoints, i.e., l_{mn} between l_m and l_n and r_{mn} between r_m and r_n. Let δ_m and $\bar{m}$ be the slope and length of the segment formed by l_m and r_m, respectively; and δ_n and $\bar{n}$ be the slope and length of the segment formed by l_n and r_n, respectively. We design multiple constraints to be applied to these two matching-corner pairs, as shown in Eq. (16).

$$\begin{cases} \text{constraint1}: & |\delta_m - \delta_n| < \lambda_\delta \\ \text{constraint2}: & |\bar{m} - \bar{n}| < \lambda_D \\ \text{constraint3}: & \left| R(I^l_{l_{mn}}) - R(I^r_{r_{mn}}) \right| < \lambda_R \end{cases} \qquad (16)$$

The intuition of Eq. (16) is that, between two matching pairs, not only the R values (Eq. (4)) of their respective midpoints should be correlated (constraint 3), but also the slope (constraint 1) and length (constraint 2) of the segments formed between these two pairs should be similar with each other as well. According to multiple constraints specified in Eq. (16), we calculate pairwise similarity between every two initial matching pairs using Eq. (17), and generate a matrix D of size Num × Num, with Num being the cardinality of T.

$$D(m,n) =$$
$$\begin{cases} \frac{s^*(l_m, r_m) + s^*(l_n, r_n)}{2} & \text{if satifying constraint in Eq.(16)} \\ 0 & \text{else} \end{cases} \qquad (17)$$

- Step 4: Final, reduced set of matching-corner pairs. Among all initial matching-corner pairs, according to Eq. (18), we search for a special pair, t, which has the strongest correlation with all other pairs.

$$t = \arg\max_{m \in [1, \text{Num}]} \left(\sum_{n=1}^{\text{Num}} D(m,n) \right) \qquad (18)$$

We then refer back to the matrix D generated in Step 3, and find all initial matching pairs that have non-zero correlation with the aforementioned special pair, t. In other words, an initial matching-corner pair will be output to the final, further reduced set as long as a non-zero value is found at the cell in D corresponding to this pair and the special pair t. Eq. (19) formally specifies this final selection step, and the resultant set T^* is the finalized, reduced set of matching-corner pairs. Note that the size of T^* is further reduced from that of T, and we explained earlier in Step 2 that T is already reduced in size compared with NCC rough match in traditional RANSAC algorithm.

$$T^* = \{(i,j)|(i,j) \in T \text{ and} D(t, l_{ij}) \neq 0\} \qquad (19)$$

Steps 1, 3, and 4 reflect our proposed heuristics with regard to the spatial consistency. Similar heuristics were presented in literature (for example [5]). It is true that heuristics described in [5] are in a more general formulation. However, our more specific formulation is exactly our unique contribution: as mentioned earlier in Section "Introduction", our algorithm is motivated by two insights on adjacent images, i.e., their small overlapping and their small difference in translation, rotation, and scaling. In other words, Eq. (9) and Eq. (16) fit better the video panoramic image stitching scenario because we take into account unique features of adjacent images to be stitched. Consequently, our methodology is more efficient (see Section "Results and Discussion" for greater details).

Image Stitching

After we obtain a finalized set of matching-corner pairs between two original images, we select one image as the reference image (the first image) and the other as the image to be aligned (the second image). We calculate the affine transformation parameters using the RANSAC algorithm. Suppose that (x,y) is a pixel in the reference image and (x',y') is the corresponding pixel in the other image. The relationships between their coordinates are illustrated in both Eq. (20) and Eq. (21), where $P = (p_0,p_1,\cdots,p_7)$ is the vector of affine transformation parameters, which can be easily calculated from T^*. Based on the values of these parameters we then map pixel coordinates in the second image into the coordinate system of the reference image. In addition, the light conditions may vary among different cameras; therefore, the panorama to be generated may be inconsistent in its overall intensity. To obtain a smooth transition in overlapping areas between images to be stitched, we utilize the weighted-sum method introduced in [14] to perform a gradual fading-in and fading-out image-stitching process to generate the final video panoramic image. Suppose that I^r is the reference image and $\widetilde{I^l}$ is the transformed source image calculated based on P, the stitched image, I^m, is generated according to Eq. (22), where G stands for pixel intensity; $\alpha = \dfrac{\hat{x}_{\max} - x}{\hat{x}_{\max} - \hat{x}_{\min}}$, $0 \leq \alpha \leq 1$, and $\hat{x}_{\max}$ and $\hat{x}_{\min}$ are maximum and minimum coordinates in the overlapping region, respectively.

$$\begin{bmatrix} e \\ f \\ g \end{bmatrix} = \begin{bmatrix} p_0 & p_1 & p_2 \\ p_3 & p_4 & p_5 \\ p_6 & p_7 & 1 \end{bmatrix} \begin{bmatrix} x' \\ y' \\ 1 \end{bmatrix} \qquad (20)$$

$$\begin{cases} x = \dfrac{e}{g} = \dfrac{p_0 x' + p_1 y' + p_2}{p_6 x' + p_7 y' + 1} \\ y = \dfrac{f}{g} = \dfrac{p_3 x' + p_4 y' + p_5}{p_6 x' + p_7 y' + 1} \end{cases} \qquad (21)$$

$$G^m(x,y) = \begin{cases} \widetilde{G^l}(x,y) & (x,y) \in \widetilde{I^l} \\ \alpha \widetilde{G^l}(x,y) + (1-\alpha) G^r(x,y) & (x,y) \in \widetilde{I^l} \cap I^r \\ G^r(x,y) & (x,y) \in I^r \end{cases} \qquad (22)$$

During the alignment and stitching of the first K frames, an initial set of optimal transformation parameters, p^{km}, can be obtained according to Eq. (23), where G stands for pixel intensity. Then, newly generated, real-time frames will reuse these optimal parameters for the stitching process. This strategy is appropriate for smooth (non-shaking) panorama generation as it can avoid unnecessary feature detection/matching. When there does exist a shaking camera, since there are more random alignments between frames, initially optimal parameters may not work for new frames. In this scenario, our solution is to calculate the similarity between overlapping regions of two images. When the similarity is lower than a predefined threshold, a new set of optimal parameters can be automatically re-calculated and applied to newly generated frames. As such, our strategy is a self-adaptive one and can well handle errors resulted from the camera shaking. In addition, the similarity between overlapping regions is regarded as a by-product of the stitching process defined in Eq. (22). As a result, only nominal time is needed for calculating the aforementioned similarity, and the algorithm efficiency is thus not much affected.

$$\begin{cases} km = \max_{1 \leq k \leq K} \dfrac{\sum_{(x,y) \in I^l_k \cap I^r_k} (\widetilde{G^l_k}(x,y) \cdot G^r_k(x,y))}{\sqrt{(\sum_{(x,y) \in I^l_k \cap I^r_k} \widetilde{G^l_k}(x,y)^2)(\sum_{(x,y) \in I^l_k \cap I^r_k} \widetilde{G^r_k}(x,y)^2)}} \\ p^{km} = (p_0^{km}, p_1^{km}, \cdots, p_7^{km}) \end{cases} \qquad (23)$$

Results and Discussion

Experimental Environment and Parameter Setup

Note that all data and images generated from our experiments, unless stated otherwise, are publicly available. No restrictions are placed on producing derivatives of original data or images except that formal acknowledgement to original inventors is required. Examples of such acknowledgement include but not limited to publication citations. The intellectual property belongs to Fuzhou University.

We used a PC (CPU E2200+2.2 GHz, 4 GB Memory, Matlab 7.0) to conduct our experiments, and the image resolution was set to 1280×720.

According to our previous experience, we set experimental parameters as follows:

- The y coordinate difference of adjacent cameras was no greater than one third of the image height, i.e., λ_H in Eq. (9) was set to $H/3$;
- λ_R in Eq. (9) was set to $0.05 \times R_{\max}$;
- The horizontal overlapping was no great than $W/3$;
- The similarity window size in Eq. (10) was set to 7×7, i.e., w was set to 3;
- The similarity threshold λ_n in Eq. (11) was set to 0.75;
- λ_δ in Eq. (16) was set to 0.5;
- The original image was segmented into regions of size 80×80, and the number of corners for each region was set to six, therefore, λ_D in Eq. (16) was set to 14 ($\cong 80/6$); and
- λ_R in Eq. (16) was set to $0.05 \times R_{\max}$.

Evaluation on Corner Selection

The experimental results are demonstrated in Figure 3. The corner-selection result from the traditional Harris detector is shown in Figure 3(b). Because the original image, Figure 3(a), was not fragmented at all, corners tended to cluster around regions with richer texture. The result from the regional corner-selection algorithm in [12] is demonstrated in Figure 3(c), where corners were evenly distributed without taking into account the texture difference among various regions. On the contrary, our algorithm (Figure 3(d)) handled well the tradeoff discussed earlier in Subsection "Improved Harris Corner Selection" (i.e., the number of corners should be proportional to region texture information and, at the same time, the corner clustering should be avoided as much as possible). Also note that different values of τ in Eq. (6) may significantly affect the corner-selection result. To be more specific, with the increasing value of τ, Figure 3(e) for example, the texture difference will be amplified and the result will be closer to that of The Harris Detector. On the other hand, if the value of τ is small, Figure 3(f) for example, the texture difference will be diminished and the result will be closer to that of the regional corner-selection algorithm in [9]. By choosing different values of τ, our corner-selection algorithm is flexible in terms of serving various purposes from users€ viewpoint.

Evaluation on Corner Matching Efficiency

Experimental results in Subsection "Evaluation on Corner Selection" have clearly indicated that our corner-selection methodology is superior to the methodology in the traditional Harris detector. Therefore, experiments in this section will be based upon our corner-selection algorithm, so that a fair comparison can be performed between different corner-matching processes. Consequently, experiments in this section do not have all possible four combinations between corner selection and corner matching methodologies. Our experimental results are demonstrated in Figures 4, 5 and Tables 1, 2, and 3. Figure 4(a) contains two original images with corners selected using our algorithm. We chose the right one-third region of the left image and the left one-third region of the right image as two regions to perform corner matching. So we had a total of 45 segmented regions $(H/80 \times W/3/80)$, and the total number of corners is $270(6 \times 45)$. The total number of matching-corner pairs from NCC rough match in the traditional RANSAC algorithm was 528 (Figure 4(b)), and the total numbers of initial and finalized matching-corner pairs from our algorithm were 456 (Figure 4(c)) and 41 (Figure 4(d)), respectively. For the purpose of analysing experimental results in a clearer manner, we further summarize these results in Table 1: the traditional RANSAC algorithm needs to calculate the extremely time-consuming NCC function for all pairwise combinations of corners $(270 \times 270 = 72,900$ in this example), and our algorithm only need to consider a small number of combinations (17,993 in this example, which is less than 25% of 72,900). As a result, our multiple-constraint corner matching was almost four times faster than NCC rough match in the traditional RANSAC algorithm $(1.81 + 0.05 = 1.86$ seconds vs. 7.20 seconds). In addition, Figure 4(d) clearly demonstrates that most of the matching pairs were correct ones. Figure 5 and Table 2 demonstrate similar results on another experiment. Due to the limited space, more experimental results can be found at: http://www.soc.southalabama.edu/huang/ImageStitching/Experiment Results.rar. Overall (average and variation) quantitative results for these additional experimental results are summarized in Table 3.

Evaluation on image stitching. We conducted two different sets of experiments to evaluate the image-stitching result. Notice that all final panoramic images in this section were obtained using the methodology discussed in Subsection "Image Stitching".

The first set of experiments on stitching evaluation. The first set of experiments was performed to compare stitching results from the traditional Harris detector, the regional corner-selection algorithm in [9], and our presented algorithm, respectively. Experimental results are demonstrated in Figures 6, 7 and Tables 4, 5. Let us first discuss results in Figure 6 and Table 4.

In Figure 6, there are two pairs of images to be stitched. Due to the rich texture information contained in the original images, the traditional Harris detector was able to obtain correct matching corners between two images on the left pair and was successful in stitching these two images. However, the traditional Harris detector was not successful in stitching images on the right pair due to not enough texture information contained in original images. The regional corner-selection algorithm had exactly opposite results: successful in stitching images on the right pair but not successful on the left pair. The example in Figure 6 may suggest that, the traditional Harris detector and the regional corner-selection algorithm are appropriate for scenarios of rich texture and for scenarios without enough texture information, respectively, but for not both. Consider the example demonstrated in Figure 6, the left pair of images has small overlapping regions but rich texture information. On the contrary, the right pair of

images has large overlapping regions but not enough texture information. On one hand, the traditional Harris detector was able to handle the left pair of images successfully but failed to deal with the right pair (Figure 6(a)). On the other hand, the regional corner-selection algorithm was able to handle the right pair but failed on the left pair (Figure 6(b)). As a comparison, our algorithm successfully handled both the left and right pairs of images because it handles well the tradeoff discussed earlier (i.e., the number of corners should be proportional to region texture information and, at the same time, the corner clustering should be avoided as much as possible). Furthermore, as shown in Table 4, a quantitative analysis indicates that our algorithm was able to obtain high intensity similarity between neighboring regions to be stitched. Therefore, the observation from Figure 6 and Table 4 is that, our algorithm is appropriate for various scenarios (rich texture information or not) and is able to obtain satisfactory intensity similarity between original images.

Figure 7 demonstrates stitching results for another pair of images. Compared with the two pairs of images in Figure 6, original images in Figure 7 contain more evenly distributed texture information. As a result, all three algorithms were able to obtain a successful stitching. However, as shown in Table 5, our algorithm is superior to the other two algorithms in terms of the quantitative analysis on the intensity similarity between original images.

The second set of experiments on stitching evaluation. The second set of experiments was performed on our algorithm alone. In fact, we implemented the entire algorithm (from corner selection to corner matching and finally to image stitching) in Visual C++2005. Results are demonstrated in Figures 8 to 12. Notice that all original images, i.e., (a), (b), and (c) in Figures 8 to 12 have small differences in translation, rotation, and scaling between each other. As discussed earlier in Section "Introduction", one motivation of our algorithm is the small difference in translation, rotation, and scaling on adjacent images to be stitched. Figure 8 is explained in detail here. We performed our proposed Harris corner selection and multiple-constraint corner matching between Figure 8(a) and Figure 8(b), and between Figure 8(b) and Figure 8(c), respectively. After we obtained two finalized sets of matching-corner pairs, we selected Figure 8(b) as the reference image and calculated the projective transformation parameters. We then mapped pixel coordinates in Figure 8(a) and Figure 8(c) into the coordinate system of Figure 8(b), respectively. Finally we performed a gradual fading-in and fading-out image-stitching process. The final result in Figure 8(d) clearly demonstrates that (i) our corner matching was accurate; (ii) we obtained a smooth transition in overlapping areas between images to be stitched; and (iii) the panorama generated satisfied human visual requirements. In addition, the preview speed of the generated panorama was around 20 frames per second, which satisfied the real-time requirement commonly found in video panorama stitching. Figures 9 to 12 exhibited convincing results from additional images. Likewise, due to the limited space in this paper, more results have been stored at the following location and can be freely downloaded: http://www.soc.southalabama.edu/huang/ImageStitching/ExperimentResults.rar. Notice that as discussed earlier in Subsection "Image Stitching", our self-adaptive strategy to obtain optimal parameter values during the image stitching can well handle errors resulted from the camera shake. Successful stitching results demonstrated in a video clip (uploaded to aforementioned extra experimental results) can justify our claim.

Limitations of the Study, Open Questions, and Future Work

1. There are a number of scenarios where our proposed methodology might be affected. In our future work, it will be worthwhile to explore how we effectively deal with such challenging scenarios. Some example cases are listed as follows.

   - What happens in case there are a lot of movements in the frame? For example, when taking videos in cars or other vehicles.
   - Besides the camera shake, can other fast movements be well handled? If so, how much overlapping between adjacent frames is considered adequate?
   - How to handle changes in lighting conditions? For example, when nearby cameras use a flash.

2. To better satisfy the real-time requirement commonly found in video panorama stitching, there is a tradeoff between obtaining more features for matching within a frame and simply capturing more frames to allow for greater overlapping. How to effectively handle such a tradeoff can be another interesting future work.

3. Other future research directions include, but are not limited to, (i) the automatic determination of the total number of corners according to the image texture information and (ii) the motion ghost challenge during the image stitching.

Conclusions

We presented an innovative algorithm to handle challenges in video panoramic image stitching, e.g., small overlapping regions, extremely time-consuming stitching processes. Our contribution can be summarized as (i) an improved, self-adaptive selection of Harris corners and (ii) a multiple-constraint corner matching. To better select high-quality corners, we fragmented original images into numerous regions and then selected corners within each region based on the normalized variance of region grayscales. This self-adaptive selection assured that corners are evenly distributed (i.e., proportional to region texture information) and, at the same time, avoided the corner clustering as much as possible. To overcome the inefficient corner matching in the traditional RANSAC algorithm, we first filtered out a large number of inappropriate corners according to their position information. We then generated an initial set of matching-corner pairs based on grayscales of adjacent regions around each corner. Finally we applied multiple constraints on every two candidate pairs to further remove incorrectly matched pairs. We were able to significantly reduce the number of iterations needed in the RANSAC algorithm, resulting in a much more efficient panorama stitching process. Experimental results demonstrated that (i) our corner matching is almost four times faster than the traditional RANSAC matching; (ii) panoramas generated from our algorithm feature a smooth transition in overlapping image areas and satisfy human visual requirements; and (iii) the preview speed of the generated panorama satisfies the real-time requirement commonly found in video panorama stitching.

Acknowledgments

The authors would like to thank Ms. Min Xiong for her help in the manuscript preparation.

Author Contributions

Conceived and designed the experiments: MZ WW BL JH. Performed the experiments: MZ WW BL JH. Analyzed the data: MZ WW BL JH. Contributed reagents/materials/analysis tools: MZ WW BL JH. Wrote the paper: MZ WW BL JH.

References

1. Hartley R, Gupta R (1997) Linear pushbroom cameras. IEEE Transactions on Pattern Analysis and Machine Intelligence 19: 963–975.
2. Chalechale A, Naghdy G, Mertins A (2005) Sketch-based image matching using angular partitioning systems. IEEE Trans on Man and Cybernetics and Part A 35: 28–41.
3. Reddy BS, Chatterji BN (1996) An FFT-based technique for translation, rotation, and scaleinvariant image registration. IEEE Transactions on Image Process 5: 1266–1271.
4. Harris CG, Stephen M (1998) A combined corner and edge detector. In: Proceedings of the 4th Alvey vision conference. Manchester, 147–151.
5. Sivic J, Zisserman A (2003) Video google: A text retrieval approach to object matching in videos. In: Proceedings of the 9th IEEE International Conference on Computer Vision. IEEE Computer Society Washington DC and USA: Computer Vision-Volume 2, 27–38.
6. Mikolajczyk K, Schmid C (2004) A performance evaluation of local descriptors. IEEE Transactions on Pattern Analysis and Machine Intelligence 27: 1615–1630.
7. Nister D, Naroditsky O, Bergen J (2006) Visual odometry for ground vehicle applications. Journal of Field Robotics 23: 3–10.
8. Zhu Q, Wu B, Xu Z (2006) Seed point selection method for triangle constrained image matching propagation. IEEE Geoscience and Remote Sensing Letters 3: 207–211.
9. Zhao W, Gong S, Liu C, Shen X (2008) A self-adaptive harris corner detection algorithm. computer engineering. Computer Engineering 34: 212–214.
10. Zhang Y, Gao G, Jia K (2009) A fast algorithm for cylindrical panoramic image based on feature points matching. Journal of Image and Graphics 14: 1188–1193.
11. Cao S, Jiang J, Zhang G, Yuan Y (2011) Multi-scale image mosaic using features from edge. Journal of Computer Research and Development 48: 1788–1793.
12. Lacey AJ, Pinitkarn N, Thacker NA (2000) An evaluation of the performance of ransac algorithms for stereo camera calibration. In: Proceedings of the 11th British Machine Vision Conference. Bristol and UK, 646–655.
13. Zitova B, Flusser J (2003) Image registration methods: a survey. Journal of Image and Vision Computing 21: 977–1000.
14. Yu H, Jin W (2009) Evolvement of research on digital image mosaics methods. Infrared Technology 31: 348–353.

8

A Method for the Evaluation of Image Quality According to the Recognition Effectiveness of Objects in the Optical Remote Sensing Image Using Machine Learning Algorithm

Tao Yuan[1,2]*, **Xinqi Zheng**[1,2], **Xuan Hu**[1,2], **Wei Zhou**[1,2], **Wei Wang**[1]

1 School of Land Sciences and Technology, China University of Geosciences, Beijing, China, **2** Key Laboratory of Land Regulation, Ministry of Land and Resources, Beijing, China

Abstract

Objective and effective image quality assessment (IQA) is directly related to the application of optical remote sensing images (ORSI). In this study, a new IQA method of standardizing the target object recognition rate (ORR) is presented to reflect quality. First, several quality degradation treatments with high-resolution ORSIs are implemented to model the ORSIs obtained in different imaging conditions; then, a machine learning algorithm is adopted for recognition experiments on a chosen target object to obtain ORRs; finally, a comparison with commonly used IQA indicators was performed to reveal their applicability and limitations. The results showed that the ORR of the original ORSI was calculated to be up to 81.95%, whereas the ORR ratios of the quality-degraded images to the original images were 65.52%, 64.58%, 71.21%, and 73.11%. The results show that these data can more accurately reflect the advantages and disadvantages of different images in object identification and information extraction when compared with conventional digital image assessment indexes. By recognizing the difference in image quality from the application effect perspective, using a machine learning algorithm to extract regional gray scale features of typical objects in the image for analysis, and quantitatively assessing quality of ORSI according to the difference, this method provides a new approach for objective ORSI assessment.

Editor: Denis G. Pelli, New York University, United States of America

Funding: This study was supported by the Fundamental Research Funds for the Central Universities (2652011299) and the National Natural Science Foundation of China (41240008, 41301118). The funders had no role in study design, data collection and analysis, decision to publish, or preparation of the manuscript.

Competing Interests: The authors have declared that no competing interests exist.

* E-mail: yuantaobj@qq.com

Introduction

Currently, remote sensing image data, mainly optical remote sensing image (ORSI) data, are a major data source to obtain spatial information. ORSI quality is closely related to its application effects, and the difference in different-quality images during object identification and information extraction is very obvious. How to accurately describe and quantitatively express the quality difference is very important to ORSI applications.

At present, ORSI quality is assessed largely by two different types of methods. The first is the subjective assessment method; that is, the ORSI quality level is judged based on some pre-specified scale standards, or the experiences of the observers on the visual effects of the images. For this method, the National Imagery Interpretability Rating Scale (NIIRS) standard is still extensively applied in remote sensing applications. This standard correlates the task needs of the user with the image quality. The second is the objective assessment method including (1) error statistics, (2) human visual system (HVS) characteristics, and (3) structural similarity (SSIM). The error statistical method obtains the differences by comparing the distorted image with the reference image through the design features, finds a number of statistical quantities, and links them to the image quality. The HVS method utilizes the fact that human eyes are inherently insensitive to some image distortion and that the nature and distribution of different noises may cause significant differences in visual effects. This method is mainly performed by simulating human subjective feeling to establish physical model fitting for human visual perception. It often uses parameters such as the distortion sensitivity, edge distortion, and sharpness to analyze image quality. The error statistical method considers a digital image as a collection of isolated image pixels, ignores the statistical correlation among local pixels, and identifies signal errors in the visual perception leading to quality aberration, which do not conform to human visual characteristics. The HVS method has several advantages, but its channel decomposition algorithm is too complex and does not sufficiently consider the correlation among pixels, leading to great differences from the real visual perception. Therefore, in certain test conditions and if there are different image distortions, its superiority over PSNR and MSE is not evident [1].

In short, image quality evaluation is an issue that has garnered worldwide interest and has made substantial progress. However, there are still many issues that should be investigated.

(1) Although the subjective assessment method represented by NIIRS has several disadvantages, such as being time-consuming, having a high cost, and requiring multiple repetitions, it is still widely used in the remote sensing (RS) field because its assessment results are closely associated with RS application effects.

(2) Objective assessment methods originating from the field of digital image or video signal processing rarely consider specific tasks and application effects of ORSI data. Commonly used assessment parameters are the variance, kurtosis, entropy, contrast, sharpness, average gradient, and edge intensity [2–3]. Based on current study results, the quantitative assessment indexes of image quality reflect some aspects of the ORSI quality under certain conditions. However, for these indicators of digital image processing, the calculation results are often significantly different from the practical application effects of object identification and information extraction of remote sensing data, or even completely opposite.

(3) These objective assessment methods utilize full-reference image quality assessment; thus, their results are relative to original reference images [4–5]. However, reference images are not used in RS practical applications; thus, these methods have limited application value in the RS field.

After the above analysis, we find that ORSI differs from methods using a common digital image. ORSI is mainly used for object identification and information extraction, and its quality is mainly reflected in the capacity of object identification and information extraction. Therefore, an ORSI quality evaluation method must be able to reflect this difference, and the evaluation results must be consistent with practical application effects. The only evaluation method now in line with this requirement is NIIRS, which is a subjective evaluation. Starting from the practical application effects of ORSI data, this study will analyze the relationship between object identification effects and image quality, and further establish an objective and quantitative ORSI quality evaluation method.

In this study, with the focus on the practical application of ORSI data, we evaluated the image quality given the ability of optical remote sensing data to recognize target objects, revealed the relationship between the commonly used objective assessment indexes and the practical application effects of RS data, and analyzed the applicability and limitations of these indexes used to represent the image data quality.

On this basis, we used a machine learning algorithm, selected sample training classifiers, and automatically identified and labeled typical objects in ORSI by programming in the Microsoft Visual C++2008 and OpenCV 2.1 environment. We then quantified the recognition results and defined the recognition rate. By standardizing the recognition rate, the practical application effects of the different qualities of ORSI can be better reflected. We propose an application-oriented ORSI quality assessment method that provides a new perspective from which to assess ORSI quality.

Materials and Methods

(1) Data Source

WorldView-2 image data are one set of commercial satellite data whose spatial resolution of the panchromatic wave band is as good as 0.5 m. Thus, the Worldview-2 panchromatic wave band data, ortho-rectified in March 2010 by Zhengzhou City, Henan Province, China, were chosen as our experimental data to study the object recognition effect. A visual inspection shows that the data have clear texture without cloud coverage and striped dead pixels.

(2) Data Pre-processing

To simulate the image data acquired by different sensors in different imaging conditions, the original image was re-acquired, its resolution was reduced, and it was used to model the image data obtained by different sensors. In order to model the image data acquired by the same sensor in different imaging conditions, three methods, including adding Gaussian noise, defocus blurring, and reducing image contrast, were used for image processing. The original image and images processed by using the above methods are shown in Fig. 1: (a) the original image, (b) a resolution-reduced image, (c) a Gaussian noise-added image, (d) a defocus-blurred image, and (e) a contrast-reduced image.

After the quality-reduction processing, we used a MATLAB program to calculate commonly used image quality indicators, such as the entropy, contrast, and sharpness, for the original image (a) and the other quality-reduced images (b), (c), (d), and (e). Then, we wrote a target recognition program based on the computer vision function library, OpenCV, selected positive and negative samples at the same positions in images (a), (b), (c), (d), and (e), utilized the Adaboost algorithm and Haar features of the target object, trained the target recognition classifier, obtained five classifiers numbered A, B, C, D, and E, respectively, and used them to examine the corresponding images (a), (b), (c), (d), and (e). After the testing program was executed, the program automatically marked the identified targets. Calculation of the ORRs of different images characterized the differences in image quality, and the image quality assessment method was presented by comparing the differences between the above identified effects and the commonly used image quality assessment parameters. The specific program execution flow is shown in Fig. 2.

(3) Selection of Recognized Objects

Automobiles are commonly distributed in urban areas and relatively easy to distinguish from their surrounding objects. More importantly, they are of a size close to the object identification limit of ORSIs (meter and sub-meter size) and have simple but unique statistical structures in an image. Thus, using automobiles as recognized objects can best reveal the effect of the image quality on object recognition and is convenient for us to determine their overall statistical quantities and recognition rates in recognition experiments.

(4) Selection of Object Recognition Methods

To compare the differences in the effects of image quality on object recognition, the methods used for object recognition should avoid human intervention and be sufficiently sensitive to the changes in image quality. The commonly used methods are the supervised classification [6], non-supervised classification, object-oriented [7–8], neural networks [9], and decision tree classifications [10]. These methods have some disadvantages, such as multiple people participation, low efficiency, and insensitivity to changes in image quality [11]; therefore, they are not suitable for evaluation of differences in image quality.

The machine learning identification method based on Haar features was chosen to identify target objects. Firstly, Haar features used for object recognition were computed by inputting positive and negative samples, then the Adaboost algorithm was used to train the classifier, and the well-trained classifier was then used to analyze the tested images and automatically identify marked targets in RS images [12]. The machine learning identification method was characterized by advantages such as little human

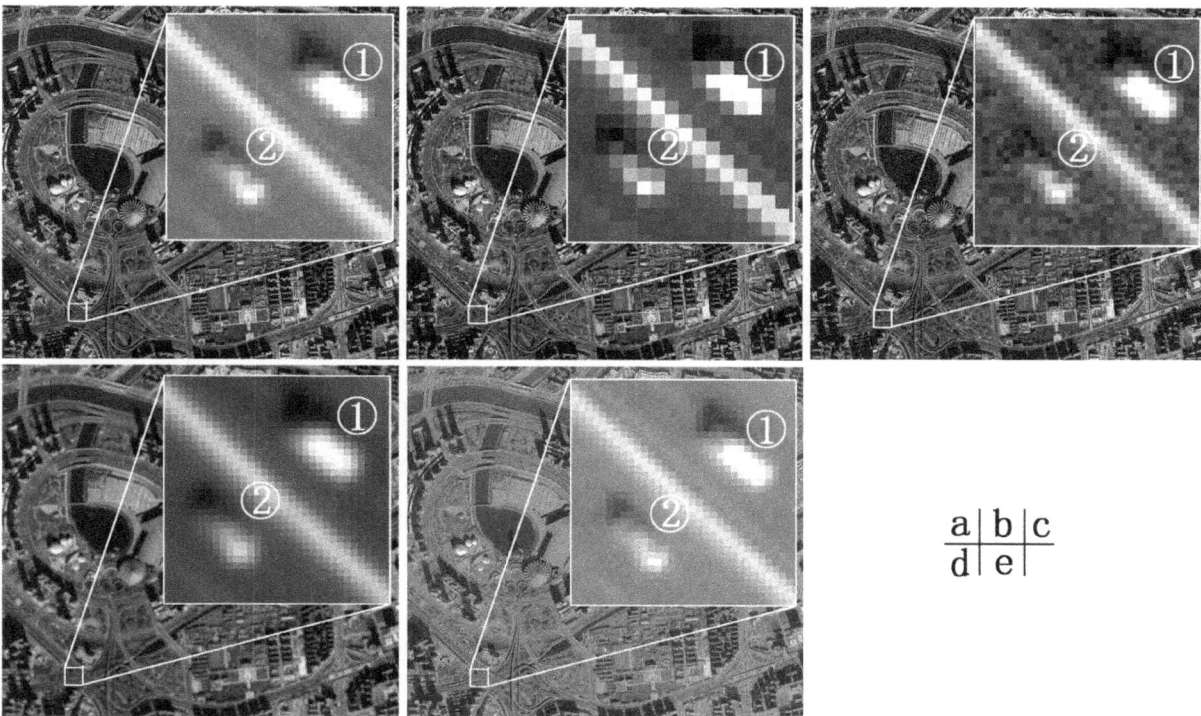

Figure 1. Comparison between the original image and the images after quality-reduction processing. (a) original image, (b) reduced-resolution image, (c) added Gaussian noise image, (d) defocus-blurred image, and (e) reduced contrast image.

intervention, fast detection speed, and sensitivity to changes in image quality. Thus, this method is suitable to reflect the changes in RS data quality.

(5) Sample Selection

The samples needed for training classifiers are intercepted from the tested images and divided into positive and negative samples. The former are the objects to be identified, i.e., the motor vehicles in the image, whereas the latter are other objects in the image that may interfere with motor vehicle identification, such as zebra crossings, bridges, shrubs, street lights, and boats in the rivers. In theory, the increase in the number of samples can improve the accuracy of the classifiers. In fact, when the number of samples is sufficiently large, the effect of improvement in classifiers is not obvious. After many comparative experiments, 500 positive samples with a pixel bitmap of the size of 32×32 and 1500

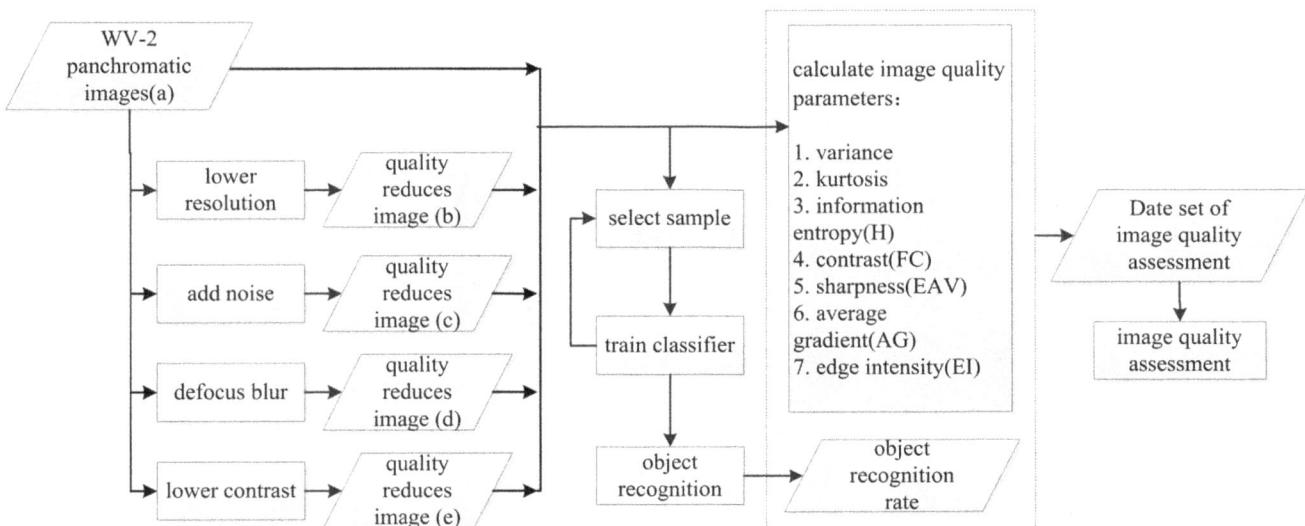

Figure 2. Flowchart of image quality assessment based on object recognition effect.

negative samples with pixel bitmap of sizes of 45×45 to 60×60 were selected for each image in this study.

(6) Classifier Training

Applying the classifier training function loads positive and negative samples and calculates their Haar-like feature values. Then, applying the Adaboost algorithm trains different weak classifiers for the training set, and combining these generated weak classifiers with weights forms a stronger final classifier [13–14]. At the end of training, classifiers A, B, C, D, and E in the XML format corresponding to images (a), (b), (c), (d), and (e), respectively, are finally obtained.

(7) Object Recognition and Mark

The vehicle recognition program written based on Microsoft Visual C++2008 and OpenCV 2.1 can directly call the previously trained classifiers A, B, C, D, and E to check the corresponding images. The search window in the recognition program is set to move for searching in the image. In order to monitor an object of unknown size in the image, it is necessary for the scanner using the search window with different size proportions to scan the picture several times and determine the parameters of the optimal search window. At the same time, a marker function is created and it will use the plus sign, "+," to mark the searched vehicle, call the cvNamedWindow function to name the image window, and call the cvShowImage function to show the resultant image examined by the classifier. The result of object recognition is shown in Fig. 3.

(8) Statistics of ORR

As the number of automobiles in the entire image is too large to accumulate statistical data of the recognition rate, 10 regions of

fixed size in the image are randomly selected, and their recognition rate and average is calculated as the vehicle recognition rate in the image. When areas where a car contrasting with the background, such as streets and squares, are selected, the correctness of the program recognition can be accurately judged. The object recognition rate (ORR) is defined as:

$$ORR = \frac{a}{m} \times 100\% \qquad (1)$$

where a is the number of vehicles correctly recognized, and m is the actual number of vehicles in the image.

(9) Calculation of Image Quality Parameters

We selected a total of seven common image quality evaluation indexes, including the variance (d), kurtosis (K), information entropy (H), contrast (FC), sharpness (EAV), average gradient (AG), and edge intensity (EI), and wrote a MATLAB program to analyze five images: (a), (b), (c), (d) and (e). Table 1 lists the various parameter formulas and their corresponding explanations.

Conclusions

(1) Analysis of ORR and Image Quality Parameters

Table 2 is obtained by comparing the image quality parameters of the original image and the degraded images with the vehicle recognition rates of corresponding images.

Nine commonly used indexes chosen in this study reveal the differences between images from the angles of the overall pixel grayscale distribution, the information entropy, and the edge sharpness. After processing, the image quality is significantly reduced, but the changes in these index values are unable to

Figure 3. Object recognition and marking effect.

Table 1. Image quality parameters and their explanations.

Name	Expression and explanation
variance (d)	$d=\dfrac{1}{m\times n}\sum_{i=1}^{m}\sum_{j=1}^{n}(f(i,j)-u)^2$ where m and n are the numbers of rows and columns in the image, respectively; f(i, j) is the grayscale value at point (i, j) in the image; u is the mean grayscale value in the image.
kurtosis (K)	$K=\dfrac{1}{m\times n}\sum_{i=1}^{m}\sum_{j=1}^{n}(f(i,j)-u)^4\Big/d^2$ where m and n are the the numbers of rows and columns in the image, respectively; f(i, j) is the grayscale value at point (i, j) in the image; u is the mean grayscale value in the image; d is the variance.
entropy (H)	$H=-\sum_{i=0}^{Max}P_i\log_2 P_i$ where Pi is the probability of a pixel grayscale value of i; max is the max grayscale value of a pixel.
contrast (FC)	$FC=\sum\delta(i,j)^2\times P_\delta(i,j)$ $\delta_{(i,j)}$ is the grayscale difference between adjacent pixels; P(i, j) is the probability of pixel distribution of the grayscale difference between adjacent pixels.
sharpness (EAV)	$EAV=\dfrac{\sum_{i=1}^{m\times n}\sum_{\alpha=1}^{8}\left|df/dx\right|}{m\times n}$ where m and n are the numbers of rows and columns in the image, respectively; α is the number of adjacent pixels; df/dx is the gradient between adjacent pixels.
average gradient (AG)	$AG=\dfrac{1}{(m-1)\times(n-1)}\sum_{i=1}^{m-1}\sum_{j=1}^{n-1}\sqrt{\dfrac{(f(i,j)-f(i+1,j))^2+(f(i,j)-f(i,j+1))^2}{2}}$ where m and n are the numbers of rows and columns in the image, respectively; f(i, j) is the grayscale value of the image at point (i, j).
edge intensity (EI)	$EI=\sum_{i}^{m}\sum_{j}^{n}\left[G_x^2(i,j)+G_y^2(i,j)\right]$ $G_x(i,j)=[f(i+1,j-1)+2f(i+1,j)+f(i+1,j+1)]-[f(i-1,j-1)+2f(i-1,j)+f(i-1,j+1)]$ $G_y(i,j)=[f(i-1,j+1)+2f(i,j+1)+f(i+1,j+1)]-[f(i-1,j-1)+2f(i,j-1)+f(i+1,j-1)]$ where m and n are the numbers of rows and columns in the image, respectively; f(i, j) is the grayscale value of the image at point (i, j).

Table 2. Statistics of ORR and image quality parameters.

Image No.	ORR (%)	Variance (d)	Kurtosis (K)	Entropy (SH)	Sharpness (EAV)	Contrast (FC)	Average gradient (AG)	Edge intensity (EI)
(a)	81.95	1864.50	5.20	6.86	11.24	29.38	0.10	75.36
(b)	65.52	2393.39	4.46	6.79	20.72	34.26	0.16	85.82
(c)	64.58	1936.22	4.81	7.13	28.47	30.34	0.13	92.83
(d)	71.21	1651.02	4.39	6.86	5.37	28.41	0.07	61.34
(e)	73.11	1310.55	6.85	6.45	9.22	23.20	0.09	60.90

better show the changes in the image quality. Images obtained by different remote sensors have different resolutions. Those indexes displaying the characteristics of the histograms of images and the image sharpness cannot intuitively show the decline in the image quality. For images obtained by the same sensor at different conditions, their resolutions are equivalent. For example, after adding Gaussian noise, all the indexes except the recognition rate of image (c) cannot intuitively reflect the decline in image quality.

(1) Assessment of Image Quality based on ORR

The vehicle identification rate in the image can better reflect the differences in the image quality; thus, if the rate acquired in the optimum condition is selected as a standard, the recognition rates of various images can be standardized, and the quality differences among different images can be compared. Because Worldview-2 satellite data are among the best commercial satellite data used in this study, the vehicle identification rate of the original image is 81.95%, close to the results of other studies [12]. Therefore, the recognition rate of 82% can be thought of as the recognition rate in the optimum condition. The ratios of the recognition rates of images (b), (c), (d), and (e) to the optimum recognition rates are 79.90%, 78.76%, 86.84%, and 89.16%.

As shown in Figure 1, image quality declines significantly after quality-degradation treatments, and the difficulty of object identification and information extraction increases. There are two vehicles labeled with ① and ②. By a general visual comparison, we can observe that vehicles in (d) and (e) are easier to recognize than those in (b) and (c). Further analysis shows that vehicle ② is better in (e) than in (d) and better in (c) than in (b). Therefore, by subjective visual perception, image quality in descending order is (a)>(e)>(d)>(c)>(b). This result is consistent with the standardized ORR data. Therefore, standardized ORR data can evaluate image quality more accurately than conventional digital image assessment indexes.

Discussion

This study proposes to conduct target object recognition on Haar-like features of typical objects by a machine learning algorithm and evaluate image quality by identifying effects. Theoretically, when ORSI resolution is low, the image is fuzzy, or image noise is significant, Haar-like features of typical objects will be unobvious and unstable, which will reduce the effectiveness of the classifier, resulting in a lower target recognition rate.

Haar-like features reflect the grayscale features in the image region. They are capable of showing the structural relationship among adjacent pixels in the image and are very sensitive to a simple graphical structure, such as an edge or segment, in a particular direction (horizontal, vertical, or diagonal). The capacity for object recognition based on Haar-like features in nature has more to do with whether the image structural characteristics of the same object in different images are obvious and have a high degree of consistency. Lowering resolution, blurring, and other techniques can weaken the structural features of the same object in the image, whereas noise can decrease the consistency of the structural features of the same object in the image, and both will cause the training difficulty in the classifier to increase and the object recognition rate to decrease. This is similar to the image quality assessment method based on structural similarity. This is a full-reference assessment method and must define a reference image and compare image structures at the same location, thus greatly limiting its applications in the RS field. The use of object recognition based on Haar-like features for image quality assessment resolves this problem well because this method is used to detect structural deformation based on both significance and consistency of the image structural characteristics of the same type of objects.

In the field of remote sensing research and application, there is not a quantitative quality-evaluating method consistent with ORSI application effects. There will be many problems if quality indicators of digital image processing are used directly. In a new perspective, this study ensures maximum objectivity of evaluation by a machine learning algorithm and classifier training on one hand and conducts target object recognition by Haar-like features of typical objects on the other hand. Actually, this study examines whether the grayscale features in the image region are obvious and stable, which has a strong relationship with the ORSI application effects and reflects data quality. As a preliminary exploration and verification, this study uses only one type of optical remote sensing image (WorldView-2 panchromatic band) and only analyzes one type of object (vehicle). Therefore, the applicability of this evaluation method may be restricted, e.g., images with no vehicles. However, this study preliminarily validates that by using a machine learning algorithm with grayscale features in the image region, the quality of remote sensing data can be reflected by object recognition effects quantitatively and objectively. This is a new idea regarding quantitative evaluation of ORSI quality, and it is theoretically feasible to create a new ORSI evaluation system for different applications through this method. In practice, we can learn from quantitative remote sensing research methods to set up a large number of targets on the ground, then collect data and conduct target recognition. Thus, under a uniform standard, we can conduct quality evaluation on different data acquired by different sensors.

Additionally, our study was only focused on an experiment with panchromatic band data, and does not refer to the assessment of multispectral data. The latter may consider the analysis of the principal components of multiple wave bands and extract them for evaluation. With image analysis technology based on machine learning continuously developing, a more comprehensive, objective, and quantitative ORSI assessment systems fit to different mission requirements can be established.

References

1. Wang Z, Bovik AC, Lu L (2002) Why is image quality assessment so difficult. IEEE International Conference on Acoustics Speech and Signal Processing: 3313–3316.
2. Zhang J, Ong SH, Le TM (2011) Kurtosis-based no-reference quality assessment of JPEG2000 images. Signal Processing: Image Communication 26(1): 13–23.
3. Seghir ZA, Hachou F (2011) Edge-region information measure based on deformed and displaced pixel for image quality assessment. Signal Processing: Image Communication 26 (8): 534–549.
4. Chen Y, Blum RS (2009) A new automated quality assessment algorithm for image fusion. Image and Vision Computing 27(10): 1421–1432.
5. Mathieu C, Patrick LC, Dominique B (2008) Objective quality assessment of color images based on a generic perceptual reduced reference. Signal Processing: Image Communication 23(4): 239–256.
6. Guo QH, Li WK, Liu DS, Chen J (2012) A Framework for Supervised Image Classification with Incomplete Training Samples. Photogrammetric Engineering & Remote Sensing 78(6): 595–604.
7. Niu X (2006) A semi-automatic framework for highway extraction and vehicle detection based on a geometric deformable model. Journal of Photogrammetry & Remote Sensing 61: 170–186.
8. Holt AC, Seto EYW, Rivard T, Gong P (2009) Object-based Detection and Classification of Vehicles from High-resolution Aerial Photography. Photogrammetric Engineering & Remote Sensing 75(7): 871–880.
9. Jin X, Davis CH (2007) Vehicle detection from high-resolution satellite imagery using morphological shared-weight neural networks. Image and Vision Computing 25(9): 1422–1431.
10. Eikvil L, Aurdal L, Koren H (2009) Classification-based vehicle detection in high-resolution satellite images. Journal of Photogrammetry and Remote Sensing 64: 65–72.
11. Sharma G, Merry CJ, Goel P, McCord MR (2006) Vehicle detection in 1-m resolution satellite and airborne imagery. International Journal of Remote Sensing 27 (34): 779–797.
12. Leitloff J, Hinz S (2010) Vehicle Detection in Very High Resolution Satellite Images of City Area. Geoscience and Remote Sensing 7(48): 2795–2806.
13. Freund Y, Schapire R (1999) A Short Introduction to Boosting. Journal of Japanese Society for Artificial Intelligence 14(5): 771–780.
14. Viola P, Jones M (2001) Rapid Object Detection using a Boosted Cascade of Simple Features. Computer Vision and Pattern Recognition 14(5): 1–9.

Author Contributions

Conceived and designed the experiments: TY XZ. Performed the experiments: WW. Analyzed the data: XH WZ. Wrote the paper: TY.

9

Iterative Nonlocal Total Variation Regularization Method for Image Restoration

Huanyu Xu, Quansen Sun*, Nan Luo, Guo Cao, Deshen Xia

School of Computer Science and Technology, Nanjing University of Science and Technology, Nanjing, Jiangsu, China

Abstract

In this paper, a Bregman iteration based total variation image restoration algorithm is proposed. Based on the Bregman iteration, the algorithm splits the original total variation problem into sub-problems that are easy to solve. Moreover, non-local regularization is introduced into the proposed algorithm, and a method to choose the non-local filter parameter locally and adaptively is proposed. Experiment results show that the proposed algorithms outperform some other regularization methods.

Editor: Xi-Nian Zuo, Institute of Psychology, Chinese Academy of Sciences, China

Funding: This work was supported by grants from Programs of the National Natural Science Foundation of China (61273251, 61003108), (http://www.nsfc.gov.cn) Doctoral Fund of Ministry of Education of China (200802880017) (http://www.cutech.edu.cn), and the Fundamental Research Funds for the Central Universities No. NUST2011ZDJH26. The funders had no role in study design, data collection and analysis, decision to publish, or preparation of the manuscript.

Competing Interests: The authors have declared that no competing interests exist.

* E-mail: qssun@126.com

Introduction

Image restoration is a classical inverse problem which has been extensively studied in various areas such as medical imaging, remote sensing and video or image coding. In this paper, we focus on the common image acquisition model: an ideal image $u \in \mathbb{R}^n$ is observed in the presence of a spatial invariance blur kennel $A \in \mathbb{R}^{n \times n}$ and a additive Gaussian white noise $n \in \mathbb{R}^n$ with zero mean and standard deviation σ. Then, the observed image $f \in \mathbb{R}^n$ is obtained by:

$$f = Au + n. \tag{1}$$

Restoring the ideal image from the observed image is ill-posed since the blur kennel matrix is always singular. A common way to solve this problem is to use regularization methods, in which regularization terms can be invited to restrict the solutions. Regularization methods generally have the form as follows:

$$\arg\min_u \ \|Au - f\|^2 + \lambda R(u), \tag{2}$$

where $\|\cdot\|$ denotes the Euclidean norm, λ is a positive regularization parameter balancing the fitting term and the regularization term, $R(u)$ is the regularization term. The total variation regularization proposed by Rudin, Osher and Fatemi [1] (also called the ROF model) is a well known regularization method in this field. The total variation norm has a piecewise smooth regularization property, thus the total variation regularization can preserve edges and discontinuities in the image. The unconstrained ROF model has the form

$$\arg\min_u \ \|Au - f\|^2 + \lambda \|u\|_{TV}, \tag{3}$$

where the term $\|u\|_{TV}$ stands for the total variation of the image. The continuous form of the total variation is defined as

$$\|u\|_{TV} = \int_\Omega |\nabla u(t)| dt. \tag{4}$$

Many numerical methods was proposed to solve (3). When A is an identity matrix, the ROF model (3) turns into a TV denoising problem, methods as Chambolle's projection method [2], semi-smooth Newton methods [3], multilevel optimization method [4] and split Bregman method [5]. When A is a blur kennel matrix, (3) turns into a TV deblurring problem, we have prime-dual optimization algorithms for TV regularization [6–9], forward backward operator splitting method [10], interior point method [11], majorization-minimization approach for image deblurring [12], Bayesian framework for TV regularization and parameter estimation [13–15], using local information and Uzawa's algorithm [16,17], using regularized locally adaptive kernel regression [18], augmented Lagrangian methods [19,20] and so on. However, the problem is far from perfectly solved, problems as edge and detail preserving [21,22], ringing effect reducing [23–25] and varied blur kernels and noise types image restoration [26,27] still need better solutions.

The purpose of this paper is to propose an effective total variation minimization algorithm for image restoration. The algorithm is based on Bregman iteration which can give significant improvement over standard models [28]. Then, we solve the proposed algorithm by alternately solving a deblurring problem and a denoising problem [29,30]. In addition, we propose a local adaptive nonlocal regularization approach to improve the restoration results.

The structure of the paper is as follows. In the next section, an iterative algorithm for total variation based image restoration is proposed, moreover we present a nonlocal regularization under the proposed algorithm framework with local adaptive filter parameter to improve the restoration results. Section Experiments shows the experimental results. Section Conculsions concludes this paper.

Methods

Iterative Approach for TV-based Image Restoration

Bregman iteration for image restoration. We first consider a general minimization problem as follows:

$$\arg\min_{u} \lambda J(u) + H(u,f), \qquad (5)$$

where λ is the regularization parameter, J is a convex nonnegative regularization functional and the fitting functional H is convex nonnegative with respect to u for fixed f. This problem is difficult to solve numerically when J is non-differentiable, and the Bregman iteration is an efficient method to solve the minimization problem.

Bregman iteration is based on the concept of "Bregman distance". The Bregman distance of a convex functional $J(\cdot)$ between points u and v is defined as:

$$D_J^p(u,v) = J(u) - J(v) - <p, u-v>, \qquad (6)$$

where $p \in \partial J$ is a sub-gradient of J at the point v. Bregman distance generally is not symmetric, so it is not a distance in the usual sense, but the Bregman distance measures the closeness of two points. $D_J^p(u,v) \geq 0$ for any u and v, and $D_J^p(u,v) \geq D_J^p(w,v)$ for all points w on the line segment connecting u and v. Using Bregman distance (6), the original minimization problem (5) can be solved by an iterative procedure:

$$\begin{cases} u^{k+1} = & \arg\min_{u} \lambda D_J^{p^k}(u,u^k) + H(u,f) \\ p^{k+1} = & p^k - \partial H(u^{k+1}, f), \end{cases} \qquad (7)$$

where $\partial H(u^{k+1}, f)$ denotes a sub-gradient of H at u^{k+1} and $\lambda > 0$. When we choose $H(u,f) = \|Au-f\|^2$ and $J(u) = \|u\|_{TV}$, (7) turns into the total variation minimization problem, then (7) can be converted into the following two step Bregman iterative scheme [28]:

$$\begin{cases} u^{k+1} = & \arg\min_{u} \lambda \|u\|_{TV} + \|Au - f^k\|^2 \qquad (8) \\ f^{k+1} = & f^k + f - Au^{k+1}. \qquad (9) \end{cases}$$

In [28], the authors mentioned that the sequence u^k weakly converges to a solution of the unconstrained form of (5), and the sequence $\|Au^k - f\|^2$ converges to zero monotonically. We can see from (8),(9) that,the Bregman iteration just turns the original problem (5) into a iteration procedure and add the noise residual

back into the degenerated image at the end of every iteration. Bregman iteration converges fast and gets better results than standard methods. Bregman iteration was widely used in varied areas of image processing [5,31–33].

General framework of the iterative algorithm. As introduced above, we use the Bregman iteration (8),(9) to bulid the main iterative framework. Rather than considering (3), we consider the problem as follows:

$$\arg\min_{u,g} \|Ag-f\|^2 + \lambda\|u\|_{TV} \text{ subject to } g=u. \qquad (10)$$

We separated the variable u in (3) into two independent variables, so we can split the original problem (3) into subproblems which are easy to solve. This problem is obviously equivalent to (3). We can replace (10) into the unconstrained form:

$$\arg\min_{u,g} \|Ag-f\|^2 + \lambda_1\|g-u\|^2 + \lambda_2\|u\|_{TV}. \qquad (11)$$

λ_1 and λ_2 are regularization parameters balancing the three terms. If the regularization parameter λ_1 is big enough, the problem (11) is close to the problem (10), and the solutions of the problems are similar. If we let $J(u,g) = \|Ag-f\|^2 + \lambda_2\|u\|_{TV}$ and $H(u,g) = \|g-u\|^2$, we can see that J and H are all convex, and then (10) is a simple application of (7). Thus, the above problem can be solved by using Bregman iteration:

$$\begin{cases} (u^{k+1}, g^{k+1}) = & \arg\min_{u,g} J(u,g) - <p_u{}^k, u-u^k> \\ & - <p_g{}^k, g-g^k> + \lambda_1\|g-u\|^2 \\ p_u{}^{k+1} = & p_u{}^k - \lambda_1(u^{k+1} - g^{k+1}) \\ p_g{}^{k+1} = & p_g{}^k - \lambda_1(g^{k+1} - u^{k+1}). \end{cases} \qquad (12)$$

Similar to [28], we can reform the above procedure into a simple two step iteration algorithm:

$$\begin{cases} (u^{k+1}, g^{k+1}) = \arg\min_{u,g}\|Ag-f\|^2 + \lambda_2\|u\|_{TV} + \lambda_1\|g-u-b^k\|^2 & (13) \\ b^{k+1} = b^k + u^{k+1} - g^{k+1}. & (14) \end{cases}$$

As we can see in (13), when λ_1 tends to infinity, the above algorithm is equal to the original Bregman iterative algorithm in [28]. we use an alternating minimization algorithm [29,30] to solve (13). We split (13) into a deblurring and a denoising subproblems. Thus, (13) can be solved by the following two step iterative formation:

$$\begin{cases} g^{k+1} = \arg\min_{g}\|Ag-f\|^2 + \lambda_1\|g-u^k-b^k\|^2 & (15) \\ u^{k+1} = \arg\min_{u} \lambda_1\|g^{k+1}-u-b^k\|^2 + \lambda_2\|u\|_{TV}. & (16) \end{cases}$$

We can see that (15) is an l^2-norm differentiable problem, we can solve it as follows:

A

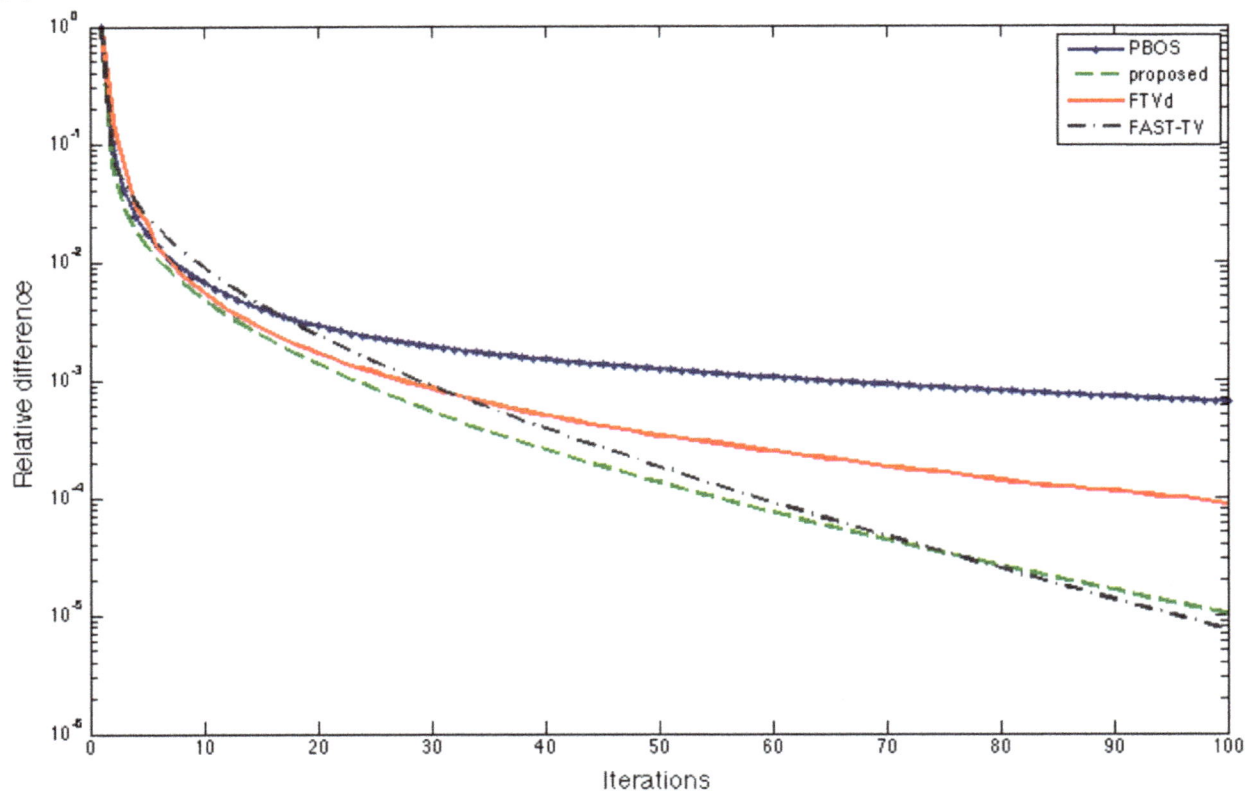

B

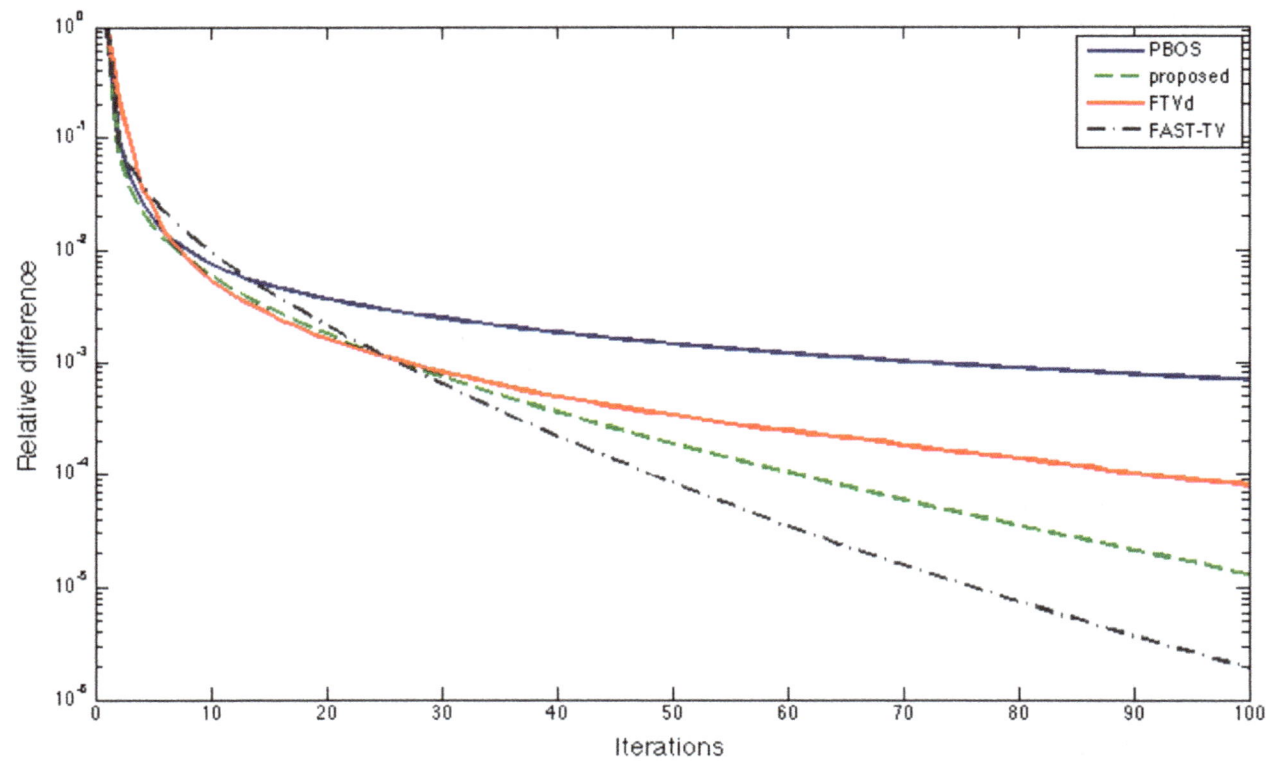

Figure 1. Convergence speed between algorithm 1, operator splitting TV, FTVd and FAST-TV. A. 7×7 Gaussian kernel with *sigma* $= 3$ and gaussian noise with $\sigma = 2$ B. 9×9 average kernel and gaussian noise with $\sigma = 3$. The figure shows the convergence speed between four methods using the Cameraman image and two different blur kennels. Axis X stands for the iteration times, Axis Y stands for the relative difference between restored images in two iterations, that is $\|u^k - u^{k-1}\|/u^k$.

$$g^{k+1} = (\frac{1}{\lambda_1} A A^T + I)^{-1} (\frac{1}{\lambda_1} A^T f + u^k + b^k), \qquad (17)$$

where I is the identity matrix and the matrix $\frac{1}{\lambda_1} A A^t + I$ is invertible. Then (15) can be solved by optimization techniques such as Gauss-Seidel, conjugate gradient or Fourier transform. As for (16), it is a exact total variation denoising problem, we can use Chambolle's projection algorithm [2], semismooth Newton method [3] or split Bregman algorithm [5] to solve this problem.

Thus, the proposed alternating Bregman iterative method for image restoration can be formed as follows:

Algorithm 1: Alternating Bregman iterative method for image restoration.

Initialize $k = 0$ and $b^0 = u_n^0 = 0$

while $\|u_n^k - u_n^{k-1}\|/\|u_n^k\| < \tau$ or $k < maxitertimes$ **do**

  for $i = 1 : n$ **do**

    $u_0^{k+1} = u_n^k$

    $g_i^{k+1} = \arg\min_g \|Ag - f\|^2 + \lambda_1 \|g - u_{i-1}^{k+1} - b^k\|^2$

    $u_i^{k+1} = \arg\min_u \lambda_1 \|g_i^{k+1} - u - b^k\|^2 + \lambda_2 \|u\|_{TV}$

  end

  $b^{k+1} = b^k + u_n^{k+1} - g_n^{k+1}$

  $k = k + 1$

end

Analysis of the proposed algorithm. First, we show some important monotonicity properties of the Bregman iteration proposed in [28].

Theorem 1 The sequence $H(u_k, f)$ obtained from the Bregman iteration is monotonically nonincreasing. And assume that there exists a minimizer $\tilde{u} \in BV(\Omega)$ of $H(., f)$ such that $J(\tilde{u}) < \infty$. Then.

$$H(u_k, f) \leq H(\tilde{u}, f) + \frac{J(\tilde{u})}{k}, \qquad (18)$$

and, in particular, u_k is a minimizing sequence.

Moreover, u_k has a weak-* convergent subsequence in $BV(\Omega)$, and the limit of each weak-* convergent subsequence is a solution of $Au = f$. If $\tilde{u}$ is the unique solution of $Au = f$, then $u_k \to \tilde{u}$ in the weak-* topology in $BV(\Omega)$.

Then, we show that the alternating minimization algorithm (15) and (16) also convergence to the solution of the sub-problem (13) [30]. Let L be the difference matrix and $NULL(\cdot)$ denotes the null space of the corresponding matrix, we obtain the following theorem.

Theorem 2 For any initial guess $u_0 \in R^{n^2}$, suppose $\{u_i\}$ is generated by (15) and (16), then u_i converges to a stationary point

of (10). And when A is a matrix of full column rank, u_i converges to a minimizer of (10).

Then, we can get the following convergence theorem of the proposed alternating Bregman iterative method.

Theorem 3 Let A be a linear operator, consider the algorithm 1. Suppose $\{u_i\}$ is a sequence generated by algorithm 1, u_i converges to a solution of the original constrained problem (3).

Proof. Let $\{u_i\}$ and $\{g_i\}$ be the sequence obtained from (13), and every u_i is the solution of the (13), moreover $H(u, g) = \|u - g\|^2 \to 0$ with the increase in iterations of the algorithm 1. Suppose in one iteration, there is u^* and g^* satisfying $u^* = g^*$, and let the true solutions of the problem (10) be $\tilde{u}$ and $\tilde{g}$, then.

$$\|u^* - g^*\| = \|\tilde{u} - \tilde{g}\| = 0. \qquad (19)$$

Due to u^* and g^* satisfy (11), u^* and g^* can enable the convex function (11) to obtain its Minimum value. Then.

$$\|Ag^* - f\|^2 + \lambda_1 \|g^* - u^*\|^2 + \lambda_2 \|u^*\|_{TV}$$
$$\leq \|A\tilde{g} - f\|^2 + \lambda_1 \|\tilde{g} - \tilde{u}\|^2 + \lambda_2 \|\tilde{u}\|_{TV}. \qquad (20)$$

Thus, we can obtain.

$$\|Ag^* - f\|^2 + \lambda_2 \|u^*\|_{TV} \leq \|A\tilde{g} - f\|^2 + \lambda_2 \|\tilde{u}\|_{TV}. \qquad (21)$$

Owning to $\tilde{u}$ and $\tilde{g}$ are the true solutions of the problem (10), this inequality implies that u^* and g^* are also the solutions of the problem (10), thus are the solutions of the original unconstrained problem (3).

Connection with other methods. We noticed that the equation (17) can be rewrite as follows:

$$g^{k+1} = u^k + b^k - (A^T A + \lambda_1 I) A^T (A(u^k + b^k) - f^k), \qquad (22)$$

thus, the proposed algorithm 1 can be interpreted as follows:

$$\begin{cases} g^{k+1} &= u^k + b^k - (A^T A + \lambda_1 I) A^T (A(u^k + b^k) - f^k) \\ u^{k+1} &= \arg\min_u \lambda_2 \|u\|_{TV} + \lambda_1 \|g - u - b^k\|^2 \\ b^{k+1} &= b^k + u^{k+1} - g^{k+1}. \end{cases} \qquad (23)$$

The preconditioned Bregmanized nonlocal regularization (PBOS) algorithm [33] can be formed as:

$$\begin{cases} g^{k+1} &= u^k - \delta A^T (A A^T +)^{-1} (A u^k - f^k) \\ u^{k+1} &= \arg\min_u \mu \|u\|_{TV} + \frac{1}{\delta} \|u - g^{k+1}\|^2 \\ f^{k+1} &= f^k + f - A u^{k+1}. \end{cases} \qquad (24)$$

the left and right pseudo inverse approximation are equal:

Figure 2. Restoration results on a 256×256 **Cameraman image degraded by a** 7×7 **Gaussian kernel with** *sigma* $=3$ **and a gaussian noise with** $\sigma = 2$. A. Original Image B. Degraded Image C. Operator Splitting TV D. ForWard E. FTVd F. FAST-TV G. NLTV+BOS H. Algorithm 1 I. Algorithm 2.

$$A^T(AA^T + \epsilon)^{-1} = (A^T A +)^{-1} A^T \qquad (25)$$

Compare these two methods, we can see the only difference between them is the way to calculate the noise and add it back to the iteration. The PBOS method calculates the undeconvolutioned noise and only add it back to the calculation of g, while the proposed method calculate the deconvolutioned noise and add it back to both the calculations of g and u, we believe that is why the proposed algorithm have a faster converge speed and better restoration results according to the experiments in section 0.

Adaptive Nonlocal Regularization

Nonlocal regularization. Recently, nonlocal methods have been extensively studied, the nonlocal means filter was first proposed by Buades et al [34]. The main idea of the nonlocal means denoising model is to denoise every pixel by averaging the other pixels with similar structures (patches) to the current one. Based on the nonlocal means filter, Kindermann et al. [35] tried to investigate the use of regularization functionals with nonlocal correlation terms for general inverse problems. Inspired by the graph Laplacian and the nonlocal means filter, Gilboa and Osher defined a variational framework based nonlocal operators [36]. In

Figure 3. Restoration results on a 256×256 **Cameraman image degraded by a** 9×9 **average kernel and a gaussian noise with** $\sigma = 2$. A. Original Image B. Degraded Image C. Operator Splitting TV D. ForWard E. FTVd F. FAST-TV G. NLTV+BOS H. Algorithm 1 I. Algorithm 2.

the following, we use the definitions of the nonlocal regularization functionals introduced in [36].

Let $\Omega \subset \mathbb{R}^2$, $x \in \Omega$, $u(x)$ is a real function $\Omega \rightarrow \mathbb{R}$ and w is a nonnegative symmetric weight function. Then the nonlocal gradient $\nabla_w u(x)$ is defined as the vector of all partial differences $\nabla_w u(x, \cdot)$ at x:

$$\nabla_w u(x,y) = (u(y) - u(x)) \sqrt{w(x,y)},$$

and the graph divergence div_w of a vector $p : \Omega \times \Omega \rightarrow \mathbb{R}$ can be defined as:

$$div_w p(x) = \int_\Omega (p(x,y) - p(y,x)) \sqrt{w(x,y)} dy,$$

the weight function is defined as the nonlocal means weight function:

$$w(x,y) = exp\left\{ -\frac{(G_a * \|f(x+\cdot) - f(y+\cdot)\|^2)(0)}{2h^2} \right\}, \quad (26)$$

where G_a is the Gaussian kernel with standard deviation a, h is the filtering parameter related to the standard variance of the noise,

Table 1. PSNR and SSIM results of the methods on five different images with a 7×7 Gaussian kernel with **sigma** $= 3$ and gaussian noise $\sigma = 2$.

Image	Blur/noise variance	PSNR	SSIM
	Operator splitting	25.81	0.701
	ForWARD	26.18	0.682
	FTVd	25.71	0.818
Cameraman	FAST-TV	26.20	0.813
	NLTV+BOS	26.00	0.830
	Algorithm 1	26.40	0.830
	Algorithm 2	27.21	0.832
	Operator splitting	23.52	0.667
	ForWARD	23.79	0.701
	FTVd	23.12	0.667
Barbara	FAST-TV	23.28	0.706
	NLTV+BOS	23.59	0.732
	Algorithm 1	23.65	0.701
	Algorithm 2	23.89	0.734
	Operator splitting	25.88	0.716
	ForWARD	25.61	0.654
	FTVd	25.49	0.762
Man	FAST-TV	25.65	0.755
	NLTV+BOS	25.99	0.780
	Algorithm 1	26.06	0.781
	Algorithm 2	26.37	0.786
	Operator splitting	27.60	0.813
	ForWARD	27.30	0.799
	FTVd	26.21	0.825
Boats	FAST-TV	26.78	0.821
	NLTV+BOS	28.01	0.843
	Algorithm 1	27.45	0.841
	Algorithm 2	28.05	0.844
	Operator splitting	27.29	0.799
	ForWARD	27.08	0.791
	FTVd	26.46	0.824
Lenna	FAST-TV	27.09	0.812
	NLTV+BOS	27.52	0.836
	Algorithm 1	27.54	0.832
	Algorithm 2	28.00	0.836
	Operator splitting	26.02	0.739
	ForWARD	25.99	0.725
	FTVd	25.40	0.779
Average	FAST-TV	25.80	0.781
	NLTV+BOS	26.22	0.804
	Algorithm 1	26.22	0.797
	Algorithm 2	26.70	0.806

Table 2. PSNR and SSIM results of the methods on five different images with a 9×9 average kernel and gaussian noise $\sigma = 3$.

Image	Blur/noise variance	PSNR	SSIM
	Operator splitting	24.68	0.691
	ForWARD	24.54	0.649
	FTVd	25.18	0.789
Cameraman	FAST-TV	24.87	0.781
	NLTV+BOS	25.16	0.755
	Algorithm 1	25.21	0.760
	Algorithm 2	25.98	0.777
	Operator splitting	23.12	0.639
	ForWARD	23.22	0.660
	FTVd	23.12	0.683
Barbara	FAST-TV	23.18	0.657
	NLTV+BOS	23.21	0.677
	Algorithm 1	23.16	0.680
	Algorithm 2	23.66	0.706
	Operator splitting	24.34	0.618
	ForWARD	23.98	0.588
	FTVd	24.12	0.701
Man	FAST-TV	24.06	0.675
	NLTV+BOS	24.51	0.702
	Algorithm 1	24.59	0.706
	Algorithm 2	24.90	0.712
	Operator splitting	25.82	0.748
	ForWARD	25.40	0.736
	FTVd	24.87	0.780
Boats	FAST-TV	25.11	0.762
	NLTV+BOS	26.15	0.783
	Algorithm 1	25.73	0.778
	Algorithm 2	26.33	0.783
	Operator splitting	26.01	0.766
	ForWARD	25.57	0.746
	FTVd	25.52	0.771
Lenna	FAST-TV	25.67	0.765
	NLTV+BOS	26.22	0.776
	Algorithm 1	26.03	0.766
	Algorithm 2	26.63	0.783
	Operator splitting	24.79	0.692
	ForWARD	24.54	0.676
	FTVd	24.56	0.730
Average	FAST-TV	24.58	0.728
	NLTV+BOS	25.05	0.739
	Algorithm 1	24.94	0.738
	Algorithm 2	25.50	0.752

and the $\cdot$ in $f(x + \cdot)$ stands for a square patch centered by point x. When the reference image f is known, the nonlocal means filter is a linear operator. The definition of the weight function (26) shows that the value of the weight is significant only when the patch around y has similar structure as the corresponding patch around x.

The nonlocal TV norm can be defined as isotropic L^1 norm of the weighted graph gradient $\nabla_w u(x)$:

$$J_{NLTV,w}(u) = \int_{\Omega} |\nabla_w u(x)| dx \qquad (27)$$

$$= \int_{\Omega} \sqrt{\int_{\Omega} (u(x) - u(y))^2 w(x,y) dy} dx. \qquad (28)$$

The main idea of the nonlocal regularization is to generalize the local gradient and divergence concepts to the nonlocal form. Then the nonlocal means filter is generalized to the variational framework.

The nonlocal means filter and the nonlocal regularization functionals can reduce noise efficiently and preserve textures and contrast of the image. Generally, it is good to choose a reference image as close as possible to the original ideal image to calculate the weights. However, the original image structures are broken in the degraded image, we can not get the precise weights between the pixels, thus the weights should be calculated from a preprocessed image [37]. In our alternating minimization framework, we get the deblurred image at the first step, then denoise the deblurred image at the second step. As the nonlocal regularization functionals are robust to the noise, and the structures of the deblurred image are close to the original ideal image, we can calculate the weights by using the deblurred image as the reference image, then apply a nonlocal denoising step to obtain the restored image.

Adaptive nonlocal parameter selection. Within the alternating Bregman iterative method, we can use g^{k+1} as the reference image to calculate the weights of the nonlocal regularization functionals, then use the weights to denoise the deblurred image g^{k+1} at every iteration. Note that the nonlocal filter parameter h is related to the standard variance of the noise, however we do not know the exact noise of the image g^{k+1}. Moreover, when we use the single filter parameter h for the whole image, there will be regions oversmoothed or undersmoothed in the restored image, because a single filter parameter h is not optimal for all the patches in the image. As the nonlocal TV norm defined in (27), we will calculate the filter parameter h adaptively using local information and get the local h for every pixel in the image.

Inspired by local regularization in [21], we define the local power as:

$$P_r(x,y) = \frac{1}{|\Omega|} \int_{\Omega} (I(x',y') - I_r(x',y'))^2 \Omega_{x,y}(x',y') dx' dy', \qquad (29)$$

and $\Omega_{x,y}(x',y') = \Omega(|x-x'|,|y-y'|)$ is a normalized smoothing window, here we use a Gaussian window. I_r is the expected image, Ω is a region to calculate the local power centered at (x,y).

Then we use the local power to calculate the local h as follows:

$$h(x,y) = \alpha \sqrt{P_r(x,y)}. \qquad (30)$$

The advantage of localizing the filter parameter h is that it can control the denoising process over image regions according to their content, the smooth regions have average w between there neighbors, texture and edge regions have big w only when the patches are similar. Besides, we do not have to know or estimate the noise condition. In this paper, we use a preprocessed oversmoothed image us as the expected image instead of the mean of the patch to get more accurate results. The oversmoothed

image is obtained by a standard TV model using a large regularization parameter.

By applying the above adaptive nonlocal regularization, the algorithm 1 can be reformed as the following algorithm, where W is the function to calculate the weights between points and us is the preprocessed oversmoothed image:

Algorithm 2: Adaptive nonlocal alternating Bregman iterative method for image restoration.

Initialize $k = 0$ and $g_0^0 = u_n^0 = 0$

while $\|u_n^k - u_n^{k-1}\|/\|u_n^k\| < \tau$ *or* $k < maxitertimes$ **do**

    for $i = 1 : n$ **do**

        $u_0^{k+1} = u_n^k$

        $g_n^{k+1} = \arg\min_g \|Ag - f\|^2 + \lambda_1 \|g - u_{n-1}^{k+1} - b^k\|^2$

        $w_n^{k+1} = W(g_n^{k+1}, us, \alpha)$

        $u_n^{k+1} = \arg\min_u \lambda_1 \|g_n^{k+1} - u - b^k\|^2 + \lambda_2 \|u\|_{NLTV/w}$

    end

    $b^{k+1} = b^k + u_n^{k+1} - g_n^{k+1}$

    $k = k + 1$

end

Experiments

In this section, we present some experimental results of the proposed alternating Bregman method and the adaptive nonlocal alternating Bregman method, and compare them with the operator splitting TV regularization [10], NLTV based BOS algorithm, FTVd algorithm [38], FAST-TV [30] and ForWaRD algorithm [39]. The ForWaRD algorithm is a hybrid Fourier-wavelet regularized deconvolution (ForWaRD) algorithm that performs noise regularization via scalar shrinkage in both the Fourier and wavelet domains.

We use the conjugate gradient method to solve the first subproblem in algorithm 1 and algorithm 2, use the Chambolle's projection algorithm to solve the second subproblem in algorithm 1 and the nonlocal version of the Chambolle's projection algorithm in algorithm 2. In algorithm 1 and algorithm 2, we set $\lambda_1 = 0.02$ and $\lambda_2 = 0.01$ by experiment results, the inner iteration times n can be set as 1 or 2 and the stopping condition is $\|u^k - u^{k-1}\|/\|u^k\| < 10^{-3}$. There are lots of work on determining the parameters in the regularization [40,41], but this work is out of the scope of this paper and we will get in to it later. In algorithm 2, we set the patch size as 5×5, searching window as 21×21 and set the Gaussian variance parameter as $\sigma = 5$ to calculate the local variance. And we set the nonlocal parameter factor as $\alpha = 2$. For the operator splitting method, the regularization parameter is set as 0.1. For the NLTV based BOS method, the regularization parameter is set as $\lambda = 0.2$, searching window is set as 21×21, the patch size is set as 5×5 and the nonlocal filter parameter is set as $h = 30$. For the FTVd algorithm, we set $\mu = 1000$. For the FAST-TV, we set α_1 as 0.003 and 0.005 according to the degeneration of the image. And the best valve of α_1. For the ForWaRD algorithm, we set the threshold as 3 * *sigma*, *sigma* is the standard deviation of the noise, and the regularization parameter is set to 1.

First, we compare the convergence speed of the proposed algorithm 1 with the preconditioned BOS algorithm, the FTVd algorithm and the FAST-TV algorithm in Figure 1. We can find that the proposed algorithm 1 converge faster than the other three methods at first, and still much faster than the preconditioned BOS algorithm and the FTVd algorithm later, close or a little bit slower than the FAST-TV at the end of the iterations. Usually, the stopping condition of the relative difference is set to 10^{-3} or 10^{-4}. Thus, the proposed algorithm 1 can reach the stopping condition with fewer iterations than other algorithms. In terms of the computation time, the FTVd algorithm is the fastest owning to its strategies and code optimization. And the proposed algorithm 1 is faster than the operator splitting algorithm and the FAST-TV algorithm. As for the nonlocal methods, convergence can not be promised after some iterations, so we compare the computation time between these methods. As the computation of the nonlocal weights, the nonlocal based algorithms cost more computing time than the non nonlocal ones. The NLTV based BOS algorithm stops with 25 steps for 180 seconds, and the preconditioned NLTV based BOS algorithm stops with 8 steps for 75 seconds, however the proposed algorithm 2 stops with 5 steps for only 47 seconds.

Next, we show some image restoration results of these methods to illustrate the effectiveness of the proposed algorithms. We use the classical Cameraman image, so as to be comparable to other image restoration works. The Cameraman image can be found at http://www.imageprocessingplace.com/root_files_V3/image_databases.htm. Figure 2 and Figure 3 show the restoration results on the Cameraman with two kind of blur kernels. We can see from the results that, the ForWard method can get a good restoration result when the image is not slightly blurred, but poor on the heavily blurred situation, besides the ForWard method can not restore edges clearly. The restoration results of the operator splitting TV method have artificial strips which affect the visual appearance of the restored images. FTVd method and FAST-TV can effectively remove noise from the degenerated images, and have higher PSNRs than the ForWard method and the operator splitting TV method, however, a lot of details are also smoothed. The results of the proposed algorithm 1 have good visual appearance, clear edges and preserved image contrast. The NLTV based BOS method (the preconditioned BOS has almost the same result) and the proposed algorithm 2 have better restoration results than the not nonlocal ones, and the proposed algorithm 2 have more details restored and a higher PSNR.

The Table 1 and Table 2 shows the restoration results on 5 different images and 2 different degradation situations. We can see that, the PSNR and SSIM of the proposed algorithms are generally higher than the methods being compared.

Conclusions

In this paper, we propose a Bregman iteration based total variation image restoration algorithm. We split the restoration problem into a three step iteration process, and these steps are all easy to solve. In addition, we propose a nonlocal regularization under the framework of the proposed algorithm using a point-wise local filter parameter, and a method to adaptively determine the filter parameter. Experiments show that the algorithm converges fast and the adaptive nonlocal regularization method can obtain better restoration results. In the future, we will consider the weights updating problem in a theoretical way and apply the proposed algorithms for other regularization problems such as compressed sensing.

Author Contributions

Conceived and designed the experiments: HX QS. Performed the experiments: HX GC. Analyzed the data: HX QS DX. Wrote the paper: HX NL GC.

References

1. Rudin L, Osher S, Fatemi E (1992) Nonlinear total variation based noise removal algorithms. Physica D: Nonlinear Phenomena 60: 259–268.
2. Chambolle A (2004) An Algorithm for Total Variation Minimization and Applications. Journal of Mathematical Imaging and Vision 20: 89–97.
3. Ng MK, Qi L, Yang Yf, Huang Ym (2007) On Semismooth Newton's Methods for Total Variation Minimization. Journal of Mathematical Imaging and Vision 27: 265–276.
4. Chan T, Chen K (2006) An optimization-based multilevel algorithm for total variation image denoising. Multiscale Model Simul 5: 615–645.
5. Goldstein T, Osher S (2009) The split Bregman method for L1 regularized problems. SIAM Journal on Imaging Sciences 2: 323–343.
6. Hintermller M, Stadler G (2006) An infeasible primal-dual algorithm for tv-based inf-convolutiontype image restoration. SIAM Journal on Scientific Computing 28: 1–23.
7. Esser E, Zhang X (2009) A general framework for a class of first order primal-dual algorithms for TV minimization. UCLA CAM Report : 1–30.
8. Carter J (2001) Dual methods for total variation-based image restoration. University of California Los Angeles.
9. Esser J (2010) Primal Dual Algorithms for Convex Models and Applications to Image Restoration, Registration and Nonlocal Inpainting. University of California Los Angeles.
10. Combettes PL, Wajs VR (2005) Signal recovery by proximal forward-backward splitting. Multiscale Modeling Simulation 4: 1168–1200.
11. Nikolova M (2006) Analysis of half-quadratic minimization methods for signal and image recovery. SIAM Journal on Scientific computing 27: 937–966.
12. Oliveira JaP, Bioucas-Dias JM, aT Figueiredo M (2009) Adaptive total variation image deblurring: A majorization-minimization approach. Signal Processing 89: 1683–1693.
13. Chantas G, Galatsanos N, Likas A, Saunders M (2008) Variational Bayesian image restoration based on a product of t-distributions image prior. IEEE transactions on image processing 17: 1795–805.
14. Babacan SD, Molina R, Katsaggelos AK (2008) Parameter estimation in TV image restoration using variational distribution approximation. IEEE transactions on image processing 17: 326–39.

15. Chantas G, Galatsanos NP, Molina R, Katsaggelos AK (2010) Variational bayesian image restoration with a product of spatially weighted total variation image priors. IEEE transactions on image processing 19: 351–62.
16. Almansa A, Ballester C, Caselles V (2008) A TV based restoration model with local constraints. Journal of Scientific Computing 34: 612–626.
17. Bertalmio M, Caselles V, Rougé B (2003) TV based image restoration with local constraints. Journal of Scientific Computing 19: 95–122.
18. Takeda H, Farsiu S, Milanfar P (2008) Deblurring using regularized locally adaptive kernel regression. IEEE transactions on image processing 17: 550–63.
19. Wu C, Tai XC (2010) Augmented Lagrangian method, dual methods, and split Bregman iteration for ROF, vectorial TV, and high order models. SIAM Journal on Imaging Sciences 3: 300–339.
20. Pang ZF, Yang YF (2011) A projected gradient algorithm based on the augmented Lagrangian strategy for image restoration and texture extraction. Image and Vision Computing 29: 117–126.
21. Gilboa G, Sochen N (2003) Texture preserving variational denoising using an adaptive fidelity term. Proc VLSM.
22. Li F, Shen C, Shen C, Zhang G (2009) Variational denoising of partly textured images. Journal of Visual Communication and Image Representation 20: 293–300.
23. Prasath VS, Singh A (2009) Ringing Artifact Reduction in Blind Image Deblurring and Denoising Problems by Regularization Methods. 2009 Seventh International Conference on Advances in Pattern Recognition : 333–336.
24. Liu H, Klomp N, Heynderickx I (2010) A perceptually relevant approach to ringing region detection. IEEE transactions on image processing 19: 1414–26.
25. Nasonov A (2010) Scale-space method of image ringing estimation. IEEE International Conference on Image Processing (ICIP) : 2793–2796.
26. Chen DQ, Zhang H, Cheng LZ (2010) Nonlocal variational model and filter algorithm to remove multiplicative noise. Optical Engineering 49: 077002.
27. Nikolova M (2004) A variational approach to remove outliers and impulse noise. Journal of Mathematical Imaging and Vision 20: 99–120.
28. Osher S, Burger M, Goldfarb D, Xu J, Yin W (2006) An Iterative Regularization Method for Total Variation-Based Image Restoration. Multiscale Modeling & Simulation 4: 460.

29. Wang Y, Yang J, Yin W, Zhang Y (2008) A New Alternating Minimization Algorithm for Total Variation Image Reconstruction. SIAM Journal on Imaging Sciences 1: 248.

30. Huang Y, Ng MK,Wen YW (2008) A Fast Total Variation Minimization Method for Image Restoration. Multiscale Modeling & Simulation 7: 774.

31. Yin W, Osher S, Goldfarb D (2008) Bregman iterative algorithms for l1-minimization with applications to compressed sensing. SIAM J Imaging Sci 1: 143—168.

32. Cai JF, Osher S, Shen Z (2009) Linearized Bregman Iterations for Frame-Based Image Deblurring. SIAM Journal on Imaging Sciences 2: 226–252.

33. Zhang X, Burger M, Bresson X, Osher S (2010) Bregmanized Nonlocal Regularization for Deconvolution and Sparse Reconstruction. SIAM Journal on Imaging Sciences 3: 253–276.

34. Buades A, Coll B, Morel JM (2005) On image denoising methods. SIAM Multiscale Modeling and Simulation 4: 490–530.

35. Kindermann S, Osher S, Jones PW (2005) Deblurring and Denoising of Images by Nonlocal Functionals. Multiscale Modeling & Simulation 4: 1091–1115.

36. Gilboa G, Osher S (2008) Nonlocal operators with applications to image processing. Multiscale Model Simul 7: 1005–1028.

37. Lou Y, Zhang X, Osher S, Bertozzi A (2009) Image Recovery via Nonlocal Operators. Journal of Scientific Computing 42: 185–197.

38. Wang Y, Yin W (2007) A fast algorithm for image deblurring with total variation regularization. CAAM Technical Report TR07–10.

39. Neelamani R, Choi H, Baraniuk R (2002) Forward: Fourier-wavelet regularized deconvolution for ill-conditioned systems. IEEE Trans on Signal Processing 52: 418–433.

40. Wen Y (2009) Adaptive Parameter Selection for Total Variation Image Deconvolution. Numerical Mathematics: Theory, Methods and Applications 2: 427–438.

41. Liao H, Li F, Ng MK (2009) Selection of regularization parameter in total variation image restoration. Journal of the Optical Society of America A 26: 2311–2320.

10

Kite Aerial Photography for Low-Cost, Ultra-high Spatial Resolution Multi-Spectral Mapping of Intertidal Landscapes

Mitch Bryson[1]*, Matthew Johnson-Roberson[2], Richard J. Murphy[1], Daniel Bongiorno[1]

1 Australian Centre for Field Robotics, The University of Sydney, Sydney, NSW, Australia, **2** The Department of Naval Architecture and Marine Engineering, University of Michigan, Ann Arbor, Michigan, United States of America

Abstract

Intertidal ecosystems have primarily been studied using field-based sampling; remote sensing offers the ability to collect data over large areas in a snapshot of time that could complement field-based sampling methods by extrapolating them into the wider spatial and temporal context. Conventional remote sensing tools (such as satellite and aircraft imaging) provide data at limited spatial and temporal resolutions and relatively high costs for small-scale environmental science and ecologically-focussed studies. In this paper, we describe a low-cost, kite-based imaging system and photogrammetric/mapping procedure that was developed for constructing high-resolution, three-dimensional, multi-spectral terrain models of intertidal rocky shores. The processing procedure uses automatic image feature detection and matching, structure-from-motion and photo-textured terrain surface reconstruction algorithms that require minimal human input and only a small number of ground control points and allow the use of cheap, consumer-grade digital cameras. The resulting maps combine imagery at visible and near-infrared wavelengths and topographic information at sub-centimeter resolutions over an intertidal shoreline 200 m long, thus enabling spatial properties of the intertidal environment to be determined across a hierarchy of spatial scales. Results of the system are presented for an intertidal rocky shore at Jervis Bay, New South Wales, Australia. Potential uses of this technique include mapping of plant (micro- and macro-algae) and animal (e.g. gastropods) assemblages at multiple spatial and temporal scales.

Editor: Simon Thrush, National Institute of Water & Atmospheric Research, New Zealand

Funding: This work was supported through the Early Career Researcher Development Scheme through the Faculty of Engineering and Information Technologies at the University of Sydney, the Australian Research Council and the New South Wales State Government. No additional external funding was received for this study. The funders had no role in study design, data collection and analysis, decision to publish, or preparation of the manuscript.

Competing Interests: The authors have declared that no competing interests exist.

* E-mail: m.bryson@acfr.usyd.edu.au

Introduction

Plant and animal assemblages that live in intertidal regions such as rocky shores are part of a complex, dynamic ecosystem, the structure and functioning of which can vary across a cascade of spatial and temporal scales [1]. Intertidal ecosystems have primarily been studied using field-based sampling e.g. [2–5] at appropriate resolutions to capture the spatial variability at which assemblages occur. When studies cover a broad area of shoreline or occur at multiple sites, extensive field-based sampling requires a large amount of logistical effort and data is often not recorded in a contiguous manner (i.e. different parts of the shoreline are sampled at different times). Remote sensing is the ideal tool to collect contiguous data over large areas in a snapshot of time, however conventional remote sensing platforms (i.e. satellites and manned-aircraft) provide data at relatively coarse spatial and temporal resolution. Current state-of-the-art commercially available high-resolution satellite imagery can provide resolutions of 2.4 m per pixel for multi-spectral imagery and 0.6 m per pixel for panchromatic imagery, at a cost of $3000-5000US per imagery scene [6]. Manned aerial photography and airborne lidar provide higher resolutions (up to 0.3 m per sample point [7]), depending on flying height, and are typically more expensive, with targeted

data collection costing in the order of tens of thousands of dollars per flight [8,9]. The low-resolution and high costs of targeted data collection (i.e. at a specific time and place) limits the effectiveness of conventional remote sensing in small-scale environmental science and ecologically-focussed studies. Furthermore this data does not provide information on topographic variability at small scales (centimeters and meters), which is known to influence the distribution of assemblages of plant and animal species [10].

In this paper we develop data collection techniques and data processing algorithms for constructing ultra-high resolution (sub-centimeter) three-dimensional (3D) multi-spectral maps of inter-tidal rock platforms using a low-cost kite-based mapping system. The objective of our work is to develop a system that provides both topographic data and multi-spectral imagery over a broad area of the intertidal environment (hundreds of meters of shoreline) at appropriate spatial scales for ecologically-focussed studies. We describe a methodology for collecting images from two consumer-grade digital cameras, one that is a standard three-channel colour camera and the other a camera that has been converted to image in near-infrared (wavelengths greater than 720 nm). The cameras are carried on the flying line of single-line kite that is flown over the target site and used to collect multiple overlapping images of the terrain. The images are then processed in a procedure that automatically extracts and matches spatial features across multiple images and uses this information to incrementally reconstruct both a 3D map of the terrain and the position of the camera as each image was captured. Results are presented from a 200 m stretch of temperate intertidal rocky shoreline at Jervis Bay, New South Wales, Australia with an average topographic ground sampling distance of 2.5 cm and an imagery resolution of 5 mm per pixel. The resulting maps contain both topographic data and multi-spectral imagery. Topographic data allows for high-resolution measurements such as the vertical position on the shore with respect to tides and the slope and aspect with respect to the sun. The multi-spectral imagery (red, green, blue and near-infrared) is used to derive a Normalised Difference Vegetation Index (NDVI) that allows for the identification of intertidal species present during imaging and the potential to provide data on variables such as chlorophyll and algal biomass. The use of readily available and cheap consumer-grade equipment and the non-technically de-manding process of collecting images in the field will hopefully allow for high-temporal resolution repeat sampling (i.e. in the order of days to weeks).

Related Work

Intertidal Ecology and Remote Sensing. The distribution of assemblages of plant and animal species in rocky intertidal and soft sedimentary intertidal systems is known to be highly variable across a cascade of spatial scales ranging from cm to km [4,11–16]. A complex interplay of bottom-up and top-down ecological processes are thought to regulate spatial distributions [17]. Remote sensing is able to provide the synoptic context in which spatial variations in assemblages can be quantified, yet the use of remote sensing has been limited by poor resolution.

Manned aircraft-based remote sensing has been used to study intertidal environments over large scales and low-spatial resolution. In [18], airborne hyperspectral imagery at a spatial resolution of 5 m was used to map the fractional coverage of macroalgae over a one kilometer-long rocky shore. In [19], airborne hyperspectral sensing was acquired at a 4 m resolution to measure sediment grain size and microphytobenthic biomass over several kilometres of an intertidal flat.

The advent of low-cost field spectroradiometers has resulted in a large body of work in field-based spectroscopy in recent years [3]. Spectroscopy in the visible and near-infrared bands has allowed for the measurement of indices relating to macrophyte cover [20] and microphytobenthic biomass [21] through the correlation of chlorophyll concentration and reflectance at different red and near-infrared bands. In [22] similar indices were derived for use with 3-band digital camera using red, green and near-infrared bands.

3D photogrammetry obtained from in-field, ground-based photography has been used to measure habitat structure and topographical complexity on rock intertidal shores. In [2] ground-based stereo-photos and photogrammetry techniques were used to measure different forms of structural complexity along repeated 30 cm transects to study the correlation to gastropod abundance in both rocky intertidal and mangrove habitats. In [5] similar field-based photogrammetry were used to measure topographic complexity along with surface temperature measurements to study the influence of habitat structure on body size and abundance of different invertebrates on an intertidal rocky shore.

Kite and Balloon Photography. Kites and balloons have been used for low-altitude aerial photography for decades, most prolifically in archaeology (see [23] for a good review) as a low-cost means of recording cultural heritage or providing a visual record of excavation activities. Balloons have typically been used when windspeeds in an area are low and kites when windspeeds are higher, providing complimentary use in a range of environmental conditions. These platforms are typically constructed from low-cost materials, are safe due to their low-weight and stability and easy to deploy in a range of conditions promoting their use in a wide range of applications such as boat-launched coral aerial photography [24], counting Antarctic penguin numbers [25] and monitoring in a humanitarian emergency [26]. In [27] the authors describe the use kite aerial photography in environmental site investigation and [28] discusses the complexities of kite aerial photography using near-infrared film in the study of vegetation and soil properties. In [29], the authors develop a kite aerial photography platform that includes real-time pan/tilt control and image stabilisation. Previous work in these applications has often focussed on taking either a single or a small number of photographs.

More recently kite aerial colour photography has been used to map and classify different vegetation types in the alpine zone using both supervised classification and unsupervised clustering of the three-band colour imagery [30]. The work in [31] and [32] used kite aerial photography to produce digital elevation models from a small collection of photographs producing topographic maps with pixel sizes of 4 cm and 5–7.5 cm over an area of 100-by-100 m and 50-by-50 m respectively. In [32] topographic maps were used to track the progress of gully erosion over a period of 2–4 years.

In relation to studies in the intertidal zone, the work in [33] used a helium blimp to take photographic stereo pairs of a section of rocky intertidal shore using both colour and near-infrared film. The photographs were geo-referenced to an 18-by-18 m plot at a resolution of 2 cm per pixel.

Unmanned Aerial Vehicles (UAVs). Unmanned Aerial Vehicles (UAVs) [34] are a relatively recent technology that have been used to produce high resolution maps owing to their ability for low-altitude flight. In [35], the authors demonstrated the use of UAVs for collecting images over rangelands with a spatial resolution of 5 cm per pixel. In the work of [36], a fixed-wing UAV was used to produce geo-referenced imagery maps with a resolution of 3.5 cm per pixel over an area of 4000-by-600 m in a weed monitoring application. Hovering UAVs (such as the Mikrokopter (http://mikrokopter.org)) have been demonstrated as a promising platform for low-altitude sensing for even higher resolution imaging. A hovering UAV was used to produce maps of Antarctic moss beds with a resolution of 1 cm per pixel over an area 100-by-40 m in [37]. In [38] the authors used hovering UAV imagery to produce 3D maps of a coastal cliff, producing pointclouds with a spatial resolution of 1–3 cm and an accuracy of 25–40 mm. UAVs are promising platforms for gathering high-resolution remotely sensed data; although the costs and technical skills required to operate these platforms are becoming lower over time, they are still relatively high, particularly when considering their use in small-scale ecological studies. Additionally, the current generation of rotary-wing UAVs are typically limited by low endurance (approx. 15–20 minutes) and are susceptible to failure in high-wind conditions, typically encountered in coastal regions.

Recent Developments in Photogrammetry and Structure-from-Motion. The process of measuring spatial properties from photographs or images is referred to as 'photogrammetry' and when large numbers of images are used, typically to reconstruct the three-dimensional spatial structure of an imaged scene, this process is referred to as 'structure-from-motion' [39]. Recent developments in structure-from-motion [40,41] have focussed on building 3D models of buildings from large collections of un-ordered, un-calibrated images. These methods utilise multi-core and parallel processing algorithms for efficiently combining images with relatively little additional information such as from Global Positioning Systems (GPS) or ground control points.

Overview of the contributions of our work in contrast to related work

In this paper we combine aspects of photogrammetry, multi-spectral remote sensing and low-cost image collection using a kite-based platform for building high-resolution imagery and terrain models in the intertidal zone. In contrast to previous work in kite aerial photography, our method fuses information from hundreds, and potentially thousands of overlapping monocular images using modern photogrammetry and bundle adjustment techniques [40,41]. This enables coverage over much larger areas and with higher spatial resolution. The photogrammetry techniques used employ algorithms for automatic calibration of the intrinsic and extrinsic parameters of the camera (i.e. focal length, lens distortion etc.) using the images themselves, allowing for the use of un-calibrated, low-cost consumer-grade cameras. The techniques also construct full, scale-less 3D models prior to the use of ground control points, providing flexibility in the way in which models are geo-referenced into a global coordinate system. Furthermore, we demonstrate a low-cost method for producing spatially-registered *multi-spectral* imagery from multiple consumer-grade cameras and validate the utility of this information for distinguishing macro-algae in the intertidal environment through the use of the NDVI. Compared to past work on intertidal mapping using traditional remote sensing platforms or field-based sampling, our approach provides data with an unprecedented level of resolution over a broad-scale environment. The data collection procedure is fast and easy to perform without technical experts in remote sensing or UAV operations.

Materials and Methods

Study Site

Experiments were performed over an intertidal rocky shore north of Greenfields beach (34.000°S, 151.249°E) along the western edge of Jervis Bay, New South Wales, Australia (see Figure 1). The site lies within a national park aquatic reserve and is host to various intertidal species and is largely populated by red, green and brown macro-algae such as *Homosira*, *Corallina*, *Ulva* and *Ralfsia* along the low shore. Various gastropods and tunicates (cunjevoi) were also present. Data collection was performed over a single, cloudless day during October 2012 at approximately 1:50 pm to coincide with a low tide of 0.16 m above the datum. All field studies involved only non-contact, non-invasive means of sampling (i.e. collection of images and field spectroscopy) and therefore did not require any specific permissions or ethics approvals.

Kite Aerial Photography Equipment

A kite aerial photography system (see Figure 2) was built that used a 2.7 m wingspan conynes-delta kite to lift a fixed, downwards-looking rig holding a consumer-grade digital camera. The conynes-delta kite was chosen for its stability and lifting capacity in a wide range of wind conditions. The camera was suspended from a Picavet rig that attached to the line of the kite approximately 10 m lower than the kite to minimise the impact of wind gust-induced motion of the camera. The Picavet provided mechanical levelling of the camera during changes in the flying angle of the kite. Two different versions of a consumer-grade digital camera (the Sony NEX-5 with a 16 mm pancake lens and a 16 MPix resolution) were used - a standard three-colour camera and one that had been converted for imaging in near-infrared wavelengths. A commercial service (LifePixel) was used to remove the internal near-infrared cut filter (installed in the majority of consumer digital camera to remove near-infrared light from images) and replace it with a high-pass colour cut filter that allowed transmission of wavelengths above 720 nm. The camera was chosen as a trade-off between the image quality achieved from a full-frame sensor digital single-lens reflex camera and the light-weight of a small-sensored compact digital camera.

The spectral response functions of both cameras were determined using the procedure described in [42,43] to provide precise information on the spectral sensitivities of the raw imagery from the red, green and blue channels of the colour camera and the near-infrared sensitivity of the converted camera (data was taken only from the red bayer channel of the images for this camera). Images of a Macbeth colour chart were captured using the raw mode of each camera under cloudless sunlight and measurements of the spectral response of each Macbeth colour panel were taken using a handheld spectroradiometer (Ocean Optics STS-VIS with an effective range from 400–800 nm), from which the response curves were estimated. Although the panels of the Macbeth colour chart are designed for use in colour calibration in the visible spectrum, it was found that the panels exhibited variations in reflectance at wavelengths around 720 nm and above, and were therefore also effective for calibrating images captured from the near-infrared converted camera. Figure 3 (a) illustrates the estimated spectral response curves for the cameras.

The red, green and blue channels of the colour camera displayed peaks at 470 nm, 530 nm and 606 nm respectively and the red channel of the near-infrared camera (which was found to have the best response of the three colour channels of this camera) had a peak response at 740 nm (shown in Table 1).

Process Overview

Figure 4 presents a flowchart illustrating the entire kite-based mapping process performed in the study. The process began by placing ground control points (visible markers) across the terrain and using the kite-based platform for collecting a series of overlapping images of the terrain taken from the air. After the images were acquired, they were processed using an automated procedure to produce 3D terrain models and geo-referenced photo-mosaics of the area covered by the images. The automated procedure extracted and matched feature points that were used by a structure-from-motion algorithm to reconstruct the poses from which the images were taken and a 3D point cloud of the imaged terrain. After geo-referencing this pointcloud using the measured ground control points, a photo-texturing algorithm was used to create the final map products. Table 2 provides an overview of the different software packages used at various stages in the data processing procedure. The following subsections describe these steps in detail.

Spatial and Spectral Ground Control Points

In order to geo-reference the 3D landscape models and maps reconstructed from the collected image data, flat 20-by-20 cm checkerboard-patterned panels were placed on dry rock surfaces in the environment before flight to act as ground control points. The position of each panel was measured using a GPS receiver. Because of the focus on low-cost equipment used in this study, ground control point panels were measured using a consumer-grade handheld GPS receiver that provided global position coordinates to an accuracy of ±2 m (rather than using a more expensive survey grade, differentially corrected GPS device that could potentially provide cm-level global accuracy). In order to supplement the low-accuracy GPS measurements, additional spatial constraints were measured between the ground control

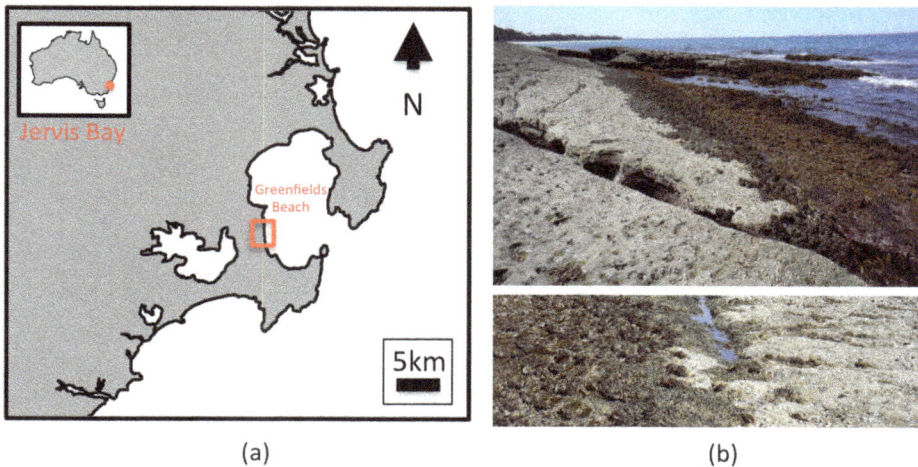

Figure 1. Study Site. (a) Intertidal rocky shore at Greenfields Beach (study site shown in red box) at Jervis Bay, New South Wales, Australia, (b) ground-based photography at the study site taken at low-tide. Google maps imagery of the site is available at: http://maps.google.com.au/?ll = -35. 085668,150.693294&spn = 0.005759,0.0109&t = h&z = 17.

points; the points were placed in triads, where each triad was arranged into an equilateral triangle pattern with an hand-measured edge length of 2 m (using a tape measure). A total of nine ground control points were placed into the environment. More details on how the control points were used are discussed in the subsection on ground control point geo-registration below.

In addition to the ground control points, ground control reference spectra were measured for various surfaces in the intertidal zone using the handheld spectroradiometer. The positions of the spectrally-measured points relative to the ground control point targets were measured so that they could be later identified in the reconstructed maps and used to validate the accuracy of the colour and near-infrared data in the maps. For each measured spectra an estimate of the expected red, green, blue and near-infrared responses as seen by the cameras was produced by multiplying the measured spectra with the estimated spectral response curves of the camera (see Figure 3 (a)) and summating over wavelength for each channel. The red and near-infrared channels were then used to produce a macrobenthos index that was compared to the index computed using the actual red and near-infrared responses measured from the camera. Figure 3 (b) illustrates examples of the reflectance spectral curves for dominant coverage types found at the Jervis Bay field site.

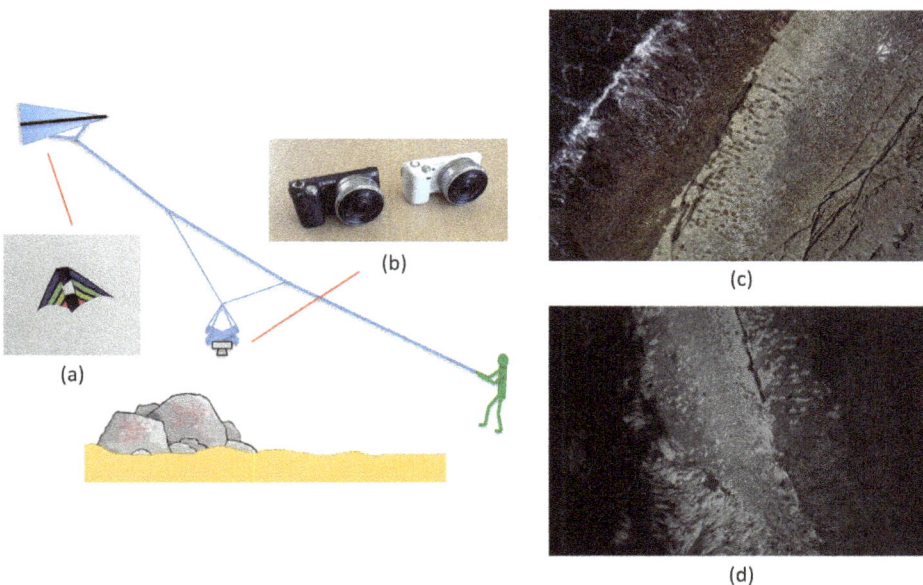

Figure 2. Kite-based image acquisition. (a) 2.7 m wingspan conynes-delta kite using to lift (b) a Picavet suspension rig that was used to attach each of the two downwards-facing Sony NEX-5 digital cameras (one colour and one near-infrared). The operator walks the kite across the intertidal zone collecting multiple, overlapping photographs. Examples of aerial images collected from an altitude of approximately 15 m are shown in colour (c) and near-infrared (d).

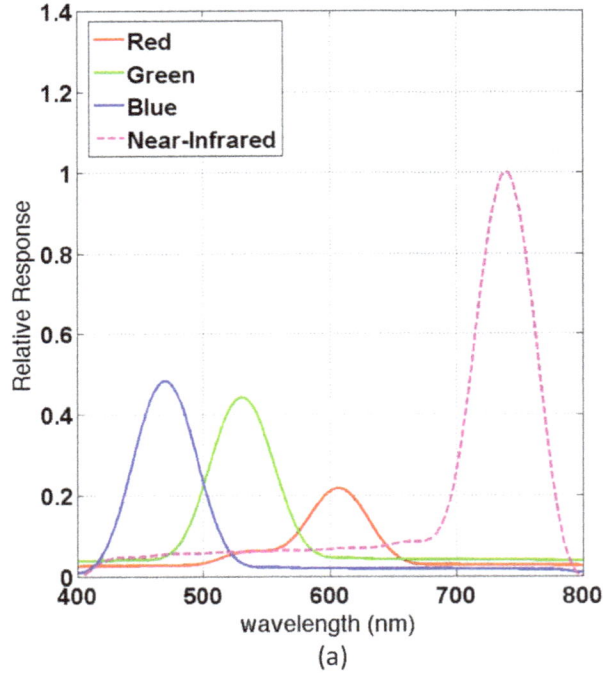

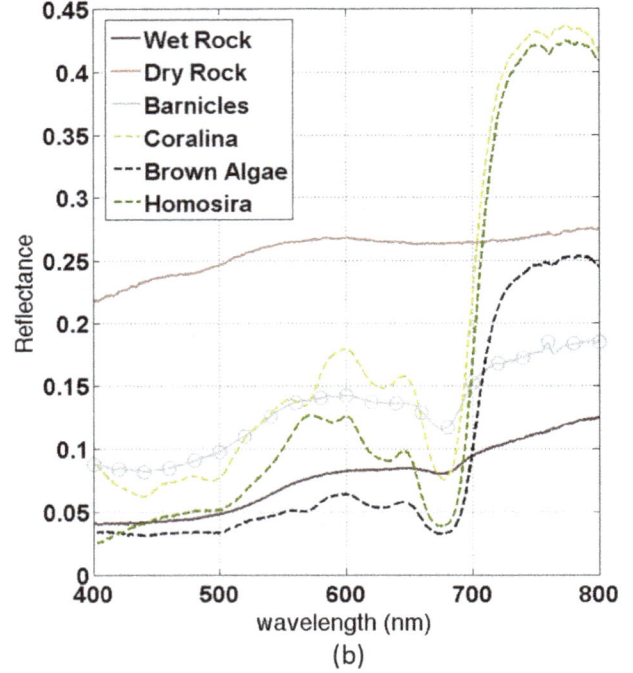

Figure 3. Spectral calibration data. (a) Spectral response functions for the colour and near-infrared converted cameras: the red, green and blue channels correspond to the three channels of the colour camera whereas the near-infrared curve corresponds to the red channel of the near-infrared converted camera, which was found to have the highest response of each of the channels for this camera. (b) Reflectance spectra for key surface coverage types in the intertidal zone measured using a handheld spectroradiometer. The reflectance spectra were used in conjunction with the camera spectral response functions to validate the measured colour of objects in the kite-based imagery.

Image Acquisition

During data collection, the kite was used to hoist the camera rig holding one camera at a time over the area of interest and the camera programmed to capture images at a frequency of approximately one shot per second using the continuous shoot mode of the camera. The aperture, exposure time and ISO were tuned by hand for each camera to the average light conditions on the day and kept constant for every image captured by a given camera. The actual values used were different for each camera (owing to the differences in the transmission properties of the internal filters of each camera). A white-balance procedure using a Spectralon target was used to balance the intensity of image data between cameras; this procedure is discussed more detail in the section below on NDVI mosaics. The kite could be flown at a variety of altitudes between approximately 10–100 m (limited by the length of the kite line) based on desired area coverage and ground spatial resolution. The desired height was achieved using distance markers on the line and by approximating the flight angle

of the kite. The kite was then slowly walked across a 200-by-30 m section of rocky shoreline and images were captured continuously using the raw mode of the camera at an altitude of approximately 15–20 m. The kite was walked in a zig-zag fashion along the shoreline rather than in a straight line in order to gather images from various perspectives with respect to the terrain. Capturing images from various perspectives was important for the functioning of the structure-from-motion algorithms discussed below, allowing for camera poses and 3D feature information to be estimated. The time taken to acquire images across the platform was approximately six minutes, after which the cameras were swapped and the process repeated to collect images in both colour and near-infrared. It was originally planned that both cameras would be flown simultaneously, however light wind conditions on the day only allowed for one camera to be flown at a time.

Figures 2 (c) and (d) illustrate example images captured by the system at an altitude of approximately 15 m from the ground, with a coverage footprint of approximately 22-by-15 m and a pixel size of approximately 4.5 mm.

Image Processing, Feature Extraction and Matching

After data collection, images were copied from each of the cameras to a desktop computer for processing. Prior to processing, images that were affected by motion blur during wind gusts or large occlusions of the terrain (for example images of people moving in the scene) were removed manually. Images were white balanced, to aid in feature extraction contrast and to provide imagery mosaics that could be easily interpreted by end-users. A total of 295 colour images and 251 near-infrared images were used. Scale-Invariant Feature Transform (SIFT) features were

Table 1. Peak response values of the colour and near-infrared converted cameras.

	Red	Green	Blue	Near-infrared
Peak Response	470 nm	530 nm	606 nm	740 nm

The red, green and blue values correspond to the three channels of the colour camera whereas the near-infrared value corresponds to the red channel of the near-infrared converted camera.

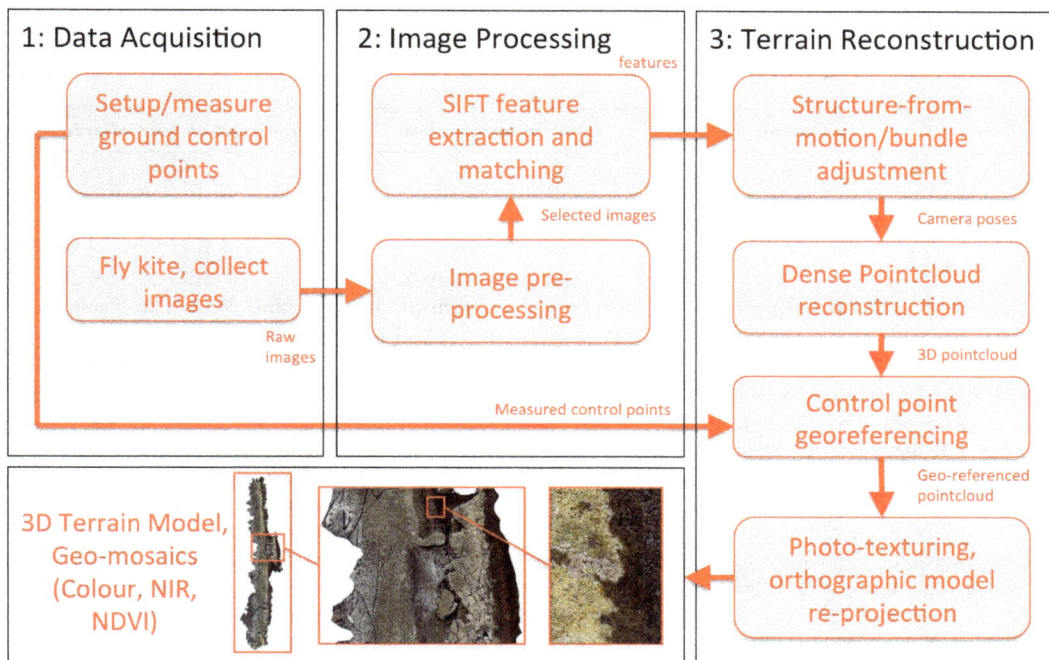

Figure 4. Overview of the kite-based mapping process. During data acquisition, ground control points are placed in the environment and images collected over the the terrain using the kite and camera. After data acquisition, images are processed to extract and match features across multiple overlapping images. These features are used to reconstruct the poses from which images were captured and a 3D pointcloud of the terrain using a structure-from-motion algorithm. The pointcloud is geo-referenced using the ground control points and a photo-texturing process is used to create 3D topographic maps and high-resolution geo-mosaics.

extracted in each image (using the implementation in [44]) and matched across all image pairs (including colour to colour, colour to near-infrared and near-infrared to near-infrared images) using a kd-tree [45] of the features present in each image. SIFT features correspond to distinctive points in the texture of surfaces captured in images and were well suited for use in the rocky intertidal environment.

Although there existed types of terrain that had a different appearance in the colour and near-infrared images, enough similarity in the properties used in the SIFT feature descriptor was present to match and register SIFT features across these two different image formats. Table 3 shows the number of matching image pairs and average feature matches per image for colour-to-

colour, colour-to-near-infrared and near-infrared-to-near-infrared image pairs. The average number of features per image matched between the colour and near-infrared images was lower than colour-to-colour or near-infrared-to-near-infrared pairs. Enough matches were found to provides a means to registering different types of image data into a common reference frame during the remainder of the processing procedure.

Robust detection of incorrect feature matches was performed using epipolar constraints between images [46]. Multi-core software implementations of these methods were developed in order to process images in parallel, speeding up the total time taken for processing. Figure 5 illustrates an example of the extracted and matched feature points between two images.

Table 2. Overview of software implementations used within the processing procedure.

Task	Software Used
Raw image conversion	DCRAW (http://www.cybercom.net/~dcoffin/dcraw/)
SIFT Feature Extraction	vlfeat (http://www.vlfeat.org/)
Feature Matching	Customised multi-core implementation using [46] and ANN (http://www.cs.umd.edu/~mount/ANN/)
Structure-from-motion, Bundle Adjustment	Bundler (http://www.cs.cornell.edu/~snavely/bundler/)
Multi-view Stereo Triangulation	PMVS2 (http://www.di.ens.fr/pmvs/)
GCP Geo-registration	Customised implementation using [49]
Photo-textured Terrain Modelling and Mosaicing	Custom implementation, software and code available: (Structured: https://github.com/mattjr/structured)

Table 3. Comparison of the number of matching image pairs and average feature matches per image for colour-to-colour, colour-to-near-infrared and near-infrared-to-near-infrared image pairs.

	colour to colour	colour to near-infrared	near-infrared to near-infrared
Number of matching image pairs	4454	3063	2630
Average number of matching features per pair	854.6	188.2	1106.2

Structure-from-motion and Pointcloud Reconstruction

A structure-from-motion/bundle adjustment software package [47] was then used to incrementally construct a 3D point feature map corresponding to the matched image feature points while simultaneously estimating camera poses and the intrinsic and extrinsic parameters for the camera. The method used was able to build a complete, scale-less 3D reconstruction (but with unknown global orientation and position) by using only the image data. The method employed a simplified model of the camera extrinsics (including focal length and lens distortion), the parameters of which were estimated from the image data and matched features themselves (as opposed to use of a pre-calibrated or metric camera). A multi-view stereo reconstruction algorithm [48] based on the correlation score of dense patches in the overlapping images was then used to produce a dense 3D point-cloud corresponding to a higher spatial resolution than by using SIFT features alone. This algorithm used the relative camera poses estimated during bundle adjustment to triangulate dense image features and robustly remove outliers from the terrain point cloud. The resulting 3D pointcloud had a spatial density that depended on the level of texture in the environment and was usually within a small factor of the image pixel size (i.e. approximately one 3D feature for every 5-by-5 pixel patch on average).

Ground Control Point Geo-registration

The ground control points that had been placed into the environment were identified in the pointcloud reconstruction and used in two ways to recover the scale, position and orientation of the final reconstruction. Firstly, the hand-measured triangle edge lengths were used to compute the absolute scale of the 3D reconstruction (i.e. the size of the model in the world) by comparing to the edge lengths of the triads of control points identified in the pointcloud reconstruction. Secondly, the GPS measured coordinates were used via Horn's method [49] to compute a transformation consisting of a translation and orientation that moved the scaled 3D model from arbitrary

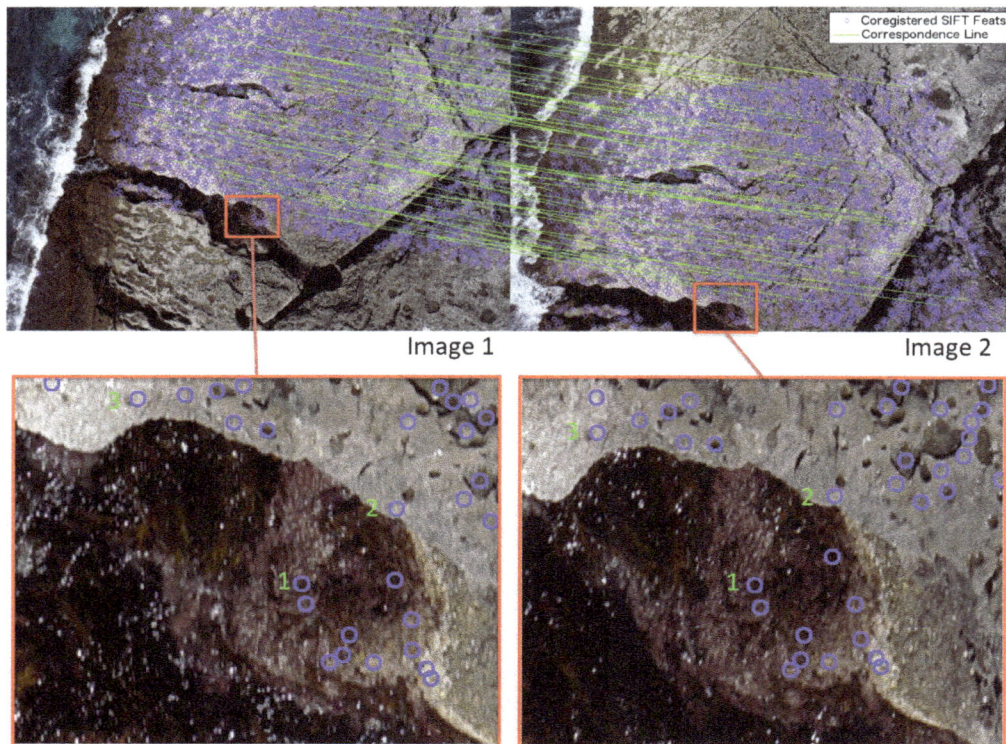

Figure 5. Example of extracted and matched SIFT image feature data. Shown are two overlapping images (image 1 and image 2) with annotated positions of SIFT features that have been matched between the two images (blue) and lines displaying the computed correspondence between points (green) (only every hundredth correspondence shown for clarity). Also shown are detailed sections of each image illustrating the feature points.

reconstruction coordinates into a geo-referenced coordinate system (i.e. latitude, longitude and altitude). Because of the low global accuracy of the GPS measured coordinates, the ground control points were not directly integrated into the structure-from-motion process.

The two different steps allowed for different information (with different levels of accuracy) to be used to compute the scale of the model and the position/orientation of the model. The scale of the landscape model was recovered accurately (owing the precision of the hand measured edge lengths) whereas the global position of the model (i.e. the absolution latitude, longitude and altitude of the model) was recovered with less accuracy (owing to the precision of the GPS measurements themselves). This distinction in the types of accuracy is important; often the local accuracy of the map (i.e. the error in the relative position of two objects in different parts of the map) is important, whereas the global accuracy of the map (i.e. the error in the absolution latitude, longitude and altitude of a given point) is of less importance. The same methods applied in our work could be used with higher-accuracy GPS ground control point measurements to achieve both high local and global accuracy in the map, but at the cost of requiring more expensive equipment.

In order to improve on the vertical accuracy of the model (owing to the poor vertical resolution of the GPS), the vertical elevation of the reconstruction was adjusted by manually extracting an outline of the water line visible in the point cloud and setting this to the known tidal datum height around the time the images were captured.

Photo-textured Terrain Model and Mosaicing

A triangulated terrain surface model was constructed from the 3D geo-referenced pointcloud using Delaunay triangulation [50]. For each face of the surface, the estimated camera poses were used to compute which images had seen the face both from the colour and near-infrared images. The images were grouped into two lists, one composed of colour images and one of near-infrared images. For each list, images were ranked based on the distance between the centerpoint of the face in the environment and the camera position from which the image was taken (and thus image resolution at this point). The best four images (those with minimum distance) from each list were then assigned to the face and the 3D coordinates of the face were projected into the 2D coordinates of each image. Using the projected coordinates, a colour value was assigned to each pixel in the final texture using a weighted averaging of these four source images according to the algorithm in [51]. The process was repeated for every face in the model to produce two sets of photographic textures, one corresponding to colour imagery and one corresponding to near-infrared imagery on the surface. The view selection and band-limited-blending of [51] allowed for only the best images of a given surface to be used in the final model, providing redundancy against images taken from poor angles.

The resulting 3D photo-textured model was visualized using a level-of-detail rendering system [51] to capture sub-centimeter details over the entire span of the map. Additionally orthographic projections of the model (imaged from directly above) were re-rendered in separate colour and near-infrared bands to produce 2D photomosaics of the entire area that were exported to a geotiff format.

Spatial and Normalised Difference Vegetation Index (NDVI) Mosaics

In addition to imagery mosaics, maps of various spatial properties including terrain elevation, slope and aspect were generated from the 3D terrain model. Slope and aspect were computed for each triangular face in the terrain model based on the normal vector of the face $\mathbf{n} = [x,y,z]^T$ (where x, y and z are the north, east and down components of the normal) using Equations 1 and 2:

$$slope = \cos^{-1}\left([0,0,1]^T \cdot \mathbf{n}\right) \quad (1)$$

$$aspect = \tan^{-1}\left(\frac{y}{x}\right) \quad (2)$$

The colour and near-infrared imagery layers were also used to compute the Normalised Difference Vegetation Index (NDVI) at each spatial point using a combination of the red and near-infrared (NIR) channels of the imagery:

$$NDVI = \frac{NIR - red}{NIR + red} \quad (3)$$

Data was taken from the raw imagery collected from each camera after a custom white balance procedure was applied to each channel. The procedure used images captured from each camera of a white Spectralon target under the same lighting conditions in which the imagery was collected using the kite. Using the measurements of the Spectralon target, for each camera a white balance gain was applied to each of the red and near-infrared channels that normalised the outputs to the ratio between the exposure times of each camera. The resulting white-balanced channels were then used to calculate the NDVI using Equation 3. This version of NDVI is closely related to the microphytobenthos index described in [21] that used precise ratios of red and near-infrared reflectance from a spectroradiometer at wavelengths of 635 and 750 nm and was found to have a linear relationship to surface algae chlorophyll concentration. In our work, the red and near-infrared channels provided by the imagery had slightly different peak response values (606 and 740 nm respectively, see Table 1) and also collected light over a wider band of wavelengths (i.e. Figure 3 (a)).

Results and Discussion

Spatial Mapping Results

Figure 6 illustrates the final photo-mosaic of the intertidal rocky shore for the colour and near-infrared photomosaic layers. The coverage achieved in each map corresponds to the locations where points on the ground were observed in at least two different camera images (and thus could be reconstructed using the structure-from-motion techniques discussed above). Coverage in the colour and near-infrared mosaics was slightly different owing to the fact that some parts of the terrain were observed only by a single type of camera. The final spatial resolution of the mosaic was 5 mm per pixel on average, with slight variations owing to variations in the flying height of the kite and the perspective from which images were captured. The detailed sections (shown in Figure 6 (c) and (d)) illustrate a division along the shoreline between the upper tidal zone, populated largely by bare rock with various grazing gastropods such as limpets and other sessile organisms such as barnacles and the mid-tidal zone populated largely by a patchwork of macroalgae and cunjevoi.

Figures 7 and 8 show the elevation, slope and aspect maps derived from the 3D terrain surface map. The final 3D terrain

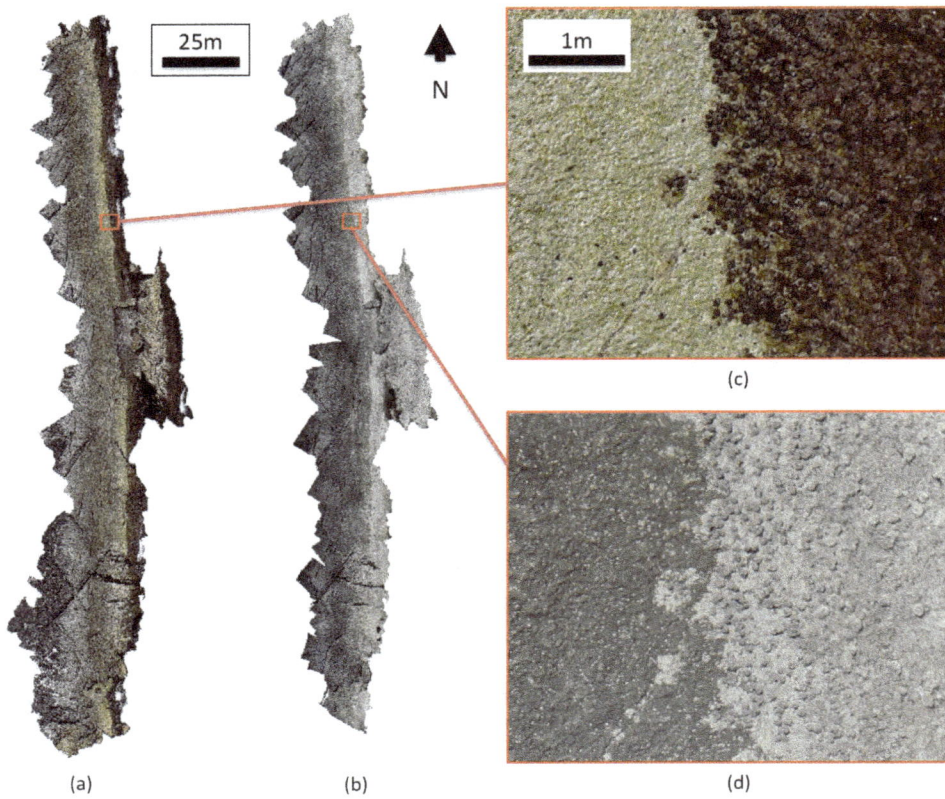

Figure 6. Colour and near-infrared photomosaic layers of the intertidal rocky shore reconstruction at Greenfields Beach. (a) Colour mosaic and (b) near-infrared mosaic for the whole shoreline. (c) and (d) show a detailed section of the colour and near-infrared mosaics illustrating the different scales achieved across the entire map.

map displayed an average 3D topographic ground sampling distance of 2.5 cm. The fine-scale structure of rock pools and crevasses is visible from the maps. Figure 9 illustrates a visualisation of the final photo-textured 3D model with various views from large-scale to fine-scale. The level-of-detail visualisation system allowed for different model scales to be visualised in a single, continuous terrain model. The highest level of detail information provided detailed structural information on individual rock crevasses with imagery information indicating the presence of different macroalgae coverage and assemblages of barnacles and limpets.

Validation of Spatial Accuracy

The residual error between the measured ground control point locations and the computed ground control point locations produced in the final maps was used as a proxy for the estimated spatial accuracy of the 3D reconstruction and photomosaics. Two different types of spatial accuracy were assessed in the maps: global accuracy (i.e. absolute position of the entire map in geo-referenced coordinates) and local/relative accuracy (i.e. the accuracy in the reconstructed relative position of two different points on the map). The expected global accuracy was assessed using the residual errors between the GPS-measured ground control point locations and the identified ground control point locations produced in the final maps. The local/relative accuracy was assessed using the residual error between the hand-measured ground control point triangle edge lengths and the edge lengths extracted from the 3D reconstructed maps. The resulting averages of the residual errors

are shown in Table 4. In both cases, the residual error was in the order of the errors associated with the external measurements themselves (i.e. GPS-based residual errors of 1.016 m compared to the accuracy of the handheld GPS, which was reported by the GPS receiver as being ±2 m and the triangle edge length residual of 0.039 m compared to the expected precision of the hand made tape measurements, which was on the order of a few centimeters).

In comparison to other studies using kite aerial photography, position errors of 0.02–0.14 cm were reported in [32], capturing images at a height of approximately 15 m and errors of 0.02–0.06 m in [33] capturing images at a height of approximately 50 m. In both of these studies, ground control points were positioned using a total-station survey and survey-grade GPS with an accuracy of 1 cm. Errors of 0.025–0.04 m were reported in [38] in aero-triangulated points computed using images from a UAV flying at approximately 50 m altitude. In this study, ground control points were measured using a Real Time Kinematic (RTK) Differential GPS system with a reported accuracy of approximately 1–4 cm. All of these studies report accuracies with respect to a geo-referenced coordinate system (i.e. global accuracies). The local/relative accuracy results reported in our work are comparable to the global accuracies of other studies, however the global accuracy of our approach is much lower (i.e. approximately 1 m compared to cm-level accuracy). This is a fundamental limitation of using low-grade (and thus inexpensive) GPS for surveying ground control points. The methods used in this paper are still applicable when survey-grade GPS measurements of ground control points are available; further analysis of

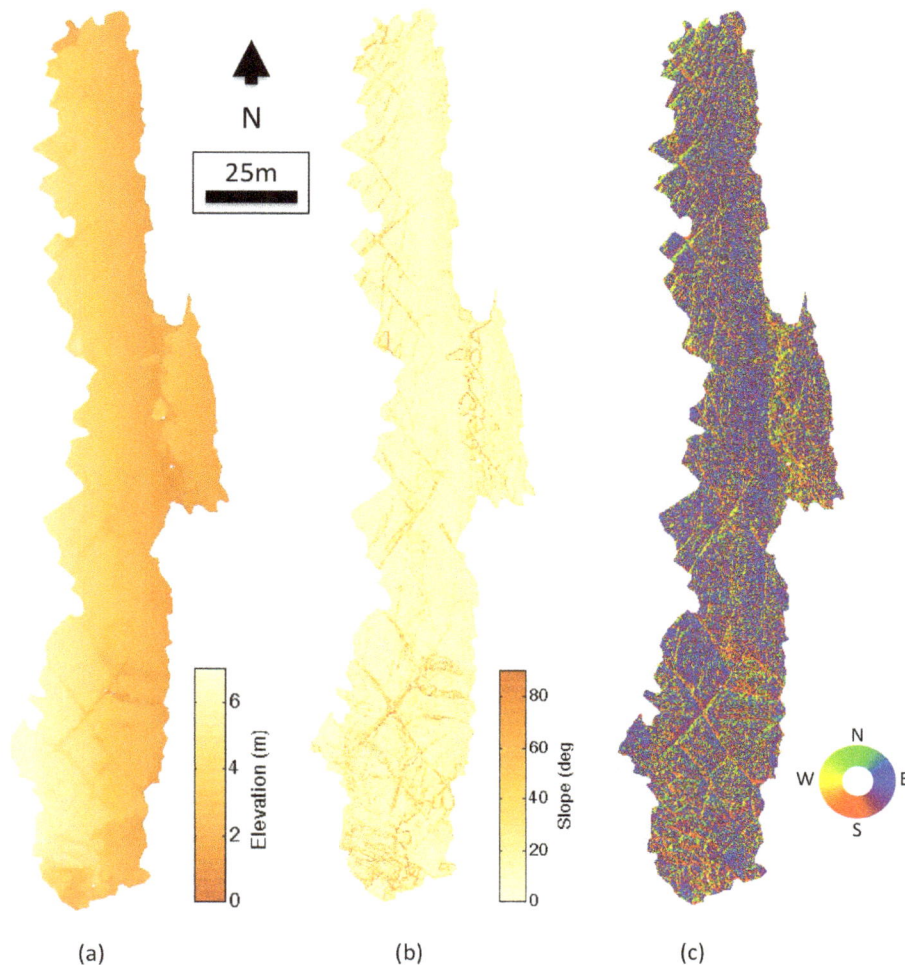

Figure 7. Elevation, slope and aspect data derived from the 3D topographic reconstruction. (a) Elevation above maximum low tide (b) slope of the terrain and (c) aspect of the terrain.

local vs. global accuracy when using these type of measurements is left to future work.

Normalised Difference Vegetation Index (NDVI) Mapping Results

Figure 10 shows the NDVI photomosaic layer computed from the custom white-balanced colour and near-infrared imagery using Equation 3. A comparison between the NDVI layer and the colour imagery layer (seen in Figures 10 (b) and (c)) of the shoreline illustrates the division between largely dry rock and an exposed mat of macroalgae and tunicates (cunjevoi). The presence of live micro- and macroalgae have been shown to correspond to NDVI values of 0.3 and 0.7 [21], bare ground and rock corresponding to values of approximately zero and water corresponding to values approaching −1.0. The NDVI mosaic allows for the inference of fine-scale spatial patterns of macro-algae to be identified from the imagery, for example within rock pools and crevasses along the upper intertidal zone.

Validation of Spectral Accuracy

The ability of the reconstruction algorithms and consumer-grade camera images to reproduce accurate colour and NDVI in the mosaic maps was assessed by comparing the reconstructed textures in the maps to the ground control reference spectra that were collected by handheld spectroradiometer. The measured spectra were multiplied through the computed spectral response function of each of the red, green, blue and near-infrared camera channels and integrated to produce and expected red, green, blue and near-infrared responses, from which a predicted NDVI value was computed for each of the dominant intertidal coverage types. These values were then compared to the measured NDVI mosaics by examining 167 manually extracted 15-by-15 pixel patches corresponding to the dominant coverage types measured in Figure 3 (b).

Figure 11 illustrates detailed views of the colour mosaic imagery showing examples of some of the dominant surface coverage types that were used in the comparison. Figure 12 illustrates the comparison for six different coverage types with error bars illustrating the variation in the manually extracted measurements. Overall there was a good correspondence between the measured and predicted NDVI values, with a large and obvious division between the macroalgae coverage types (with high NDVI) and the non-algal dominated surfaces (with NDVI close to zero). The 'wet rock' coverage type, displayed a difference between the predicted and measured NDVI outside of the error bars of the different measured samples. This was thought to be due to variations in the

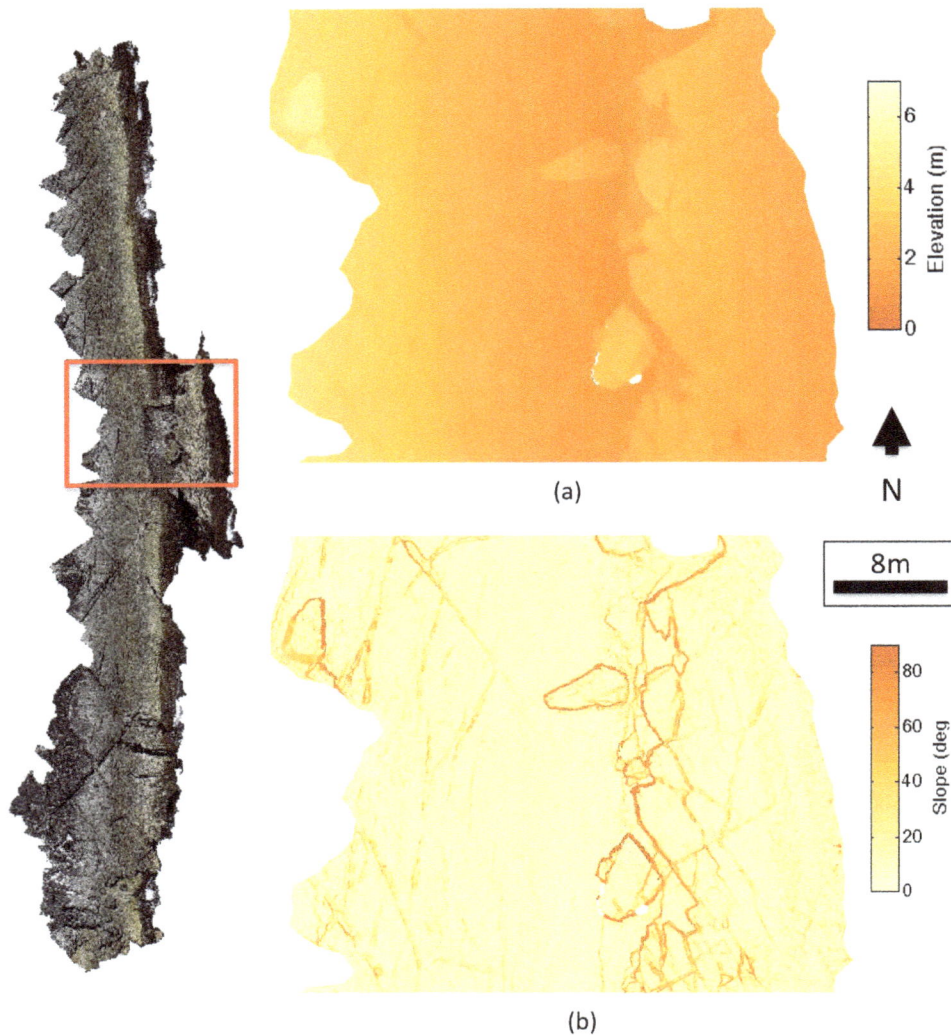

Figure 8. Detailed views of elevation and slope data derived from the 3D topographic reconstruction. (a) Elevation above maximum low tide and (b) slope of the terrain.

microalgal coverage on different wet rock surfaces in the intertidal zone that were not observable by eye during field sample collection and resulted in inaccuracies in the collected data.

Discussion

Potential Uses in Intertidal Ecology

The results presented here demonstrate the ability of our technique for collecting relevant spatial variables (i.e. elevation, slope, aspect) and spectral variables (such as NDVI) at fine spatial scales and over a broad area in the intertidal environment. The spatial variables are potentially useful for computing water emersion times, habitat structural complexity and seasonal sun exposure at different positions along the shoreline, while spectral

indices such as NDVI are potentially useful for evaluating algal biomass. Similar variables have been used as part of studies into body size and abundance of invertebrates on rocky shores [2,5], variations in microalgal biomass versus surface light availability on soft intertidal sediments [52], spatial distributions of intertidal macrobenthos versus food availability and sediment size [19] and for quantitating fractional macroalgae coverage [18]. Our low-cost method allows for collections of relevant spatial and spectral data across a cascade of fine-spatial scales, in a non-destructive manner, while preserving a permanent electronic record that can be used for comparison against repeated surveys, and thus could be an extremely useful tool within these type of studies.

Traditional remote sensing methods or field-based spectroscopy typically utilise high-spectral resolution measurements to identify and classify different plant (i.e. micro- and macroalgae) and animal (i.e. gastropods and other sessile fauna) species. Unlike these methods, the imagery provided by our technique is able to capture fine-scale details in the structure, texture and pattern in the imagery (see for example Figure 11), owing to the high-spatial

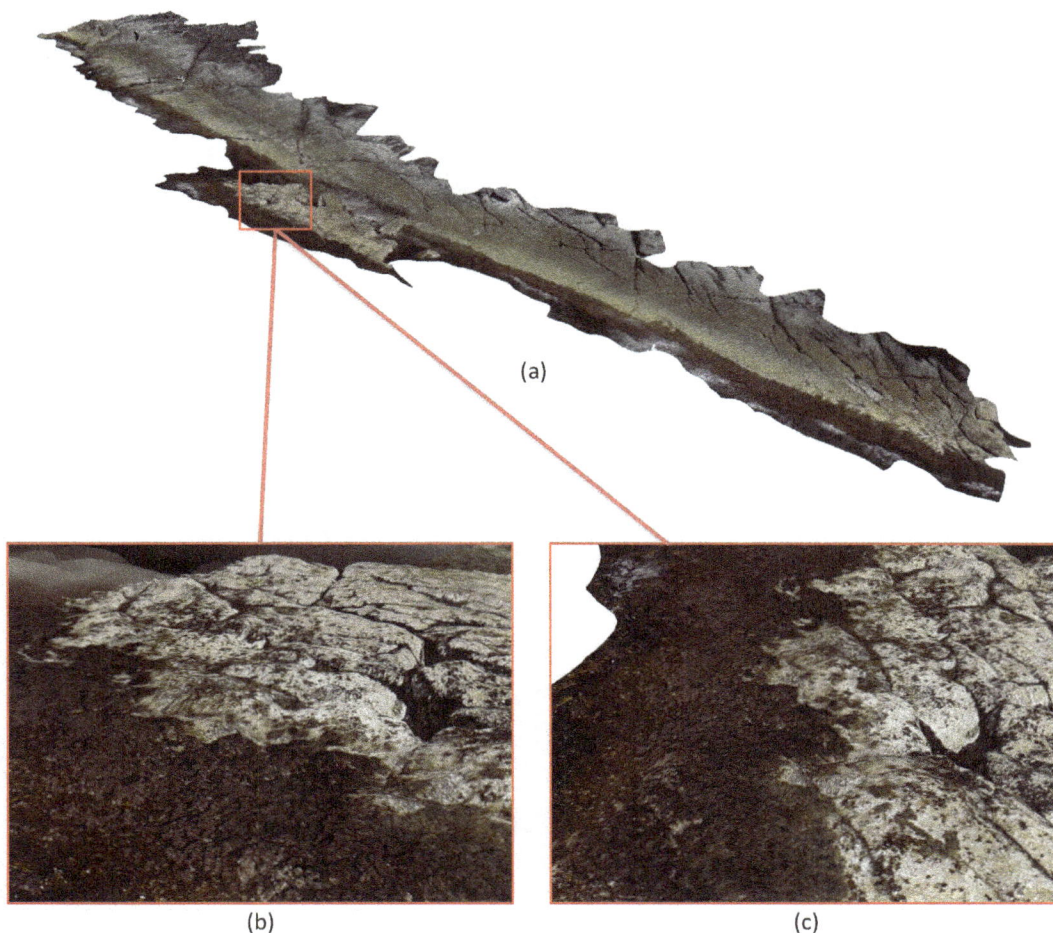

Figure 9. 3D photo-textured rocky shore reconstruction. (a) 3D oblique view of the shoreline, (b) and (c) Detailed oblique views of rock platform section from different viewing angles. The level-of-detail visualisation system allows for different model scales to be visualised in a single, continuous terrain model.

resolution, which could be used instead to automatically classify objects in the intertidal zone through the use of methods in semi-supervised machine learning and object-based image analysis [53]. For example, the colour and spectra information of cunjevoi or algae such as *Homosira* produce a uniquely identifiable texture and pattern in high-resolution imagery, owing to their physical structures, that could be used as classification features that would not be detected via traditional, lower-resolution remote sensing methods.

Potential Uses in Monitoring and Environmental Impact Assessment

To reliably detect human and environmental impacts on plant and animal populations it is necessary to sample populations at several times and at appropriate spatial resolutions before and after an impact (see [10]). Often, it is not possible to do this either for logistical or cost constraints, thus reducing the reliability of the detection of change. Our method provides the potential for distinguishing natural changes such as seasonality from human impacts such as construction, pollution or climate change in intertidal environments in a statistically robust way; the low-cost and logistical simplicity of our data collection procedure lends itself to frequent data collection, and by collecting data at fine-scales over broad areas and with full coverage of the landscape, potentially allows for precise spatial registration of data collected over multiple surveys.

Application to Studying Intertidal Mudflats

The method presented here is not limited to rocky intertidal substrata but could also be used for various applications on

Table 4. Measures of both global and local-scale (fine-scale) spatial map errors from Ground Control Point (GCP) residual errors.

Average GCP Residual (Global-scale Error)	Average GCP Triangle Edge-length Residual (Local-scale Error)
1.016 m	0.039 m

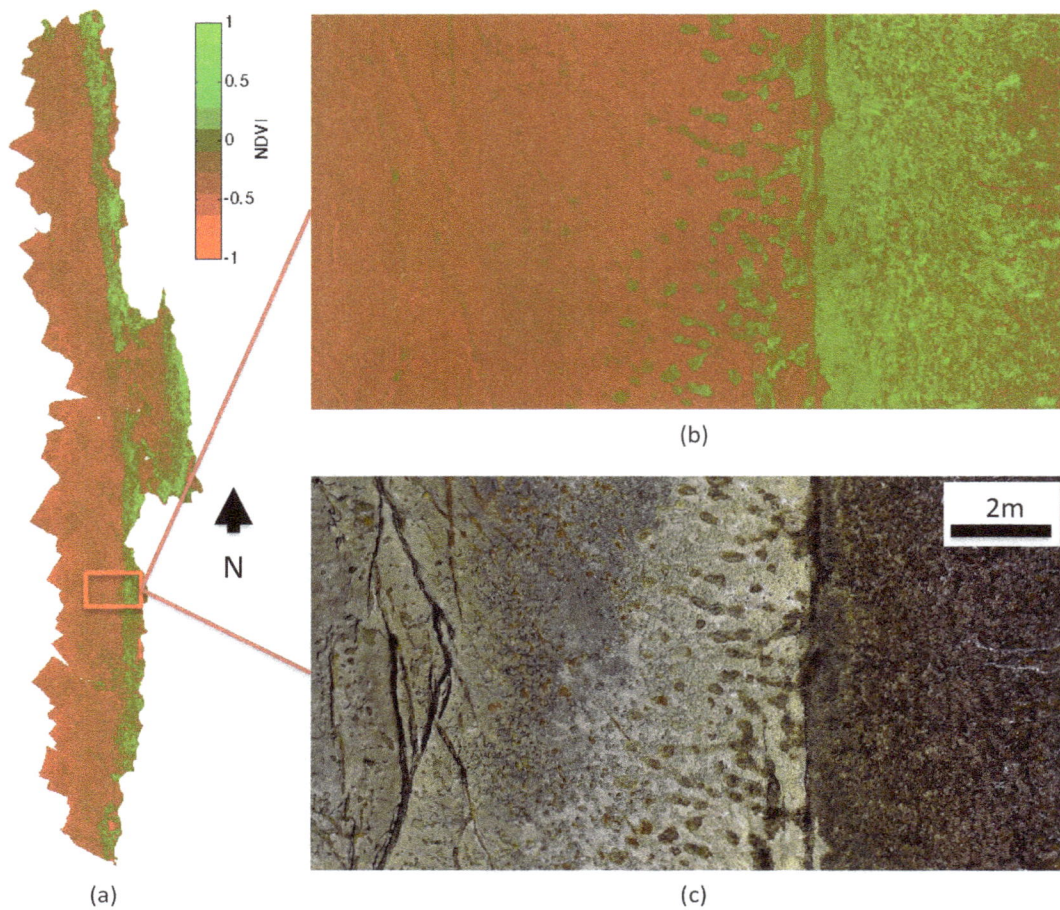

Figure 10. Normalised Difference Vegetation Index (NDVI) maps derived from raw colour and near-infrared imagery. (a) NDVI map for entire shoreline, (b) detailed view of NDVI and (c) corresponding colour imagery of detailed section.

intertidal mud flats. Conditions on intertidal mud flats can change over an interval of a few minutes - large changes in sediment properties can occur including dewatering of the sediment and the migration of microphytobenthos to the surface [54]. These changes can have significant impacts upon sediment stability over the course of a single tidal cycle. Due the dynamic nature of intertidal mudflats combined with the relatively slow pace of conventional field sampling make it impossible to make measurements of sediment properties across space that are truly independent of changes in time. Our method, by enabling large

Figure 11. Example mosaic imagery highlighting some of the dominant surface types compared in this study.

Kite Aerial Photography for Low-Cost, Ultra-high Spatial Resolution Multi-Spectral Mapping of Intertidal...

105

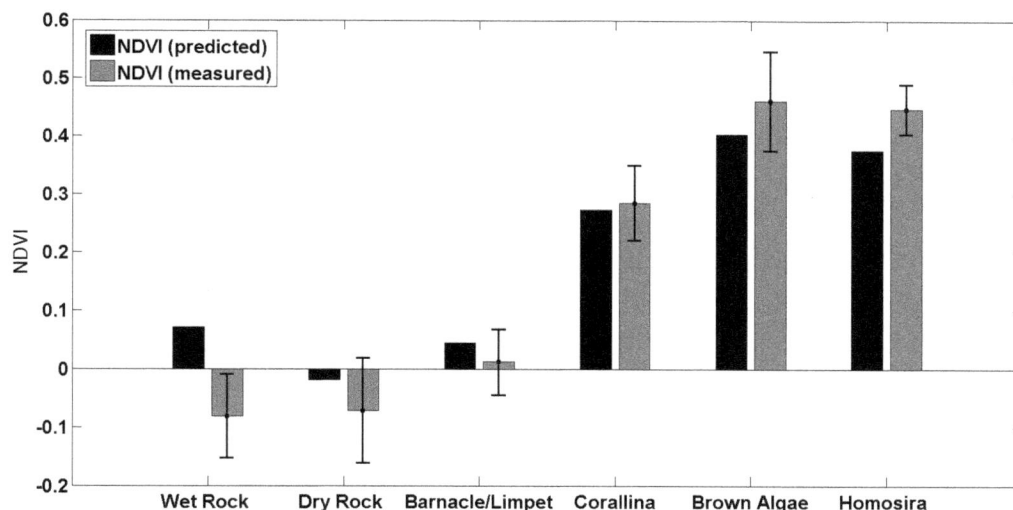

Figure 12. Comparison of predicted and measured Normalised Difference Vegetation Index (NDVI) values for different intertidal zone coverage types. Predicted NDVI values were derived using knowledge of coverage type reflectance spectra and camera spectral response functions (see Figure 3). Measured NDVI values correspond to manually extracted image patches taken from the reconstructed NDVI mosaic maps (error bars show the standard deviation from multiple sample points).

amounts of data to be acquired in a snapshot of time, at specific times in the tidal cycle, enables independent measurements to be made. Applications include the determination of impacts of structures on sediment properties or the effects of spillage of contaminants e.g. algicides, pesticides or fertiliser [55] on phytobenthos.

Conclusions and Future Work

In this paper, we have described a photogrammetric/mapping procedure that was developed for constructing high-resolution, 3D, photo-textured terrain models of intertidal areas using multiple low-altitude images collected from a consumer-grade digital camera suspended by a kite platform. Dynamic intertidal ecosystems by their nature can change rapidly at the scale of minutes to years making it almost impossible to acquire data that describe changes that occur spatially independently of temporal changes using field-based sampling. The methods presented acquire colour and topographic information across a hierarchy of spatial scales in a very small time interval, enabling changes in spatial distributions of assemblages to be determined independently of temporal changes, and at resolutions not achievable by traditional remote sensing platforms (such as satellites and manned-aircraft).

Ongoing and future work is focusing on three main areas. The first area is to develop methods for registering map data collected across multiple surveys. In some cases, the visual features (i.e. SIFT) used to register images within a survey will be suitable for this process, however the stability of these features is known to diminish when the time between images increases owing to small-scale changes in the imaged surface. Instead, current work is focussing on using robust means for multi-survey registration commonly used in remote sensing and medical imaging, such as

mutual information [56,57], to align multi-temporal datasets. The second area of future work involves the investigation of object based image analysis and machine learning algorithms for semi-supervised classification of dominant coverage types, such as different macroalgae, in the intertidal zone. This type of automated analysis could complement the mapping and reconstruction algorithms presented in this paper, providing useful information for ecologists and marine park managers without the need for laborious manual interpretation of the large quantity of image data. The third area of future work is focussing on using tethered kites and balloons as a potential platform for carrying low-cost, hyperspectral imagers or spectroradiometers to provide higher spectral resolution information than provided by the consumer-grade cameras presented in this study. Accurate spectroradiometers are becoming more accessible as technology improves, and the combination of high resolution photography and single-point spectroradiometer measurements, via cross-sensor calibration and the use of the photogrammetry techniques discussed in this paper, could provide a means to producing spatially-registered, airborne hyperspectral maps, at a lower-cost than via manned aircraft imaging systems.

Acknowledgments

Thanks to Dr. Nathan Knott from the New South Wales Department of Primary Industries for his discussions and assistance during fieldwork.

Author Contributions

Conceived and designed the experiments: MB MJR RM. Performed the experiments: MB MJR. Analyzed the data: MB MJR DB. Contributed reagents/materials/analysis tools: MB MJR DB. Wrote the paper: MB RM.

References

1. Underwood A (2000) Experimental ecology of rocky intertidal habitats: what are we learning? Journal of Experimental Marine Biology and Ecology 250: 51–76.

2. Beck M (1998) Comparison of the Measurement and Effects of Habitat Structure on Gastropods in Rocky Intertidal and Mangrove Habitats. Marine Ecology Progress Series 169: 165–178.

3. Forster R, Jesus B (2006) Field Spectroscopy of Estuarine Intertidal Habitats. International Journal of Remote Sensing 27: 3657–3669.

4. Murphy R, Tolhurst T, Chapman M, Underwood A (2008) Spatial Variation of Chlorophyll on Estuarine Mudflats determined by Field-based Remote Sensing. Marine Ecology Progress Series 365: 45–55.

5. Meager J, Schlacher T, Green M (2011) Topographic Complexity and Landscape Temperature Patterns create a Dynamic Habitat Structure on a Rocky Intertidal Shore. Marine Ecology Progress Series 428: 1–12.

6. Wang K, Franklin S, Guo X, Cattet M (2010) Remote Sensing of Ecology, Biodiversity and Con- servation: A Review from the Perspective of Remote Sensing Specialists. Sensors 10: 9647–9667.

7. Xharde R, Long B, Forbes D (2006) Accuracy and Limitations of Airborne LiDAR Surveys in Coastal Environments. In: IEEE Geoscience and Remote Sensing Symposium.

8. Knudby A, LeDrew E, Newman C (2007) Progress in the use of Remote Sensing for Coral Reef Biodiversity Studies. Progress in Physical Geography 31: 421–434.

9. Mumby P, Green E, Edwards A, Clark C (1999) The cost-effectiveness of remote sensing for tropical coastal resources assessment and management. Journal of Environmental Management 55: 157–166.

10. Underwood A (1994) On beyond BACI: Sampling designs that might reliably detect environmental disturbances. Ecological Applications 4: 3–15.

11. Decho A, Fleeger J (1988) Microscale dispersion of meibenthic copepods in response to food- resource patchiness. Journal of Experimental Marine Biology and Ecology 118: 229–243.

12. Denny M, Helmuth B, Leonard G, Harley C, Hunt L, et al. (2004) Quantifying scale in ecology: Lessons from a wave-swept shore. Ecological Monographs 74: 513–532.

13. Saburova M, Polikarpov I, Burkovsky I (1995) Spatial structure of an intertidal sandflat micro- phytobenthic community as related to different spatial scales. Marine Ecology Progress Series 129: 229–239.

14. Sandulli R, Pinckney J (1999) Patch sizes and spatial patterns of meiobenthic copepods and bentic microalgae in sandy sediments: a microscale approach. Journal of Sea Research 41: 179–187.

15. Sun B, Fleeger J, Carney R (1993) Sediment microtopography and the small-scale spectial distribution of meiofauna. Journal of Experimental Marine Biology and Ecology 167: 73–90.

16. Underwood A, Chapman M (1996) Scales of spatial pattern of distribution of intertidal inverte- brates. Oecologia 107: 212–224.

17. Thompson R, Norton T, Hawkins S (2004) Physical stress and biological control regulate the producer-consumer balance in intertidal biofilms. Ecology 85: 1372–1382.

18. Bajjouk T, Populus J, Guillaumont B (1998) Quantification of Subpixel Cover Fractions using Principle Component Analysis and a Linear Programming Method: Application to the Coastal Zone of Roscoff (France). Remote Sensing of Environment 64: 153–165.

19. van der Wal D, Herman P, Forster R, Ysebaert T, Rossi F, et al. (2008) Distribution and Dynamics of Intertidal Macrobenthos Predicted from Remote Sensing: Response to Microphytobenthos and Environment. Marine Ecology Progress Series 367: 57–72.

20. Kromkamp J, Morris E, Forster R, Honeywill C, Hagerthey S, et al. (2006) Relationship of Inter- tidal Surface Sediment Chlorophyll Concentration to Hyperspectral Reflectance and Chlorophyll Fluorescence. Estuaries and Coasts 29: 183–196.

21. Meleder V, Barille L, Launeau P, Carrere V, Rince Y (2003) Spectrometric Constraint in Analysis of Benthic Diatom Biomass using Monospecific Cultures. Remote Sensing of Environment 88: 386–400.

22. Murphy R, Underwood A, Jackson A (2009) Field-based Remote Sensing of Intertidal Epilithic Chlorophyll: Techniques using Specialised and Conventional Digital Cameras. Journal of Experi- mental Marine Biology and Ecology 380: 68–76.

23. Verhoeven G (2009) Providing an Archaeological Bird's-eye View - an Overall Picture of Ground-based Means to Execute Low-altitude Aerial Photography in Archaeology. Archaeological Prospection 16: 233–249.

24. Scoffin T (1982) Reef Aerial Photography from a Kite. Coral Reefs 1: 67–69.

25. Fraser W, Carlson J, Duley P, Holm E, Patterson D (1999) Using kite-based aerial photography for conducting Adelie penguin censuses in Antarctica. Waterbirds 23: 435–440.

26. Sklaver B, Manangan A, Bullard S, Svanberg A, Handzel T (2006) Rapid Imagery through Kite Aerial Photography in a Complex Humanitarian Emergency. International Journal of Remote Sensing 27: 4709–4714.

27. Aber J, Sobieski R, Distler D, Nowak M (1999) Kite Aerial Photography for Environmental Site Investigations in Kansas. Kansas Academy of Science 102: 57–67.

28. Aber J, Aber S, Leffler B (2001) Challange of Infrared Kite Aerial Photography. Trans of the Kansas Academy of Science 104: 18–27.

29. Murray J, Neal M, Labrosse F (2013) Development and Deployment of an Intelligent Kite Aerial Photography Platform (iKAPP) for Site Surveying and Image Acquisition. Journal of Field Robotics 30: 288–307.

30. Wundram D, Loffler J (2008) High-resolution Spatial Analysis of Mountain Landscapes using a Low-altitude Remote Sensing Approach. International Journal of Remote Sensing 29: 961–974.

31. Smith M, Chandler J, Rose J (2009) High spatial resolution data acquisition for the geosciences: kite aerial photography. Earth Surface Processes and Landforms 34: 155–161.

32. Marzolff I, Poesen J (2009) The Potential of 3D gully Monitoring with GIS using High-resolution Aerial Photography and a Digital Photogrammetry System. Geomorphology 111: 48–60.

33. Guichard F, Bourget E, Agnard J (2000) High-resolution remote sensing of intertidal ecosystems: A low-cost technique to link scale-dependent patterns and processes. Limnology and Oceanography 45: 328–338.

34. Anderson K, Gaston K (2013) Lightweight unmanned aerial vehicles will revolutionize spatial ecology. Frontiers in Ecology and the Environment 11: 138–146.

35. Rango A, Laliberte A, Steele C, Herrick J, Bestelmeyer B, et al. (2008) Using Unmanned Aerial Vehicles for Rangelands: Current Applications and Future Potentials. Environmental Practice 8: 159–168.

36. Bryson M, Reid A, Ramos F, Sukkarieh S (2010) Airborne Vision-Based Mapping and Classification of Large Farmland Environments. Journal of Field Robotics 27: 632–655.

37. Turner D, Lucieer A, Watson C (2012) An Automated Technique for Generating Georectified Mosaics from Ultra-High Resolution Unmanned Aerial Vehicle (UAV) Imagery, Based on Structure from Motion (SfM) Point Clouds. Remote Sensing 4: 1392–1410.

38. Harwin S, Lucieer A (2012) Assessing the Accuracy of Georeferenced Point Clouds Produced via Multi-View Stereopsis from Unmanned Aerial Vehicle (UAV) Imagery. Remote Sensing 4: 1573–1599.

39. Hartley R, Zisserman A (2003) Multiple View Geometry in Computer Vision. Cambridge University Press.

40. Frahm J, Pollefeys M, Lazebnik S, Gallup D, Clipp B, et al. (2010) Fast robust large-scale mapping from video and internet photo collections. ISPRS Journal of Photogrammetry and Remote Sensing 65: 538–549.

41. Agarwal S, Snavely N, Seitz S, Szeliski R (2010) Bundle Adjustment in the Large. In: European Conference on Computer Vision.

42. Pike T (2011) Using digital cameras to investigate animal colouration: estimating sensor sensitivity functions. Behavioral Ecology and Sociobiology 65: 849–858.

43. Finlayson G, Hordley S, Hubel P (1998) Recovering device sensitivities with quadratic program- ming. In: IS and T/SID sixth color imaging conference: color science, systems and applications. pp. 90–95.

44. Vedaldi A, Fulkerson B (2008) VLFeat: An open and portable library of computer vision algo- rithms (http://www.vlfeat.org/).

45. Beis J, Lowe D (2003) Shape indexing using approximate nearest-neighbor search in high- dimensional spaces. In: Computer Vision and Pattern Recognition. pp. 1000–1006.

46. Torr P, Murray D (1997) The Development and Comparison of Robust Methods for Estimating the Fundamental Matrix. International Journal of Computer Vision 24: 271–300.

47. Snavely N, Seitz S, Szeliski R (2008) Modeling the world from Internet photo collections. International Journal of Computer Vision 80: 189–210.

48. Furukawa Y, Ponce J (2010) Accurate, Dense, and Robust Multi-View Stereopsis. IEEE Trans on Pattern Analysis and Machine Intelligence 32: 1362–1376.

49. Horn BKP (1987) Closed-form solution of absolute orientation using unit quaternions. Journal of the Optical Society of America 4: 629–642.

50. Barber C, Dobkin D, Huhdanpaa H (1996) The Quickhull algorithm for convex hulls. ACM Trans on Mathematical Software 22: 469–483.

51. Johnson-Roberson M, Pizarro O, Williams S, Mahon I (2010) Generation and Visualization of Large-scale Three-dimensional Reconstructions from Under-water Robotic Surveys. Journal of Field Robotics 27: 21–51.

52. Murphy R, Underwood A, Tolhurst T, Chapman M (2008) Field-based Remote-sensing for Ex- perimental Intertidal Ecology: Case Studies using Hyperspatial and Hyperspectral Data for New South Wales (Australia). Remote Sensing of Environment 112: 3353–3365.

53. Blaschke T (2010) Object based image analysis for remote sensing. ISPRS Journal of Photogram-metry and Remote Sensing 65: 2–16.

54. Perkins R, Honeywill C, Consalvey M, Austin H, Tolhurst T, et al. (2003) Changes in microphy- tobenthic chlorophyll a and EPS resulting from sediment compaction due to dewatering: opposing patterns in concentration and content. Continental Shelf Research 23: 55–58.

55. Murphy R, Tolhurst T (2009) Effects of experimental manipulation of algae and fauna on the properties of intertidal soft sediments. Journal of Experimental Marine Biology and Ecology 379: 77–84.

56. Kita Y (2008) A study of change detection from satellite images using joint intensity histogram. In: 19th International Conference on Pattern Recognition.

57. Sabuncu MR, Ramadge P (2008) Using Spanning Graphs for Efficient Image Registration. IEEE Transactions on Image Processing 17: 788–797.

11

Study of Burn Scar Extraction Automatically Based on Level Set Method using Remote Sensing Data

Yang Liu[1,2], Qin Dai[1]*, JianBo Liu[1], ShiBin Liu[1], Jin Yang[1]

1 Institute of Remote Sensing and Digital Earth, Chinese Academy of Sciences, Beijinng, China, **2** University of Chinese Academy of Sciences, Beijing, China

Abstract

Burn scar extraction using remote sensing data is an efficient way to precisely evaluate burn area and measure vegetation recovery. Traditional burn scar extraction methodologies have no well effect on burn scar image with blurred and irregular edges. To address these issues, this paper proposes an automatic method to extract burn scar based on Level Set Method (LSM). This method utilizes the advantages of the different features in remote sensing images, as well as considers the practical needs of extracting the burn scar rapidly and automatically. This approach integrates Change Vector Analysis (CVA), Normalized Difference Vegetation Index (NDVI) and the Normalized Burn Ratio (NBR) to obtain difference image and modifies conventional Level Set Method Chan-Vese (C-V) model with a new initial curve which results from a binary image applying K-means method on fitting errors of two near-infrared band images. Landsat 5 TM and Landsat 8 OLI data sets are used to validate the proposed method. Comparison with conventional C-V model, OSTU algorithm, Fuzzy C-mean (FCM) algorithm are made to show that the proposed approach can extract the outline curve of fire burn scar effectively and exactly. The method has higher extraction accuracy and less algorithm complexity than that of the conventional C-V model.

Editor: Marie Jose Goumans, Leiden University Medical Center, Netherlands

Funding: This work was supported in part by (1).135 Strategy Planning of Institute of Remote Sensing and Digital Earth, CAS; (2). The project of Youth Innovation Promotion Association, CAS. The funders had no role in study design, data collection and analysis, decision to publish, or preparation of the manuscript.

Competing Interests: The authors have declared that no competing interests exist.

* E-mail: qdai@ceode.ac.cn

Introduction

Burn scar refers to areas that are destroyed by forest fire, grass fire and controlled burning and have not yet recovered. Forest fire is one of the most severe natural hazards. It impacts ecology structure, atmospheric systems, as well as having detremental effects on living environment. For these reasons, in order to decrease the effect of forest fire, how to detect active fire and evaluate burn area rapidly with high accuracy has to be settled urgently [1]. Detecting and assessing the spatial extent and distribution of burn scar can support forestry services to process efficient vegetation recovery and post-fire management. In general, burn area has the properties of large size and spatial variability. Remote sensing has the advantages of wide viewing angles, multi-spectral imaging and multi-temporal revisit. It has become a primary tool for extracting burn scar. The destroyed landscapes caused by forest fire are observed by satellites from space at different scales is becoming the focus of researchers across the global. Over the past several decades, a wide variety of satellite datasets have been used to generate burn scar, such as the National Oceanic and Atmospheric Administration Advanced Very High Resolution Radiometer (NOAA/AVHRR), SPOT VEGETATION, Along-Track Scanning Radiometer (ATSR), Moderate Resolution Imaging Spectroradiometer (MODIS), Landsat Thematic Mapper (TM) and the Enhanced TM plus (ETM+) [2–4]. Besides, the Landsat 8 satellite launched on February 11, 2013. It will replace Landsat 5 to acquire valuable data and imagery to be used in agriculture, education, business and politics. Furthermore, it becomes another data source for extracting burn scar.

Today, many techniques have been developed to derive burn scar information from remote sensing data: fixed thresholding algorithm with multi-spectral images or indices computed from pre-fire and post-fire images and post-classification techniques using multi-temporal data [5]. These methodologies have been commonly applied to specific fire events. However, very few of them can comprehensively adapt to as many aspects as possible. The fixed thresholding algorithm discriminates burn scar from neighbouring objects with empirically derived thresholds [6]. Researchers have defined a set of fixed thresholds to extract burn scar. This method has advantages of simplicity and processing speed, whereas the limitation is that adequate thresholds related to various reasons are difficult to choose. In order to overcome this limitation, researchers have developed automatic threshold methods based on mean value and Standard Deviation (SD). Fernandez et al. (1997) propose a method to obtain thresholds automatically which is defined as mean+2*SD from a NDVI differencing image within local window [7]. Barbosa et al. (1999) present the V13T threshold method as the mean-SD for each pixel with long temporal series [8]. However, Vafeidis and Darke (2005) find that the above methods based on SD would not get well results for different cases. Therefore, the fixed thresholds algorithms still face the challenges to choose the optimal thresholds to extract burn scar in an automatic fashion [9].

Post classification techniques use image data and indices derived from multi-temporal images to make image classification and compare the post-classification image to extract burn scar. For example, X. Cao et al. (2009) combine the GEMI-B and SVMs classifier to extract burn scar in grassland areas with high accuracy (a). St. Maxime: July 12, 2003

(b). St. Maxime: August 13, 2003

(c). Texas: April 2, 2011

(d). Texas: April 18, 2011

(e). Marker Fire: July, 19, 2013

(f). Marker Fire: August, 4, 2013

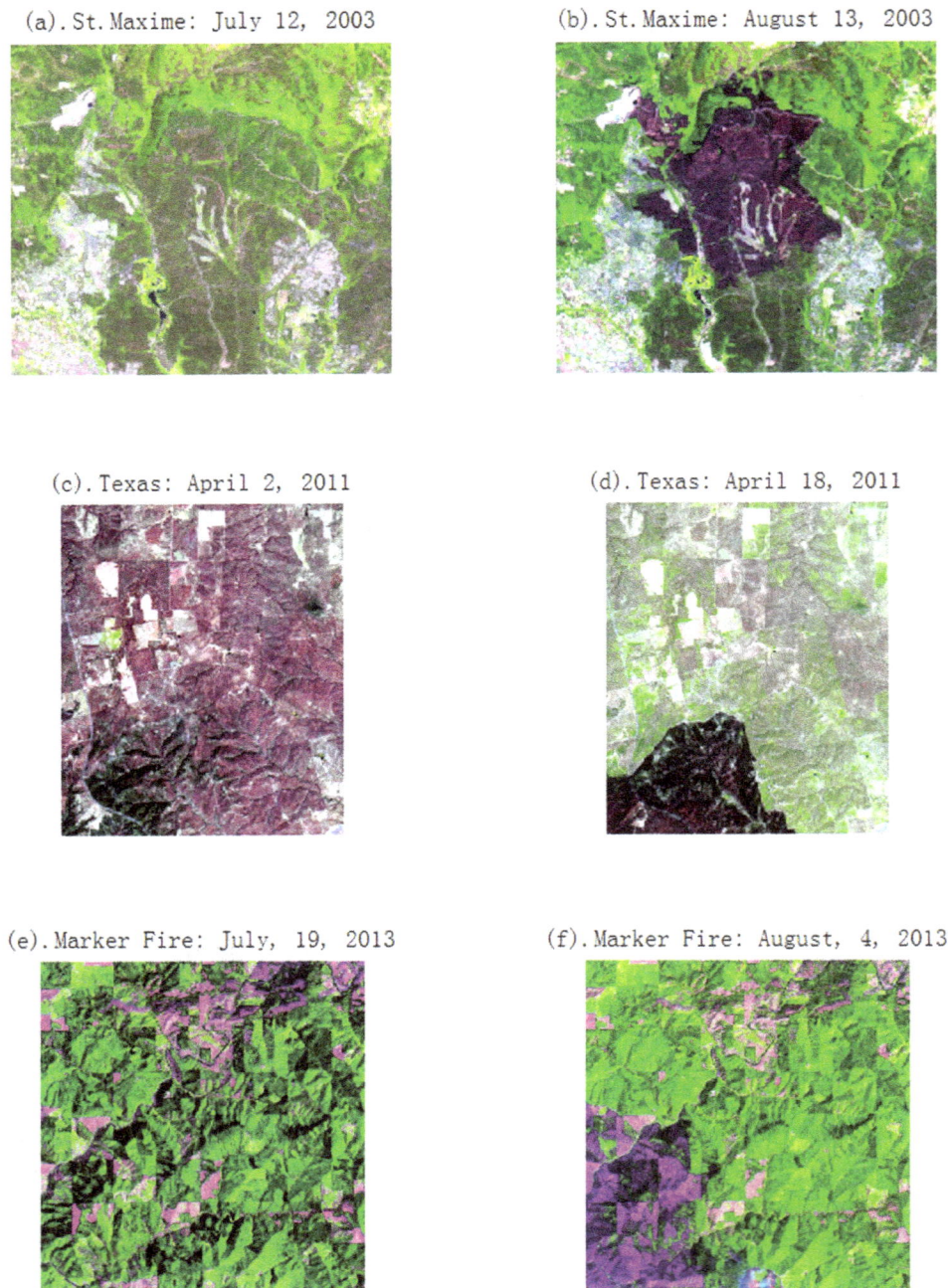

Figure 1. False color images of three data sets. (a). St. Maxime image of July 12, 2003; (b). St. Maxime image of August 13, 2003; (c). Texas image of April 2, 2011; (d). Texas image of April 18, 2011; (e). Marker Fire image of July, 19, 2013; (f). Marker Fire image of August, 4, 2013.

[10]. Pereira *et al.* (1999) use classification and decision trees to segment NOAA-AVHRR imageries into burn surface, unburn surface, and clouds [11]. The result of this method matches well with the ground truth. However, extraction accuracy of algorithms above depends on the training data of classification. In fact, the ground truth samples data are difficult to obtain.

Despite the simplicity and widespread use of the above mentioned algorithms, there exists a major drawback: lack of an automatic and unsupervised technique to extract burn scar. The objective of this study is to develop and test an unsupervised method to automatically extract burn scar without predefined

thresholds. The proposed method includes two-step process. First, compare the bi-temporal images which are taken over the same geographical area at different dates. "difference image" is produced by the comparison between pre-fire and post-fire images. Methods of obtaining difference image is generally divided into three categories: (1) simple algebraic method, for example: image differencing, image ratioing, regression analysis, etc [12]. (2) based on transformation: Principal Component Analysis (PCA), Change Vector Analysis (CVA), etc. (3) based on image features: texture, gradient, vegetation indices and so on. The methods mentioned above have made a difference on the specific aspect.

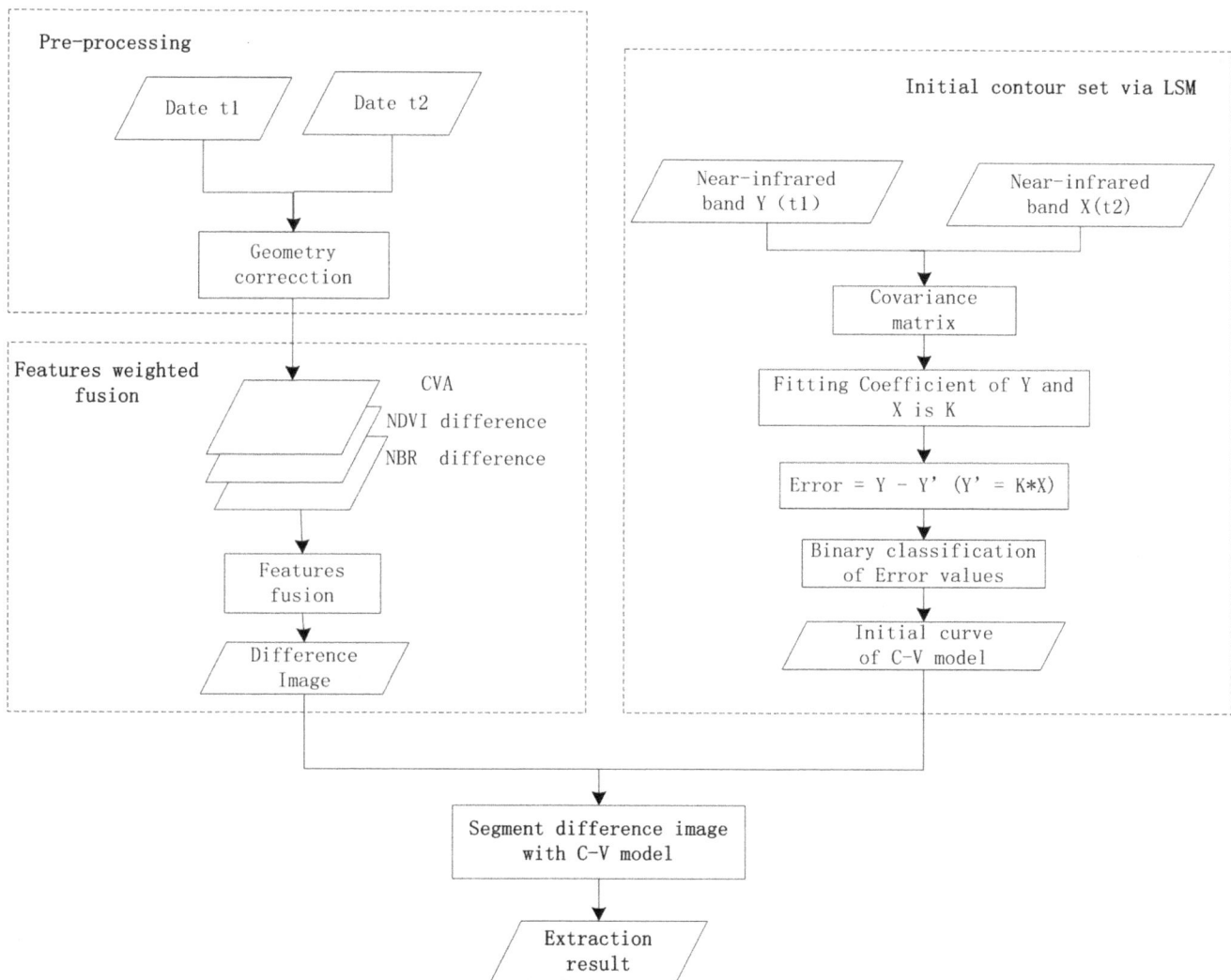

Figure 2. Procedure of fire burn scar extraction based on LSM.

However, these ones only utilize single feature to get the result, sometimes leading the extraction results are inaccurate and incomplete. Consequently, this paper proposes a method that creates the difference image via combining weighted features.

After the first step, separating the burn scar from difference image is a significant job. According to the characteristics of burn scar, this paper focuses on developing an automatic procedure to extract burn scar based on Level Set Method (LSM) without pre-defined information. LSM was proposed by Osher and Sethian in 1988, which was applied on the hydrodynamics problems [13]. LSM expresses loop curve as zeros level set of three dimensional continuous surface. It transforms the process of solving curve function into a partial differential equation of zeros level set. LSM takes it easy to follow shapes that changes topology, such as splitting, merging and developing holes. Then it has played an important part in wide fields such as: physics, materials and computer vision, etc. LSM consists two parts based on edge and region. Caselles *et al.* (1993) put forward the Geodesic Active Contour (GAC) model which is based on edge [14]. Kimmel (2003) modifies the GAC model with joining the direction information of edge [15]. Nevertheless, when the imagery is

vague, edge-based model does not perform very well. Moreover, if the objects' shapes within the image are sunken, the curve may not shrink inward leading the phenomenon that the final position of curve does not coincide with the object's real boundary. On the contrary, the model integrating global gray information behaves well, which can deal with the imagery with blurred and sunken borders. Mumford and Shah (1989) come up with the Mumford-Shah model to segment image by minimizing a energy funciton [16]. Chan and Vese (2001) utilize LSM to refine the Mumford-Shah model to simplify the process of solution, called C-V model [17]. C-V model uses the gray-value feature not gradient to partition the imagery into background and object parts. Remote sensing imagery of burn scar has various shapes caused by wind or terrain, so the boundary may be blurred and sunken. Thus, C-V model is adapted to extract burn scar. Furthermore, setting initial contour makes some effect on iteration numbers when applying C-V model to segment images. How to set initial curves of C-V model influences the speed and accuracy of extraction. There are many researchers have presented strategies to set effective initial contour. Hichiri (2013) proposes to consider the result of Support Vector Machine classification [18]. Cao *et al.* (2007) uses binary

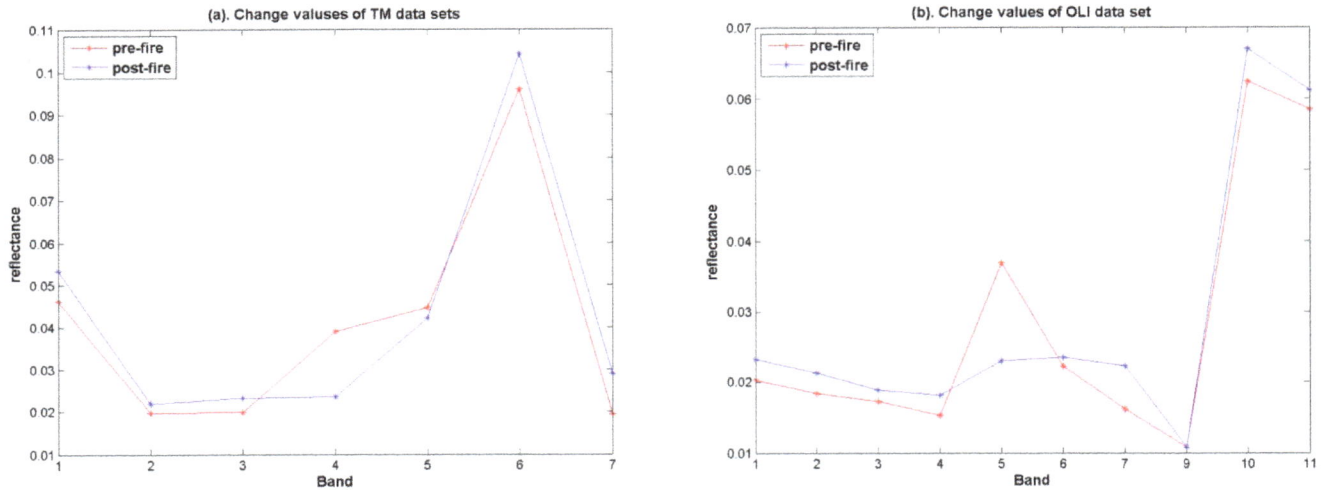

Figure 3. Change values of different bands of TM and OLI data between pre-fire and post-fire. (a). TM data sets; (b). OLI data sets.

imagery of interested objects as an initial curve via selecting areas interactively and morphological algorithm [19]. This paper also presents new method to modify the initial curve of LSM to reduce the time of consuming and improve the precision of burn scar extraction.

In summary, the proposed method can overcome the disadvantages of traditional methods with selecting empirical thresholds and depending on training set. The novel method takes advantage of LSM and multiple features to extract burn scar quickly and automatically.

Materials and Methods

Study Area and Data

(1) The first data set used in the experiments consists of two co-registered images (356×317 pixels) taken near St. Maxime, France, located between the longitudes of $6°32'15.98''E \sim 6°40'23.45''E$, and latitudes of $43°28'14.87''N \sim 43°33'32.83''N$, on the Mediterranean coast by Landsat 5 Thematic Mapper (TM) in the July 12 and August 13, 2003, respectively. During the two acquired dates, a fast-moving forest fire occurred on July 28, 2003, destroyed nearly 16,000 acres of woodland. Fig. 1(a) and (b) show false color images composited with 5,4,3 bands, respectively.

(2) The second data set are two co-registered images of 293×354 pixels acquired by Landsat 5 Thematic Mapper (TM) on the April 2, 2011 and April 18, 2011, respectively. This fire occurred on April 9, 2011 in the central Texas, USA located between the longitudes of $100°28'31.38''W \sim 100°34'16.55''W$, and the latitudes of $32°5'38.61''N \sim 32°59'53.76''N$. The false color images composited by 5,4,3 bands are showed in Fig. 1(c) and (d).

(3) The area located in 15 miles Northeast of Goldendate, Washington is selected as third data set. The data set is captured by Operational Land Imager (OLI) on Landsat 8 before and after the Mile 28 Marker fire. Mile 28 Marker fire occurred in late-July, 2013 and destroyed 10,220 acres in Washington. The subset of this data set with 396×415 pixels locates between $123°28'17.64''W \sim 123°37'4.73''W$, and the latitudes of $42°48'32.90''N \sim 42°55'18.99''N$. Fig. 1(e) and (f) show the false color images composited by 6,5,4 bands taken on July, 19, 2013 and August, 4, 2013.

(4) The experiment environment is Matlab R2010b platform based on windows 7 system with 8G memory.

In the view of Fig. 1(a)–(f) images, it can be seen that the shapes of burn scar are variable and the boundary may be sunken even blurred. Traditional methods sometimes have difficulty in discerning the burn scar exactly and completely. In the Results section, the proposed method can be validated that it can extract complete and clear burn scar.

Burn Scar Extraction Automatically Based on Level Set Method

The proposed technique flowchart is shown in the Fig. 2. The whole process is divided into three parts: First, integrate the weighted spectral features CVA, NDVI and NBR difference of the Date t_1 and Date t_2 which are corrected geometrically to get the difference image. "Features weighted fusion" sub-section introduces this step. Then, analyze the linear fitting error of two near-infrared band images which are got at t_1 and t_2 respectively and refine the initial curves through binary classification result acquired by K-means method. The second step is described in details in the sub-section "Initial contour set of LSM". Finally, make the difference image and initial contour gained above as the input and the output is extraction result of burn scar. The sub-section "Segment difference image with C-V model" presents how to process the last step of proposed method.

Features Weighted Fusion

Traditional extraction method based on LSM segments the difference image obtained by CVA methodology to extract burn

Table 1. D values produced by CVA and Fusion features method.

	CVA	Fusion Methods
TM data (St. Maxim)	1.2565	1.8350
OLI data (Marker Fire)	0.9193	1.5354

CVA represents Change Vector Analysis.

area information. Considering the fact that NDVI and NBR indices change obviously because fire wrecks the vegetation cover of forest [4]. NBR has been widely used to discriminate burn area from unburn area [20]. NDVI is claimed to solve the confusion between classes in the remote sensing images better than NBR [21]. This paper aims at integrating CVA, NDVI difference and NBR difference to get the final difference image. Like this, the difference image can enhance the recognition characteristics of burn scar.

The principle of CVA is describing the difference of single feature across the different bands between two dates. The difference represents the change of pixels in individual band. The results calculated by CVA include magnitude and direction parts. Basically, a change vector can be described with variables of each band, and magnitude component expressed the amount of the change as the direction component. Assuming that BV_{date1} and BV_{date2} are pre-fire and post-fire geometrically coregistered remote sensing images with $N \times N$ pixels and n bands, which are taken from the same area at different dates. CM_{pixel} represents the pixel value of the difference image by applying CVA algorithm. CM_{pixel} is given by (1)[22]:

$$CM_{pixel} = \sum_{k=1}^{n} \left[BV_{ijkdate2} - BV_{ijkdate1} \right]^2 \qquad (1)$$

where CM_{pixel} describes spectral difference of two different remote sensing images. The larger the amount of CM_{pixel} is, the higher the probability of change occurs. $BV_{ijkdate1}$ and $BV_{ijkdate2}$ are the kth band value of pixel located at (i,j) corresponding to Date t_1 and Date t_2 images respectively. $k = 1,2,3,\ldots,n$, n is the number of remote sensing image bands.

The NDVI is calculated as follows:

$$\text{NDVI} = \frac{\rho_{NIR} - \rho_R}{\rho_{NIR} + \rho_R} \qquad (2)$$

where ρ_{NIR} and ρ_R stand for the reflectance acquired in the near-infrared and visible (red) bands respectively. Digital number values of remote sensing images are used instead of reflectance in this paper. The NDVI difference is given by:

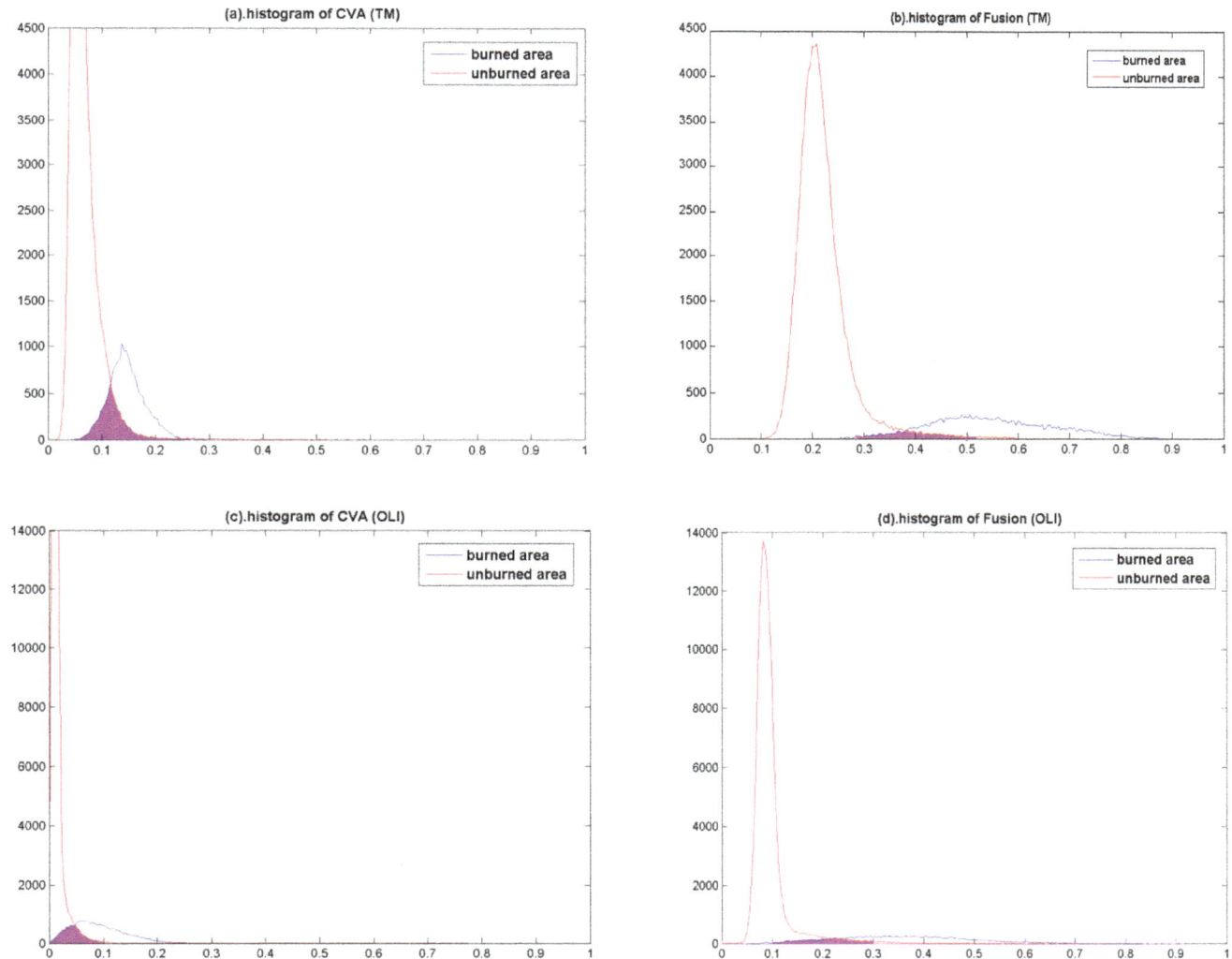

Figure 4. Histogram of burn and unburn class for two data sets. (a). Histogram of CVA :St. Maxime; (b). Histogram of fusion:St. Maxime; (c). Histogram of CVA:Marker Fire; (d). Histogram of fusion:Marker Fire.

(a).St.Maxime:Initial curve of TCV

(b).St.Maxime:Initial curve of Proposed Algorithm

(c).Texas:Initial curve of TCV

(d).Texas:Initial curve of Proposed Algorithmm

(e).Marker Fire:Initial curve of TCV

(f).Marker Fire:Initial curve of ProposedAlgorithm

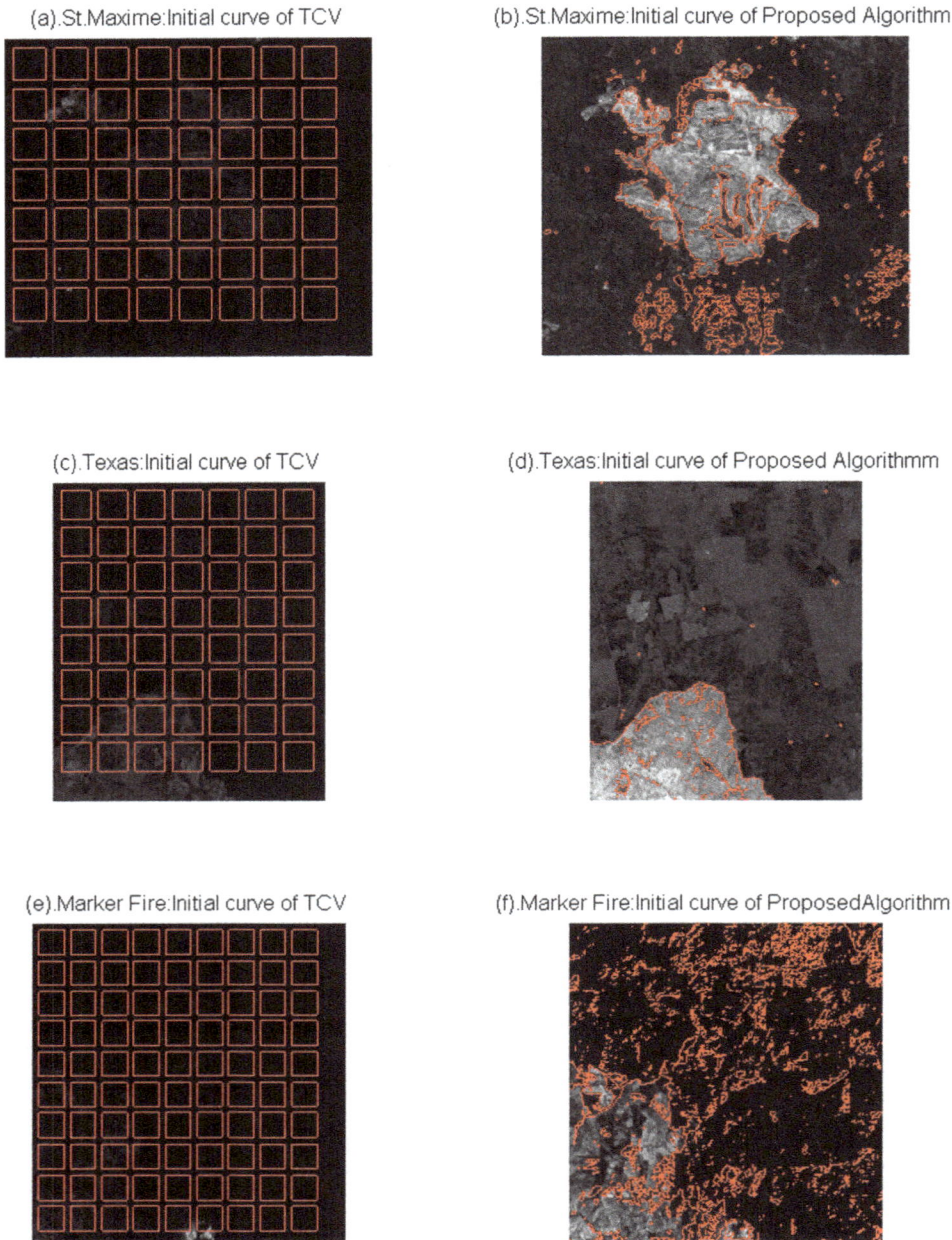

Figure 5. Comparison of different initial curves of three data sets. (a). St. Maxime:Initial curve of TCV; (b). St. Maxime:Initial curve of Proposed Algorithm; (c). Texas: Initial curve of TCV; (d). Texas:Initial curve of Proposed Algorithm; (e). Marker Fire: Initial curve of TCV; (f). Marker Fire:Initial curve of Proposed Algorithm.

$$\text{NDVI}_{dif} = \text{NDVI}_{pre-fire} - \text{NDVI}_{post-fire} \qquad (3)$$

NBR is proposed by Key and Benson in 2004 [23]. They replace the red reflectance in the NDVI with the mid-infrared reflectance value. The mid-infrared reflectance is sensitive to water of vegetation and the lignose content of non-photosynthetic vegetation. The index is given below [24]:

$$\text{NBR} = \frac{\rho_{NIR} - \rho_{MIR}}{\rho_{NIR} + \rho_{MIR}} \qquad (4)$$

where ρ_{NIR}, ρ_{MIR} are reflectance in the near-infrared and mid-infrared bands, respectively. The NBR difference is described below:

$$\text{NBR}_{dif} = \text{NBR}_{pre-fire} - \text{NBR}_{post-fire} \qquad (5)$$

In order to utilize the advantages of multiple features, this paper uses Chi Square Transformation (CST) algorithm to weighted fuse the normalized difference image obtained from CVA, NDVI and NBR algorithms to create difference image between pre-fire and post-fire data. The weight values are obtained by the standard

Table 2. Compositing of different methods.

Extraction Algorithm	Difference image	Segmentation algorithm
Proposed algorithm	Fusion results	C-V model with modified initial curve
OSTU algorithm	Fusion results	OSTU algorithm
FCM algorithm	Fusion results	FCM algorithm
TCV algorithm	CVA	C-V model with arbitrary shapes as initial curve

C-V model: Chan-Vese model; FCM: Fuzzy C-mean algorithm; TCV: Traditional C-V model algorithm with arbitrary shapes as the initial curves.

deviation of each difference image component [25].

$$\mathrm{DI} = \frac{D_{CVA}}{\sigma_C} + \frac{D_{NDVI}}{\sigma_N} + \frac{D_{NBR}}{\sigma_B} \qquad (6)$$

In (6), DI is the weighted difference Image, D_{CVA} is the difference image acquired according to CVA algorithm, D_{NDVI} is the difference image calculated by NDVI feature and D_{NBR} is the difference image obtained by NBR feature. Normalize the DI which is the final difference image by fusing the distinct difference images, where the coefficients σ_C, σ_N and σ_B are the standard deviation of three difference images, respectively.

After the fire, the reflectance of burn scar area will have a change at different bands. The change values of selected bands data of TM and OLI images within burn scar area are shown in Fig. 3.

Fig. 3 shows that reflectance values of all bands have changed after the fire. The change values of red, near-infrared and mid-infrared bands are obvious, especially the near-infrared band. NDVI is calculated by combining red and near-infrared bands data. NBR combines near-infrared band and mid-infrared band to estimate the vegetation area effected by fire. CVA represents the difference between pre-fire and post-fire images of all bands in global. At the same time, NDVI and NBR difference can emphasize the difference of red, near-infrared and mid-infrared bands in local. Thus, in order to enhance the burn scar reflectance features, the weighted fusion algorithm is a better choice.

To illustrate the fusion results, this paper uses the normalized distances D calculated with mean and standard values of burn and unburn area to compare the CVA image and weighted fusion difference image [9].

$$D = \frac{|\mu_U - \mu_B|}{\sigma_U + \sigma_B} \qquad (7)$$

where μ_B and μ_U are the means values of burn and unburn areas, σ_B and σ_U represent the standard deviation of burn and unburn areas, respectively.

The D values calculated by CVA and Fusion methods are shown in Table 1.

According to the Table 1, the D values of difference image acquired via weighted fusion method are higher 0.5785 and 0.6161 than CVA method for TM and OLI data sets, respectively. The higher D values demonstrate the method can make better discrimination between burn and unburn areas than others. [9] mentions that the larger D values can indicate low degree of histogram overlap between burn and unburn classes. The histograms of burn and unburn area for TM and OLI data sets are shown in Fig. 4.

Fig. 4 presents that the overlap area between burn and unburn classes produced by weighted fusion features is smaller than one produced by CVA algorithms for TM and OLI data sets. This comparison outcomes can illustrate the fusion algorithm can describe the characteristics of the burn class in the difference image better. Thus, this step provides a guarantee for segment algorithm to discriminate the burn scar from unburn area exactly.

Initial Contour Set of LSM

Because of the high correlation of pre-fire and post-fire remote sensing images taken by the same sensor from the same area, identification of the burn area can be made by least square method for two acquisitions images. Burn scar belongs to the change information which are within the part with large fitting error values. In order to decide the threshold to separate the Error into "change" and "no change", K-means algorithm is adopted to analyze the Error values. Then, the binary classification result is as the initial curve. The initial contour set by this method can locate appropriately the boundary of real burn scar, and improve the sufficiency of initial curve.

Suppose Y is the kth band of date t_1 of m dimensional vector, X is the kth band of date t_2 of m dimensional vector.

St.Maxime

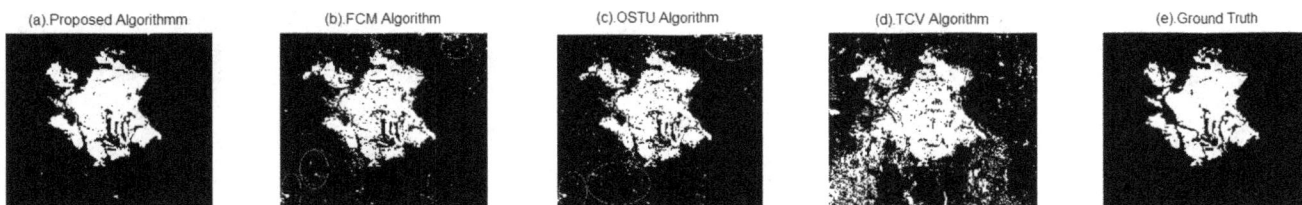

(a).Proposed Algorithmm (b).FCM Algorithm (c).OSTU Algorithm (d).TCV Algorithm (e).Ground Truth

Figure 6. Extraction results of different methods of St. Maxime. (a). Proposed algorithm; (b). FCM algorithm; (c). OSTU algorithm; (d). TCV algorithm; (e). Ground Truth.

Texas

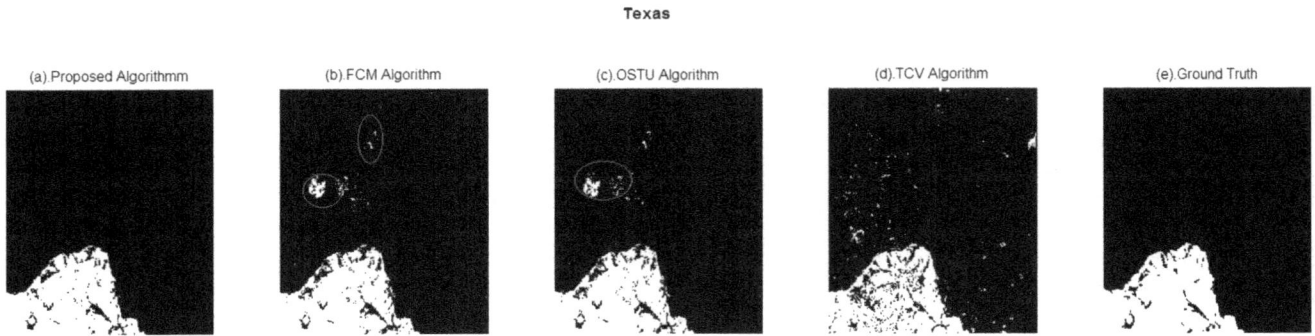

Figure 7. Extraction results of different methods of Texas. (a). Proposed algorithm; (b). FCM algorithm; (c). OSTU algorithm; (d). TCV algorithm; (e). Ground Truth.

$$Y^* = KX \tag{8}$$

where K is calculated by least square method, which satisfies the equitation $min(\frac{1}{m}\sum_{i=1}^{m}||Y-Y^*||^2)$. According to this constrained condition the following fitting coefficient can be got:

$$K = \Sigma_{21}\Sigma_{11}^{-1} \tag{9}$$

where Σ_{21} indicates covariance between Y and X, Σ_{11} is variance of X.

Error value between Y and Y^* in the following equation is calculated by Mahalanobis distance,

$$Error = (Y-Y^*)^T\Sigma_{\varepsilon}(Y-Y^*) \tag{10}$$

In (10), Σ_{ε} represents the variance of $(Y-Y^*)$, the area with lager *Error* value is defined as the change areas.

Segment Difference Image with C-V Model

The objective of Mumford-Shah segment method is to find a curve C which can minimize the fitting energy to divide the image into two non overlapping parts. The fitting energy function is:

$$F^{MS}(I,C) = \int_{\Omega}|X-I|^2dxdy + \lambda\int_{\Omega\backslash C}|\nabla I|^2dxdy + \mu|C| \tag{11}$$

where μ and λ are the parameters selected by users to fit a particular class of images, X is the image to be segmented, and Ω is the defined domain of that image. I is the average of the region separated by curve C.

In fact, it is not easy to minimize the fitting energy function because it is a non-convex surface. Chan and Vese raise to exploit LSM to solve the aforementioned problem. The C-V model simplifies the Mumford-Shah method and the Heaviside function is introduced. This algorithm divides the image into inner region Ω_1 and external region Ω_2 by the curve C. The average values of Ω_1 and Ω_2 can reflect the difference values of object and background. In this case, the simplified energy function is as follows:

$$\begin{aligned}F^{MS}(\Phi,c_1,c_2) = \quad &\int_{\Omega}|X-c_1|^2H(\Phi(x,y))dxdy \\ &+ \int_{\Omega}|X-c_2|^2(1-H(\Phi(x,y)))dxdy \\ &+ \mu\int_{\Omega}\delta_0|\nabla H(\Phi(x,y))|dxdy\end{aligned} \tag{12}$$

In (12), $H(\Phi(x,y))$ is Heaviside function, $\Phi(x,y)$ is level set function, $X(x,y)$ is the image defined in the Ω, δ_0 is Dirac delta function $\delta_0(z) = (d/dz)H(z)$. Heaviside function is given by:

$$H(z) = \begin{pmatrix}1, & z>0 \\ 0, & z\leq 0\end{pmatrix} \tag{13}$$

In practice, Heaviside is replaced by the following regularized format generally:

Marker Fire

Figure 8. Extraction results of different methods of Marker Fire. (a). Proposed algorithm; (b). FCM algorithm; (c). OSTU algorithm; (d). TCV algorithm; (e). Ground Truth.

Table 3. St. Maxim: Comparison of different methods in precision of burn scar detection.

Method	Missed alarm rate (%)	False alarm rate (%)	Right alarm rate (%)	Kappa	Iteration
Proposed algorithm	0.90	1.40	97.70	0.9684	349
TCV algorithm	1.59	14.09	84.32	0.7563	2777
OSTU algorithm	1.56	2.95	95.49	0.9372	/
FCM algorithm	2.92	1.58	95.50	0.9374	/

$$H_\varepsilon(z) = \frac{1}{2}\left(1 + \frac{2}{\pi}arctan(\frac{z}{\varepsilon})\right) \tag{14}$$

$$\delta_\varepsilon(z) = \frac{d}{dz}H_\varepsilon(z) \tag{15}$$

In (12), c_1 and c_2 are given by:

$$c_1 = \frac{\int_\Omega X(x,y)H(\Phi(x,y))dxdy}{\int_\Omega H(\Phi(x,y))dxdy} \tag{16}$$

$$c_2 = \frac{\int_\Omega X(x,y)(1 - H(\Phi(x,y)))dxdy}{\int_\Omega (1 - H(\Phi(x,y)))dxdy} \tag{17}$$

In (16) and (17), c_1 and c_2 are the average values of inner and external contour C respectively. After the c_1 and c_2 are fixed, Euler-Lagrange equations and the gradient-descent method are used to derive Φ which minimizes the fitting energy function (12):

$$\frac{d\Phi}{dt} = \delta_\varepsilon[\mu div\frac{\nabla\Phi}{|\nabla\Phi|} - (X - c_1)^2 + (X - c_2)^2] \tag{18}$$

μ is regularization parameter set by user, t is an artificial time, the solution of (18) is calculated by spatial finite difference method.

Results

Set the Initial Contour with Least Square Fitting

The bi-temporal images are observed from same sensor and same area at different time. Due to the evident change of near-infrared band of remote sensing images, it is chosen as the fitting images with least square fitting algorithm. The initial contour of traditional C-V model is arbitrary shapes. Here we choose rectangles filled in the whole image as traditional initial curve. This initial curve is as shown in Fig. 5(a)(c)(e). The initial curve proposed by this paper is as shown in Fig. 5(b)(d)(f). In the view of Fig. 5, the proposed initial curve matches the ground truth a lot, which indicates that the proposed algorithm is more reliable.

Extraction Results and Analysis

Get difference image of bi-temporal remote sensing images and modify C-V initial curve according to sections above, and then utilize the C-V model to segment the difference image with a new initial contour to extract burn scar.

To assess the validity of the proposed algorithm, different algorithms are compared with the proposed method: OSTU algorithm, Fuzzy C-mean algorithm (FCM), Traditional C-V model algorithm with rectangles as the initial curves (TCV model). These algorithms are shown in the Table 2. It can be seen from Table 2, the input difference images of OSTU algorithm, FCM algorithm and proposed algorithm are the same. Like this, the extraction results can show the different segment effect of OSTU, FCM algorithms and proposed method. The proposed algorithm has two improvements compared with traditional C-V model, which are methods of obtaining difference image and setting the initial curve. Thus in order to illustrate the effectiveness of the proposed algorithm, comparing it with the traditional C-V algorithm based on two improved points. The results figures of above methods are shown in Fig. 6, 7 and 8.

In order to evaluate the availability of the proposed method, this section aims at analysing quantitatively the extraction results with following measures: (1). Missed alarm rate: number of burn pixels classified as unburn pixels; (2). False alarm rate: number of unburn pixels classified as burn pixels; (3). Right alarm rate: the number of burn pixels and unburn pixels classified correctly and (4). Kappa coefficient. Besides, the paper also considers the iteration numbers of proposed algorithm and TCV algorithm. The overall analysis statistics of three data sets as shown in Table 3, 4 and 5.

The detailed analysis of three data sets with different shapes and size are shown as following:

(1) St. Maxime.

In Fig. 6(d), the segment result of TCV algorithm contains a lot of false information, and simply describes the rough boundary of the burn scar. The proposed method is shown in Fig. 6(a), texture of burn scar is more clear, false information is less, and maps with ground truth very well. The information inside the red circles in Fig. 6(b) and Fig. 6(c) show more false boundaries compared with proposed algorithm. Besides, the extraction results of Fig. 6(b) and Fig. 6(c) are scattered. Compared with Fig. 6(e), the proposed method can produce more real extraction results by getting more complete results and reducing false outline.

As seen in Table 3, the kappa value of the proposed method is higher than the other methods obviously for the selected data set. The commission is just 0.90%, kappa is 0.9684 and the iteration numbers is 349 of the proposed method. The extraction results of OSTU and FCM algorithms are almost the same. Their kappa coefficients are 0.93 nearly with a lower result compared with proposed method. The kappa coefficients of TCV is 0.7563 which is the lower than the other methods. The curve of novel approach can achieve the optimal location quickly because newly developed initial curve setting is close to ground truth outline. From a qualitative point of view, the proposed method can extract the higher accuracy results with least iteration numbers than the other methods.

(2) Texas.

Table 4. Texas: Comparison of different methods in precision of burn scar extraction.

Method	Missed alarm rate (%)	False alarm rate (%)	Right alarm rate (%)	Kappa	Iteration
Proposed algorithm	1.06	0.32	98.62	0.9817	196
TCV algorithm	1.91	1.12	96.97	0.9596	3297
OSTU algorithm	1.43	0.77	97.80	0.9707	/
FCM algorithm	1.37	0.82	97.81	0.9708	/

Comparing the Fig. 7(a)–(e), Fig. 7(a) has a close result mapped with ground truth. There is no fire happened in the area inside the red circle of Fig. 7(b) and (c) actually. TCV algorithm results look scattered compared with proposed method.

In the view of Table 4, in terms of kappa coefficients and iteration numbers, proposed methods produce 0.9817 and 196, respectively. Comparing this result with the ones obtained by TCV algorithm, the proposed method shows clear boundary and less false information. This result is corresponding with visually comparing results. OSUT and FCM produce the same result with kappa coefficient is 0.97.

(3) Marker fire.

Fig. 8 illustrates the efficiency of proposed method visually, Fig. 8(a) is confirmed very close to Fig. 8(e) and its outline is clear and complete. By contrast, Fig. 8(d) and Fig. 8(b) have more false information than Fig. 8(a). The OSTU algorithm performs not well with incomplete burn scar results.

Table 5 points out the proposed method can get better extraction results compared with the other ones. The validation results obtained by proposed algorithm are equal to 0.65%, 2.47%, 96.88%, and 0.9589, 890, respectively. TCV algorithm needs to iterate 5800 to meet convergence. However, proposed method decreases the performing time. The results derived by OSTU algorithm is the lowest of all, and the FCM and TCV have the nearly extraction results.

Discussion and Conclusion

Burn scar area data are important for research on forest recovery and environment change. Multi-temporal remote sensing data is perfect source for burn scar extraction. The existing methods have some limitations. The fixed threshold method depends on the experience. Usually, the adequate threshold is difficult to obtain under various cases. In practice, training set of supervised method cannot satisfy all kinds of burn scar due to their diversity on seasons and spatial extent. Therefore, relying on subjectivity, the training set needs to be increased or adjusted continuously, even interact with human. In order to overcome the limitations mentioned above. This paper proposes an automatic and unsupervised method to extract burn scar in forestry area.

The contributions in this paper are: (1) Weighted fuse the CVA, NDVI and NBR difference to create difference image which makes the burn scar outstanding and contains more rich and accurate information when extracting them. (2) Modify initial contour setting by using least square method between near-infrared images of two dates considering the vegetation cover is destroyed by forest fire. (3) Utilize C-V model to extract burn scar. The C-V model can separate the burn pixels from unburn pixels by handling the topology changes of curve automatically. This step can overcome the disadvantage that the blurred and irregular image boundary is difficult to extract. Besides, it makes the details information of burn scar more clear.

This method has been validated by comparing with TCV, OSTU, FCM methods using Landsat 5 TM data set and Landsat 8 OLI data set. The comparison results demonstrate the proposed approach can achieve better extraction results than the others. Although the three burn scar mentioned above have different shapes and are taken from different data source, the proposed method gets perfect results as well. At the same time, new optimal initial curve reduces the iteration numbers and speeds the extraction of burn scar. The experimental results illustrate that this newly developed algorithm can realize the automatic processing of burn scar extraction with higher accuracy.

This method is suited to be applied on the burn scar extraction of forestry area, especially covered with vegetation. The prerequisites of proposed method are that the two dates images are covered without cloud and taken at the same season. Due to cloud cover, season change may confound extraction results of proposed method. So in the future, in order to extract burn area more conveniently and exactly. This paper would take cloud and time interval factors into account. Through experiments analysis, the results indicate that this method can be easily adapted and applied to extract burn scar from medium resolution satellite images without cloud after the fire occurred. High resolution satellite image is difficult to acquire and has more complex properties than medium ones. Further, the proposed method would get hold of different resolution satellite images into consideration to realize the burn scar extraction of real time satellite data automatically and timely. Once the fire hazards happen, it is always difficult to get the sample data of the disaster areas in a short time for supervised

Table 5. Marker Fire: Comparison of different methods in precision of burn scar extraction.

Method	Missed alarm rate (%)	False alarm rate (%)	Right alarm rate (%)	Kappa	Iteration
Proposed algorithm	0.65	2.47	96.88	0.9589	890
TCV algorithm	2.61	2.79	94.60	0.9389	5800
OSTU algorithm	10.66	0.0164	89.32	0.8746	/
FCM algorithm	3.27	1.54	95.19	0.9374	/

methods. As a better alternative method, the proposed method can obtain extraction results timely and accurately to help the forestry department compute the area destroyed by fire and process vegetation recovery quickly.

Acknowledgments

Thanks to USGS website for providing free Landsat 5 TM and Landsat 8 OLI data set free. We thank Dr. David Hudson from Geoscience Australia and Australian National University for his help in revising this paper. We also appreciate the editor and anonymous reviewers for their helpful comments.

Author Contributions

Conceived and designed the experiments: QD YL JBL. Performed the experiments: YL QD SBL. Analyzed the data: YL QD JY. Contributed reagents/materials/analysis tools: YL QD. Wrote the paper: YL QD.

References

1. Sedano F, Kempeneers P, Miguel JS, Strobl P, Vogt P (2013) Towards a pan-european burnt scar mapping methodology based on single date medium resolution optical remote sensing data. International Journal of Applied Earth Observation and Geoinformation 20: 52–59.
2. Sedano F, Kempeneers P, Strobl P, McInerney D, San Miguel J (2012) Increasing spatial detail of burned scar maps using irs-awifs data for mediterranean europe. Remote Sensing 4: 726–744.
3. Gong P, Pu R, Li Z, Scarborough J, Clinton N, et al. (2006) An integrated approach to wildland fire mapping of california, usa using noaa/avhrr data. Photogrammetric engineering and remote sensing 72: 139.
4. Chuvieco E, Martin MP, Palacios A (2002) Assessment of different spectral indices in the red-near-infrared spectral domain for burned land discrimination. International Journal of Remote Sensing 23: 5103–5110.
5. Silva J, Sá AC, Pereira J (2005) Comparison of burned area estimates derived from spot-vegetation and landsat etm+ data in africa: Inuence of spatial pattern and vegetation type. Remote sensing of environment 96: 188–201.
6. Li Z, Kaufman YJ, Ichoku C, Fraser R, Trishchenko A, et al. (2001) A review of avhrr-based active fire detection algorithms: Principles, limitations, and recommendations. Global and regional vegetation fire monitoring from space, planning and coordinated international effort : 199–225.
7. Fernández A, Illera P, Casanova JL (1997) Automatic mapping of surfaces affected by forest fires in spain using avhrr ndvi composite image data. Remote sensing of environment 60: 153–162.
8. Barbosa P, Gregoire J, Pereira J (1999) An algorithm for extracting burned areas from time series of avhrr gac data applied at a continental scale - an overview. Remote Sensing of Environment 69: 253–263.
9. Vafeidis A, Drake N (2005) A two-step method for estimating the extent of burnt areas with the use of coarse-resolution data. International Journal of Remote Sensing 26: 2441–2459.
10. Cao X, Chen J, Matsushita B, Imura H, Wang L (2009) An automatic method for burn scar mapping using support vector machines. International Journal of Remote Sensing 30: 577–594.
11. Pereira J (1999) A comparative evaluation of noaa/avhrr vegetation indexes for burned surface detection and mapping. Geoscience and Remote Sensing, IEEE Transactions on 37: 217–226.
12. Singh A (1989) Review article digital change detection techniques using remotely-sensed data. International journal of remote sensing 10: 989–1003.
13. Osher S, Sethian JA (1988) Fronts propagating with curvature-dependent speed: algorithms based on hamilton-jacobi formulations. Journal of computational physics 79: 12–49.
14. Caselles V, Catté F, Coll T, Dibos F (1993) A geometric model for active contours in image processing. Numerische mathematik 66: 1–31.
15. Kimmel R (2003) Fast edge integration. In: Geometric Level Set Methods in Imaging, Vision, and Graphics, Springer. 59–77.
16. Mumford D, Shah J (1989) Optimal approximations by piecewise smooth functions and associated variational problems. Communications on pure and applied mathematics 42: 577–685.
17. Chan T, Vese L (2001) Active contours without edges. Image Processing, IEEE Transactions on 10: 266–277.
18. Hichri H, Bazi Y, Alajlan N, Malek S (2013) Interactive segmentation for change detection in multispectral remote-sensing images. Geoscience and Remote Sensing Letters, IEEE 10: 298–302.
19. Biao C, Qi L (2007) Ultrasound image segmentation method based on mathematical morphology and level set. China Measurement Technology 33: 114–117.
20. Kontoes C, Poilve H, Florsch G, Keramitsoglou I, Paralikidis S (2009) A comparative analysis of a fixed thresholding vs. a classification tree approach for operational burn scar detection and mapping. International Journal of Applied Earth Observation and Geoinformation 11: 299–316.
21. Chuvieco E, Englefield P, Trishchenko AP, Luo Y (2008) Generation of long time series of burn area maps of the boreal forest from noaa-avhrr composite data. Remote Sensing of Environment 112: 2381–2396.
22. Zhao Y (2003) The principles and methods of application and analysis of remote sensing. Beijing: Science Press.
23. van Wagtendonk JW, Root RR, Key CH (2004) Comparison of {AVIRIS} and landsat etm+ detection capabilities for burn severity. Remote Sensing of Environment 92: 397–408.
24. Miller JD, Thode AE (2007) Quantifying burn severity in a heterogeneous landscape with a relative version of the delta normalized burn ratio (dnbr). Remote Sensing of Environment 109: 66–80.
25. D'Addabbo A, Satalino G, Pasquariello G, Blonda P (2004) Three different unsupervised methods for change detection: an application. In: Geoscience and Remote Sensing Symposium, 2004. IGARSS'04. Proceedings. 2004 IEEE International. IEEE, volume 3, 1980–1983.

12

Spatio-Temporal Variability of Aquatic Vegetation in Taihu Lake over the Past 30 Years

Dehua Zhao*, Meiting Lv, Hao Jiang, Ying Cai, Delin Xu, Shuqing An

Department of Biological Science and Technology, Nanjing University, Nanjing, P R China

Abstract

It is often difficult to track the spatio-temporal variability of vegetation distribution in lakes because of the technological limitations associated with mapping using traditional field surveys as well as the lack of a unified field survey protocol. Using a series of Landsat remote sensing images (i.e. MSS, TM and ETM+), we mapped the composition and distribution area of emergent, floating-leaf and submerged macrophytes in Taihu Lake, China, at approximate five-year intervals over the past 30 years in order to quantify the spatio-temporal dynamics of the aquatic vegetation. Our results indicated that the total area of aquatic vegetation increased from 187.5 km^2 in 1981 to 485.0 km^2 in 2005 and then suddenly decreased to 341.3 km^2 in 2010. Similarly, submerged vegetation increased from 127.0 km^2 in 1981 to 366.5 km^2 in 2005, and then decreased to 163.3 km^2. Floating-leaf vegetation increased continuously through the study period in both area occupied (12.9 km^2 in 1981 to 146.2 km^2 in 2010) and percentage of the total vegetation (6.88% in 1981 to 42.8% in 2010). In terms of spatial distribution, the aquatic vegetation in Taihu Lake has spread gradually from the East Bay to the surrounding areas. The proportion of vegetation in the East Bay relative to that in the entire lake has decreased continuously from 62.3% in 1981, to 31.1% in 2005 and then to 21.8% in 2010. Our findings have suggested that drastic changes have taken place over the past 30 years in the spatial pattern of aquatic vegetation as well as both its relative composition and the amount of area it occupies.

Editor: Jose Luis Balcazar, Catalan Institute for Water Research (ICRA), Spain

Funding: This research was supported by the National Natural Science Foundation of China (31000226) and Environmental Monitoring Research Foundation of Jiangsu Province (1014). The funders had no role in study design, data collection and analysis, decision to publish, or preparation of the manuscript.

Competing Interests: The authors have declared that no competing interests exist.

* E-mail: dhzhao@nju.edu.cn

Introduction

With the rapid development of the Chinese economy since the "reform and opening-up" policy was implemented in 1978, human activities have placed a growing stress on the country's freshwater lakes, which are extremely vulnerable to natural and anthropogenic disturbance [1–3]. As a result, most freshwater lake ecosystems in China have experienced drastic changes during this period [4,5]. Because of the important ecological and socioeconomic functions of aquatic macrophytes, such as stabilization of sediments, purification of water, slowing of water currents and maintenance of fishery production [6–10], examining the temporal dynamics of these organisms following implementation of the 1978 policy changes in China can provide valuable information concerning the mechanisms driving shifts in distribution of aquatic macrophytes that can be used to better manage inland waters.

Taihu Lake, the third-largest freshwater lake in China, is located in the core of the Yangtze Delta within the lower reaches of the Yangtze River Basin, one of the most developed areas in China. Since the 1978 policy changes, the Taihu Lake catchment has experienced rapid socio-economic development. Currently, the catchment contains 3.7% of the Chinese population and 11.6% of its Gross Domestic Product (GDP) within an area of 36,900 km^2 that accounts for only 0.4% of China's total land area. Concurrent with the rapid socio-economic development, the aquatic ecosystem of Taihu Lake has degraded appallingly [11–13], with the degradation being widely attributed to eutrophica-tion and human activities such as flood control projects and wetland reclamation [11], and the distribution and community structure of aquatic macrophytes have clearly changed [14,15].

Although several field inventories conducted since the 1960s have provided data on community structure in Taihu Lake, most of these inventories have provided little or only approximate information on distributional ranges due to technological limitations in addition to the substantial variability in distribution area both within a year and among years [14,15]. Because continuous data sets containing information on exact distributions of different aquatic vegetation types are lacking, the ability to monitor the dynamics of aquatic vegetation and identify the driving forces behind changes in its distribution is restricted. Moreover, disturbances resulting from human activities [15,16] inhibit the ability to clearly identify the particular role of eutrophication in the temporal succession process, further limiting our understanding of the forces driving succession.

We divided Taihu Lake into six sections according to the relative influences of human activities and environmental factors in order to better describe the spatio-temporal variability. By reconstructing the distribution of aquatic vegetation types in each of the six sections from 1981 to 2010 using a series of Landsat images (i.e. ETM+, TM and MSS images) and field validation inventories from 2009 and 2010, this study sought to track the spatio-temporal variability of aquatic vegetation distribution in

Taihu Lake since the "reform and opening-up" policy was implemented in 1978.

Materials and Methods

2.1 Study Area

Our study area included the entirety of Taihu Lake as well as the surrounding area within 500 meters of the lake boundary, where most of the emergent vegetation was distributed. Taihu Lake has an average depth of 1.9 m and occupies a surface area of 2,425 km². The Taihu Lake catchment plays an important role in China's political economy, with the GDP per capita in this area exceeding US$8,000 after 2007, more than three times the country's average. Per unit of land, the catchment's economic yield is 57 times the national average [11]. Indirect human activities such as flood control projects and pollution inputs, as well as direct human activities such as wetland reclamation and pen-fish-farming have transformed the formerly healthy aquatic ecosystem of Taihu Lake [16].

Because of the substantial spatial variation in the aquatic environments of Taihu Lake, we divided the lake into six sections (Fig. 1): I, Meiliang Bay and Zhushan Bay, the most polluted area in the lake [17]; II, Gonghu Bay, through which large amounts of water have been flushed into the lake from the Yangtze River since 2001 [18,19]; III, the eastern coastal areas that represent the traditional distribution area of aquatic vegetation; IV, the east bay, where most of the activities related to fisheries production are focused; V, the central area and the west coastal areas, which occupy 58.8% of the lake; VI, the southeast section, which connects the central area and the east bay.

We used the spectral characteristics of the remotely-sensed images to classify the aquatic vegetation of Taihu Lake into three types: emergent vegetation, floating-leaf and floating vegetation, and submerged vegetation [20,21]. To investigate the temporal dynamics of aquatic vegetation distribution in Taihu Lake over the past 30 years, we mapped the distributions at approximate five-

Table 1. Dates of the Landsat ETM+, TM and MSS images used in this study.

Years	Sensors	Date-1	Date-2	Date-3
2010	ETM+	3/13	8/20	9/21
2009	TM	1/13	8/25	9/10
2005	ETM+	3/31	6/19	9/7
2000	ETM+	3/18	8/8	10/12
1995	TM	2/24	8/3	8/19
1989	TM	1/14	7/17	10/21
1984	MSS	12/16 (1983)	8/4	9/5
1981	MSS	2/4	7/16	9/8

(Due to the absence of high quality MSS images in the winter of 1984, the MSS image dated 16 December 1983 was used instead).

year intervals. Considering the availability of clear remote sensing images, the final study years were chosen to be 1981, 1984, 1989, 1995, 2000, 2005, 2009 and 2010 (Table 1). The 2009 data were used to evaluate the method developed in this study to identify aquatic vegetation types.

2.2 Field Surveys

Field surveys were conducted on 14–15 September 2009 and 27 September 2010 to gather data for model building and accuracy assessment of the classified maps, with 783 samples collected for open water or aquatic vegetation (Fig. 1). An additional 182 samples of reed (emergent) vegetation or terrestrial areas (e.g., shoreline roads and buildings such as docks, businesses and factories) were obtained from a 1:50,000 land use and land cover map due to logistical difficulties in maneuvering a boat in the dense reed vegetation. A total of 426 and 539 training or validation samples were collected for 2009 and 2010, respectively. The field survey procedures have been described in detail by Zhao et al. [20].

2.3 Image Processing

The procedure used in this study was similar to that implemented by Zhao et al. [20] and thus is only briefly summarized here. Images used in this study had no more than 10% cloud cover in accordance with the recommended standard for aquatic remote sensing [22]. Prior to atmospheric correction, cloud-contaminated pixels were removed from all images using interactive interpretation. Atmospheric corrections were applied to the images using the cosine approximation model (COST; Chavez, 1996) using ERDAS IMAGINE 9.2 (Leica Geosystems Geospatial Imaging, LIC), and geometric correction was carried out for all the images using second-order polynomials with an accuracy higher than 0.5 pixel.

We used combinations of winter (January through March when the biomass of aquatic vegetation was lowest) and summer (June through October when the biomass of aquatic vegetation was highest) images from each study year for aquatic vegetation identification. For each study year, three clear Landsat images were selected (one from winter and two from summer) and formed into two pairs in which the winter image was paired with each summer image. Thus, a total of sixteen pairs were used in this study (Table 1). The ground truth samples from 2009 and 2010 were used to evaluate the accuracy of the final aquatic vegetation classifications derived from the 2009 and 2010 image pairs.

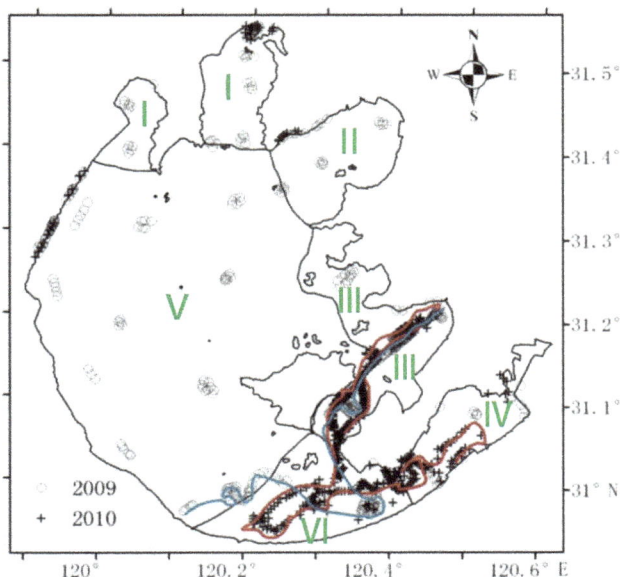

Figure 1. The study area showing the six sections of Taihu Lake and the distribution of 965 training samples (426 in 2009 and 539 in 2010) of emergent, floating-leaf and submerged macrophytes and other compositional types.

2.4 Aquatic Vegetation Identification

To identify emergent, floating-leaf and submerged vegetation, we developed classification tree (CT) models for direct application between images from different dates and sensors [20,21]. Considering the differences in both wavelength range and the spectral response curve among images from different sensors (i.e., ETM+, TM and MSS), we developed classification model structures manually for emergent, floating-leaf and submerged vegetation and then obtained quantitative thresholds for specific images from CT analysis of the data in our statistical software package (PASW-Statistics v. 18). We selected three spectral indices to identify emergent, floating-leaf and submerged vegetation: the Normalized Difference Vegetation Index (NDVI) [23], Modified Normalized Difference Water Index (MNDWI) [24], and average reflectance of the blue, green and red bands from the remote sensing image (AVE123) [25]. Summer (−s) and winter (−w) values, as well as the difference between the summer and winter values (−s−w), were calculated for each of these indices for use in the models.

The basic classification model structures for identification of emergent, floating-leaf and submerged vegetation are shown in Fig. 2. Three spectral indices, MNDWI-(w), NDVI-(s) and NDWIF-(s), and one spatial parameter, distance to boundary (DB), were used for identification of emergent vegetation; two spectral indices, MNDWI-(w) and NDVI-(s), were used to identify floating-leaf vegetation; and four spectral indices, MNDWI-(w), NDVI-(s), AVE123-(s-w) and AVE-(s), were used to identify submerged vegetation.

2.5 Robustness of CT Models

CT models developed for 2009 image pairs (Fig. 2) had an overall accuracy of 92.5%, with classification accuracies of 89.1%, 92.9% and 88.8% for emergent, floating-leaf and submerged

vegetation, respectively. When the CT models were applied to the image pairs of 2010, overall accuracy was 91.7%, with classification accuracies of 91.4%, 89.2% and 87.9% for emergent, floating-leaf and submerged vegetation, respectively (Table 2). These results suggested that the CT analysis could be used to effectively identify the aquatic vegetation in Taihu Lake. Therefore, the CT models were used to map the distribution of aquatic vegetation at different times during the past 30 years.

Results

3.1 Spatio-temporal Dynamics of Distribution Area

We used the CT models developed in this study to map emergent, floating-leaf and submerged vegetation in 1981, 1984, 1989, 1995, 2000, 2005 and 2010 (Fig. 3). Substantial changes in aquatic vegetation distribution have taken place over the past 30 years. From 1981 to 2005, the area of aquatic vegetation gradually increased, from 187.5 km^2 in 1981 to 485.0 km^2 in 2005– a 159% increase (Fig. 4). However, compared with 2005, the area of aquatic vegetation decreased suddenly in 2010 to 341.3 km^2.

In addition to distribution area, the spatial pattern of aquatic vegetation has also experienced substantial temporal variation. Over the past 30 years, aquatic vegetation has spread gradually from the East Bay (section IV) to sections III and VI. In section IV, aquatic vegetation area increased from 116.6 km^2 in 1981 to 150.7 km^2 in 2005 and then suddenly decreased to 74.5 km^2 in 2010, but the proportion of vegetation in section IV relative to that in the entire lake decreased continuously from 62.3% in 1981, to 31.1% in 2005 and then to 21.8% in 2010. In sections III and VI, aquatic vegetation area increased, respectively, from 10.1 and 13.3 km^2 in 1981 (5.4% and 8.3% of the total area) to 115.7 and 145.6 km^2 in 2005 (23.9% and 30.0% of the total area), and then decreased to 94.5 and 99.2 km^2 in 2010 (27.6% and 29.0% of the total area).

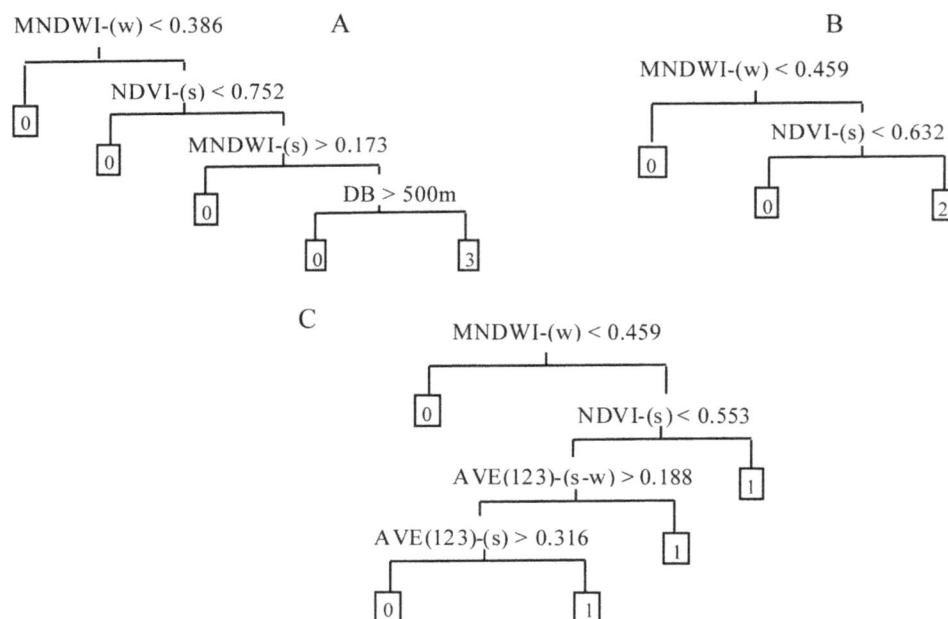

Figure 2. Classification tree models established for (A) emergent vegetation, (B) floating-leaf vegetation and (C) submerged vegetation. The numbers 3, 2 and 1 in the end nodes of the classification trees represent emergent, floating-leaf and submerged vegetation, respectively, whereas 0 represents other types. Variables used are the Modified Normalized Difference Water Index (MNDWI), the Normalized Difference Vegetation Index (NDVI), the average reflectance of the blue, green and red image bands (AVE123) and the distance to the lake bank (DB). Variables were calculated by season (s = summer, w = winter) or differences among seasonal values (e.g., s-w).

Table 2. Confusion matrix of the CT models developed in this paper as applied to 2009 and 2010 data, respectively (in number of field samples).

			Prediction					
			Emergent vegetation	Floating-leaf vegetation	Submerged vegetation	Other types	Classification accuracy (%)	Overall accuracy (%)
2009	Truth	Emergent vegetation	49	4	1	1	89.1	92.5
		Floating-leaf vegetation	4	91	3	0	92.9	
		Submerged vegetation	0	5	103	8	88.8	
		Other types	0	1	5	151	96.2	
2010	Truth	Emergent vegetation	74	7	0	0	91.4	91.7
		Floating-leaf vegetation	5	132	7	4	89.2	
		Submerged vegetation	0	6	102	8	87.9	
		Other types	0	0	8	186	95.9	

3.2 Composition of Emergent, Floating-leaf and Submerged Vegetation

Drastic changes have occurred to the areas occupied by all the vegetation types (emergent, floating-leaf and submerged) over the past 30 years, and the temporal dynamics differed by vegetation type (Fig.4). The area occupied by emergent vegetation gradually decreased from 47.7 km^2 in 1981 to 10.6 km^2 in 2005 (a 77.8% decrease) and then increased to 31.8 km^2 in 2010 (a 300% increase over 2005). The temporal dynamics of emergent vegetation in section IV followed this pattern, decreasing from 26.6 km^2 in 1981 to 4.2 km^2 in 2005 and then increasing to 12.8 km^2 in 2010. The area occupied by floating-leaf vegetation increased continually through time, from 12.9 km^2 in 1981 to 146.2 km^2 in 2010, a 10.3-fold increase. This continuous increase

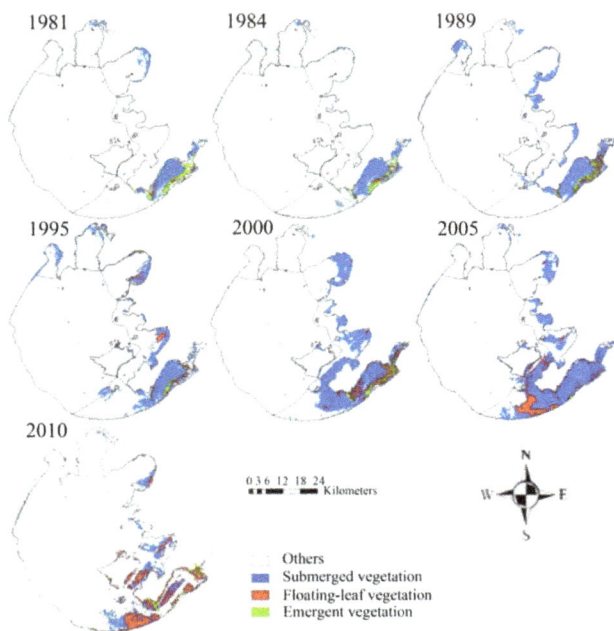

Figure 3. Maps showing the distribution of emergent, floating-leaf and submerged vegetation in Taihu Lake in 1981, 1984, 1989, 1995, 2000, 2005 and 2010.

was primarily a result of the increases in sections III, IV and VI, where floating-leaf vegetation increased 33.6, 29.6 and 46.5 km^2, respectively, in 2010 over that in 1981. Submerged vegetation, on the other hand, gradually increased in area from 127.0 km^2 in 1981 to 366.5 km^2 in 2005 (a 189% increase), then decreased suddenly in 2010 (163.3 km^2).

Substantial changes were also observed in the spatial pattern for each vegetation type. In 1981, emergent, floating-leaf and submerged vegetation were primarily distributed in section IV, where they comprised 56.3%, 81.2% and 62.6%, respectively, of the total area occupied by these vegetation types within the entire lake. After 1981, emergent, floating-leaf and submerged vegetation gradually expanded outward from section IV to the adjacent sections. By 2010, the proportion of emergent, floating-leaf and submerged vegetation in section IV had decreased to 39.2%, 27.5% and 13.2%, respectively, of the total area occupied by these types.

Additionally, the relative abundances of the three aquatic vegetation types changed substantially over the 30 years of this study. Emergent vegetation decreased from 25.4% of total aquatic vegetation in 1981 to only 2.18% in 2005, then rose slightly to 9.31% in 2010; this was consistent with the emergent vegetation in section IV, which decreased from 22.9% of total aquatic vegetation within this section in 1981 to 2.76% in 2005, then rose slightly to 17.2% in 2010. Floating-leaf vegetation increased as a percentage of total aquatic vegetation over the study period, from 6.88% in 1981 to 22.3% in 2005 and then to 42.8% in 2010. The percentage of submerged vegetation increased from 67.7% in 1981 to 75.6% in 2005 and then decreased to 47.9% in 2010.

Discussion

4.1 Driving Forces

The high spatio-temporal variability of aquatic macrophyte distribution in Taihu Lake over the past 30 years can be attributed to both direct and indirect influences of human activities [11]. Because consistency is lacking with regard to quantitative information on the spatio-temporal vegetation dynamics in relation to human activities, we are limited to a qualitative discussion of the forces driving the observed changes.

4.1.1 Direct influences of human activities. Human activities such as planting and conservation or harvesting and removal have affected the distribution area of aquatic vegetation directly. The incentives for planting and conserving aquatic

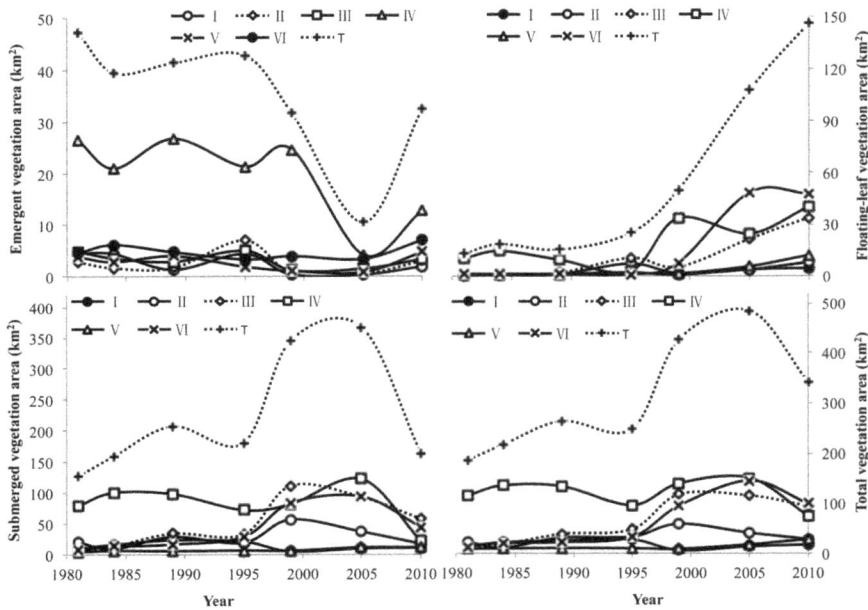

Figure 4. The temporal dynamics of emergent, floating-leaf, and submerged vegetation as well as those of total aquatic vegetation area (the sum of emergent, floating-leaf, submerged vegetation) in the six sections of Taihu Lake (see Fig. 1) between 1981 and 2010. T is the total area of the lake (i.e. the sum of the six sections).

vegetation in the lake came primarily from fishery production activities that can be quantified by pen-fish area. Pen-fish farming activities in Taihu Lake began in the early 1980s and reached a peak in 2000 (108.1 km^2, accounting for more than 80% of the East Bay of Taihu Lake), after which the pen-fishing area began to decrease (Fig. 5). In 2009 and 2010, the area open to pen-fishing decreased suddenly to only about one-fourth what it was in 2000.

In addition to pen-fishing activities, numerous water conservation projects in recent years have slightly increased the lake area occupied by vegetation through planting and restoration of aquatic macrophytes [26]. Direct removal of aquatic vegetation in Taihu Lake has resulted primarily from four types of human activities: (1) wetland reclamation for farming and building construction, which took place mostly prior to 1985 [14,15]; (2) flood-control projects that built concrete embankments instead of maintaining the original wetland buffer zones [11]; (3) large-scale dredging that was implemented in section IV and the surrounding area in recent years [26,27]; and (4) harvesting of aquatic vegetation to be used as food for fish. These direct influences probably had the greatest impact on the distribution of aquatic vegetation in section IV as well as the distribution of emergent vegetation within the entire lake [28].

4.1.2 Indirect influences of human activities. Indirect impacts of human activities on the aquatic vegetation of Taihu Lake were more complex than direct planting or removal activities. First, the high sedimentation rate and swampy conditions in section IV and the area surrounding section VI were likely responsible for the increase in distribution area of aquatic vegetation in section VI before 2005, which was mitigated by the implementation of dredging after 2007 [14,28,29]. Second, water level and its seasonal pattern, determined largely by flood-control projects and artificial regulation, influenced the distribution area and composition of aquatic vegetation in the entire lake [14,15,30]. Third, eutrophication caused primarily by exogenous input from surrounding rivers and fishery production activities probably acted as one of the most important factors affecting the

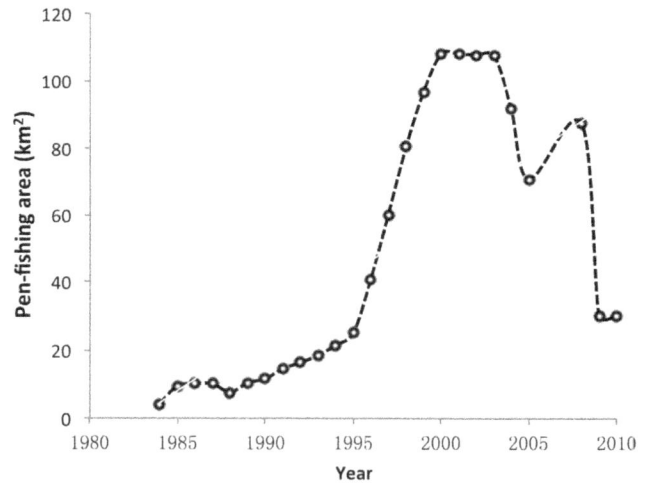

Figure 5. Temporal dynamics of area open to pen-fishing in Taihu Lake in the past 30 years.

temporal variability of both area and composition of aquatic vegetation in sections I, III and V [15,31,32]. Fourth, the action of flushing water into the lake from the Yangtze River through section II decreased water clarity and thus decreased submerged vegetation in section II [18,19,33].

4.2 Implications for Management

Currently, eutrophication is one of the main aquatic environmental problems in Taihu Lake [34]. In order to remove excess nutrients and recover the degraded aquatic ecosystem of Taihu Lake, numerous costly water conservation projects that include planting and restoration of aquatic macrophytes have been implemented in recent years, particularly after the water supply crisis in Wuxi City that was induced by the 2007 blue-green algal bloom [11,35,36]. Our results indicated that aquatic macrophytes are distributed over a large area in Taihu Lake and experienced substantial changes in their distributions over the past 30 years (i.e. between 187.5 and 485.0 km^2). Compared to the expensive water conservation projects, it is probably more economical to protect existing communities of aquatic macrophytes [30]. Therefore, future study should be focused on the effective management of the hundreds of square kilometers of aquatic macrophytes in Taihu Lake. However, we cannot over-rely on the purification function of aquatic vegetation for the recovery of the aquatic ecosystems in Taihu Lake because the aquatic ecosystem can shift suddenly from a clear-water plant-dominated state to a turbid algal-dominated state if the pollutants increase further in the lake [37]. Because it is much more difficult to promote a shift in aquatic ecosystems from the turbid algal-dominated state to the clear-water plant-dominated state than it is to instigate a shift in the opposite direction [37–39], preventing the initial shift from the clear-water plant-dominated state to the turbid algal-dominated state is probably one of the most important current goals for effective lake management.

Due to the different ecological and socioeconomic functions performed by different species and the different aquatic vegetation types such as submerged and floating-leaf vegetation [14,40], the artificial regulation of aquatic vegetation composition is also very important for the management of Taihu Lake. Field studies have indicated that the dominant species in the lake have changed from *Potamogeton maackianus*, *Hydrilla verticillata*, *Vallisneria spiralis* and *Zizania caduiftora*, to *Elodea muttalli*, *Potamogeton malaianus* and *Nymphoides peltata*. The area covered by *Zizania caduiftora* and *Phragmites communis* has decreased drastically in response to human activities - plantation reclamation in particular [14,15,40]. Our findings were consistent with these field investigations: the area of emergent vegetation gradually decreased, while the ratio of floating-leaf vegetation to submerged vegetation gradually increased over the past 30 years. Current management strategies in Taihu Lake such as the large-scale dredging carried out in section IV and the surrounding area after 2007 [28,29], the decrease in eutrophication [31] and the decrease in pen-fishing area [32] are probably beneficial to restoring Taihu Lake to its original submerged-dominant status, whereas water level management activities that decrease the water level in the rainy season and increase the water level in the dry season will likely result in the opposite scenario [30].

Author Contributions

Conceived and designed the experiments: DZ SA. Performed the experiments: DZ DX HJ YC ML. Analyzed the data: DZ. Wrote the paper: DZ HJ.

References

1. Gullström M, Lundén B, Bodin M, Kangwe J, Ohman MC, et al. (2006) Assessment of changes in the seagrass-dominated submerged vegetation of tropical Chwaka Bay (Zanzibar) using satellite remote sensing. Estuar Coast Shelf Sci 67: 399–408.
2. Chambers P, Lacoul P, Murphy K, Thomaz S (2008) Global diversity of aquatic macrophytes in freshwater. Hydrobiologia 595: 9–26.
3. Wu G, de Leeuw J, Skidmore AK, Prins HHT, Liu Y (2007) Concurrent monitoring of vessels and water turbidity enhances the strength of evidence in remotely sensed dredging impact assessment. Water Res 41: 3271–3280.
4. Jin X (2008) The key scientific problems in lake eutrophication studies. Acta Scientiae Circumstantiae 28: 21–23.
5. Yang G, Ma R, Zhang L, Jiang J, Yao S, et al. (2010) Lake status, major problems and protection strategy in China. J Lake Sci 22: 799–810.
6. Orth RJ, Moore KA (1983) Chesapeake Bay: an unprecedented decline in submerged aquatic vegetation. Science 222: 51–53.
7. Jackson JBC, Kirby MX, Berger WH, Bjorndal KA, Botsford LW, et al. (2001) Historical overfishing and the recent collapse of coastal ecosystems. Science 293: 629.
8. Brisson J, Chazarenc F (2009) Maximizing pollutant removal in constructed wetlands: Should we pay more attention to macrophyte species selection? Sci Tot Environ 407: 3923–3930.
9. van der Heide T, van Nes EH, van Katwijk MM, Olff H, Smolders AJP (2011) Positive feedbacks in seagrass ecosystems–Evidence from large-scale empirical data. PLoS one 6: e16504.
10. Wang Q, Cheng S (2010) Review on phytoremediation of heavy metal polluted water by macrophytes. Environ Sci Technol 33: 96–102.
11. An S, Wang RR (2009) The human-induced driver on the development of Lake Taihu: In Lee, Xuhui ed. Lectures on China's Environment, Yale School of Forestry and Environmental Studies.
12. Le CF, Li Y, Zha Y, Sun D, Huang C, et al. (2009) A four-band semi-analytical model for estimating chlorophyll a in highly turbid lakes: The case of Taihu Lake, China. Remote Sens Environ 113: 1175–1182.
13. Zhao DH, Cai Y, Jiang H, Xu DL, Zhang WG, et al. (2011) Estimation of water clarity in Taihu Lake and surrounding rivers using Landsat imagery. Adv Water Resour 34: 165–173.
14. Gu X, Zhang S, Bai X, Hu W, Hu Y, et al. (2005) Evolution of community structure of aquatic macrophytes in East Taihu Lake and its wetlands. Acta Ecol Sinica 25: 1541–1548.
15. Liu WL, Hu WP, Chen YG, Gu XH, Hu ZX, et al. (2007) Temporal and spatial variation of aquatic macrophytes in west Taihu lake. Acta Ecol Sinica 27: 159–170.
16. Ma RH, Duan HT, Gu XH, Zhang SX (2008) Detecting aquatic vegetation changes in Taihu Lake, China using multi-temporal satellite imagery. Sensors 8: 3988–4005.
17. Deng JC, Chen Q, Zhai SJ, Yang XC, Han HJ, et al. (2008) Spatial distribution characteristics and environmental effect of N and P in water body of Taihu Lake. Environ Sci 29: 3382–3386.
18. Li Y, Acharya K, Yu Z (2011) Modeling impacts of Yangtze River water transfer on water ages in Lake Taihu, China. Ecol Eng 37: 325–334.
19. Hu W, Zhai S, Zhu Z, Han H (2008) Impacts of the Yangtze River water transfer on the restoration of Lake Taihu. Ecol Eng 34: 30–49.
20. Zhao DH, Jiang H, Yang TW, Cai Y, Xu DL, et al. (2012) Remote sensing of aquatic vegetation distribution in Taihu Lake using an improved classification tree with modified thresholds. J Environ Manage 95: 98–107.
21. Jiang H, Zhao D, Cai Y, An S (2012) A method for application of classification tree models to map aquatic vegetation using remotely sensed images from different sensors and dates. Sensors 12: 12437–12454.
22. Kloiber SM, Brezonik PL, Bauer ME (2002) Application of Landsat imagery to regional-scale assessments of lake clarity. Water Res 36: 4330–4340.
23. Rouse JW, Haas RH, Schell JA, Deering DW, Harlan JC (1974) Monitoring the vernal advancement of retrogradation of natural vegetation; Greenbelt, MD, USA, 309–317. Texas A & M University, Remote Sensing Center.
24. Xu H (2006) Modification of normalised difference water index (NDWI) to enhance open water features in remotely sensed imagery. Int J Remote Sens 27: 3025–3033.
25. Härmä P, Vepsäläinen J, Hannonen T, Pyhälahti T, Kämäri J, et al. (2001) Detection of water quality using simulated satellite data and semi-empirical algorithms in Finland. Sci Tot Environ 268: 107–121.
26. Yu HQ, Wang JC, Deng JT, Ao HY, Fang T (2009) Effect of dredging and re-establishment of aquatic macrophytes on organic matter in sediments of West Wuli Hu of Taihu. J Agro-Environ Sci 28: 1903–1907.
27. Lv ZL (2012) Practice and thoughts on comprehensive treatment of water pollution in Taihu Lake. J Hohai Univ 40: 123–128.
28. Qin B, Hu W, Chen W (2004) Process and mechanism of environmental changes of the Taihu Lake: Science Press, Beijing.
29. Wu Q, Hu Y, Li W, Chen K, Pan J (2000) Tendency of swampiness of East Taihu Lake and its causes. Acta Scientiae Circumstantiae 20: 275–279.
30. Zhao D, Jiang H, Cai Y, An S (2012) Artificial regulation of water level and its effect on aquatic macrophyte distribution in Taihu Lake. PLoS ONE 7: e44836.
31. Jin X, Yan C, Xu Q (2007) The community features of aquatic plants and its influence factors of lakeside zone in the north of Lake Taihu. J Lake Sci 19: 151–157.
32. Qing BQ (2009) Progress and prospect on the eco-environmental research of Lake Taihu. J Lake Sci 21: 445–455.
33. Xie P (2008) Historical development of cyanobacteria with bloom disaster in Lake Taihu :why did water pollution incident occur in Gonghu Waterworks in 2007? After 30 years, can we succeed to rescue Lake Taihu from bloom disaster? Beijing: Science Press.
34. Qin B (2009) Lake eutrophication: Control countermeasures and recycling exploitation. Ecol Eng 35: 1569–1573.

35. Yang M, Yu J, Li Z, Guo Z, Burch M, et al. (2008) Taihu Lake not to blame for Wuxi's woes. Science 319: 158.

36. Guo L (2007) Doing battle with the green monster of Taihu Lake. Science 317: 1166.

37. Scheffer M, Carpenter S, Foley JA, Folke C, Walker B (2001) Catastrophic shifts in ecosystems. Nature 413: 591–596.

38. Contamin R, Ellison AM (2009) Indicators of regime shifts in ecological systems: what do we need to know and when do we need to know it. Ecol Appl 19: 799–816.

39. Zimmer KD, Hanson MA, Herwig BR, Konsti ML (2009) Thresholds and stability of alternative regimes in shallow prairie–parkland lakes of central North America. Ecosystems 12: 843–852.

40. He J, Gu XH, Liu GF (2008) Aquatic macrophytes in East Lake Taihu and its interaction with water environment. J Lake Sci 20: 790–795.

13

Using LiDAR Data to Measure the 3D Green Biomass of Beijing Urban Forest in China

Cheng He[1], Matteo Convertino[2,3], Zhongke Feng[4], Siyu Zhang[1]*

1 Nanjing Forest Police College, Nanjing, China, **2** School of Public Health-Division of Environmental Health Sciences, University of Minnesota Twin-Cities, Minnesota, United States of America, **3** Institute on the Environment, University of Minnesota Twin-Cities, Minnesota, United States of America, **4** Institute of GIS, RS&GPS, College of Forestry, Beijing Forestry University, Beijing, China

Abstract

The purpose of the paper is to find a new approach to measure 3D green biomass of urban forest and to testify its precision. In this study, the 3D green biomass could be acquired on basis of a remote sensing inversion model in which each standing wood was first scanned by Terrestrial Laser Scanner to catch its point cloud data, then the point cloud picture was opened in a digital mapping data acquisition system to get the elevation in an independent coordinate, and at last the individual volume captured was associated with the remote sensing image in SPOT5(System Probatoired'Observation dela Tarre)by means of such tools as SPSS (Statistical Product and Service Solutions), GIS (Geographic Information System), RS (Remote Sensing) and spatial analysis software (FARO SCENE and Geomagic studio11). The results showed that the 3D green biomass of Beijing urban forest was 399.1295 million m^3, of which coniferous was 28.7871 million m^3 and broad-leaf was 370.3424 million m^3. The accuracy of 3D green biomass was over 85%, comparison with the values from 235 field sample data in a typical sampling way. This suggested that the precision done by the 3D forest green biomass based on the image in SPOT5 could meet requirements. This represents an improvement over the conventional method because it not only provides a basis to evalue indices of Beijing urban greenings, but also introduces a new technique to assess 3D green biomass in other cities.

Editor: Matteo Convertino, University of Florida, United States of America

Funding: This work was supported by the state forestry administration project "948" (2013-4-65); National Project "863" (2009AA12Z327, 2008AA121305-4); Beijing Natural Science Foundation (09D0297); National Natural Science Foundations of China (30872038, 30671696). The funders had no role in study design, data collection and analysis, decision to publish, or preparation of the manuscript.

Competing Interests: The authors have declared that no competing interests exist.

* E-mail: siyu85878817@163.com

Introduction

It is hard to estimate the amount of urban green space due to its characteristics of diverse structure and scattered distribution [1,2]. Therefore, 3D green biomass could be vividly defined as a 3D volume of the stems and leaves of all plants growing in the region [3], which can not only more accurately reflect the proportion of all vegetations in the region than such traditional 2D indicators as forest area and coverage, but also provide some ecological efficiency and green indexes suitable for the ecological assessment of the urban landscape, while playing an important role in planning the city and building the forestry discipline [4,5] .

In general, two methods can be used for the 3D Green Biomass estimation: the ground survey and the estimating with the remote sensing technology [6–9]. Actually, the ground survey is difficult to be done on a large scale even if the value can get a high accuracy because the green biomass can be acquired by the 3D volume measured by each tree's crown width and diameter at the breast height so that the systems need continuous field tests to be improved [10–21]. The remote sensing technology has been widely used in vegetation classification, forest fire monitoring and 3D green biomass measuring. Lv et al. [22] calculated the 3D volume by the crown height and width, how the crown width and height on the aerial photo was first measured. Cheng et al. [24] acquired the 3D volume by using the screen tracking vectorization

by means of GIS, who first made some field investigations to get the data of leaf area and vegetation coverage, and then combined the data with some high resolution images (IKONOS). Zhou et al. [25] succeeded in making an estimation of the 3D green biomass of Shanghai urban city forest by classifying the species on the aerial photos with high resolution, then simulating the stereo quantity by the plane quantity on the computer. Compared with the traditional ground work, the remote sensing techniques mentioned above have made greater improvements with such less cost as manpower, material and time, leading a fast calculation of 3D green biomass on a large scale. However, although the approaches can mitigate some problems, their precisions are not guaranteed because they are computered by the crown volume based on an appropriate formula suitable for the crown shape. Additionally, there are so timely and limited calculations that it is difficult to be widely promoted. Therefore, it is essential to find a more precise and generalized approach capable of achieving the 3D green biomass by means of remote sensing retrieval method today when the ecological environment is more and more important.

The spatial distribution of leaf area determines resource capture and canopy exchanges with the atmosphere. It is generally tedious and time-consuming to measure the spatial distribution of leaf area, even when 3D digital techniques are employed [21–26]. Many tree models, like light models, therefore choose individual canopies as a volume filled with leaf area. Simple shapes like ellipsoids or frustums have been extensively used to model tree shape [26–31]. More sophisticated parametric envelopes have been proposed by Cescatti (1997) to extend the range of modeled canopy shapes, and non-parametric envelopes like polygonal envelopes are expected to fit any tree shape [32]. However, all the envelopes showed that different shape models for the same tree may lead to large differences in crown volume [33–36]. None of these methods for tree crown volume estimation has been evaluated by comparison with direct measurements. Moreover, neither method accounts for the fractal nature of plants, because only one value of crown volume is computed (i.e., at the observation scale) and changes in crown volume with measurement scale are ignored.

On the other hand, airborne laser scanners can be used to acquire vertical and horizontal forest structure in detail as scanning targets with laser pulses. In particular, such vertical measurements enable the prediction of forest biomass and carbon storage. Furthermore, laser sensors can be used to accurately measure topographical information, the physical properties of a forest and other information. Therefore, ALS(Auto Scanner Laser System)has been recognized as a more efficient and precise instrument than field surveys and optical remote sensing techniques [37–43]. Since the early to mid 1980s, several studies using full waveform sensors have been performed for forest inventory, merchantable timber volume estimation [44], and forest canopy characterization [45–47]. Recently, several researchers have applied discretely emitted laser pulses for the individual- and stand-level tree height estimation [47–53]and height-based timber volume estimates [44–46].However, there is currently no effective approach in the methods mentioned above to resolve the problems on how to calculate the canopy volume accurately and quickly, especially for the volume on a large scale.

In this paper, the 3D green biomass of Beijing urban forest was calculated and analyzed based on the remote sensing retrieval model. This approach, in which to obtain the point cloud data of the crown, 30 different trees in size of each species from over 30 common tree species in Beijing urban area(like *arborvitae, cedar, pine, cypress, ginkgo, poplar, sycamore, willow tree, Sophora japonica, Ailanthus altissima, Koelreuteria, ash, maple, cork oak* and other about ten common tree species)were chosen for scanning with laser scanner FARO(FARO develops and markets portable CMMs (coordinate measuring machines) and 3D imaging devices to solve dimensional metrology problems.), was designed to calculate the 3D green biomass of single wood by CASS (CAD AID SURVEY SYSTEM) software(CASS mapping equipment in the South in AutoCAD 2004 to develop a new generation of digital terrain cadastral mapping software)which has been patented in China in 2011 [24–28], associated with the remote sensing image in SPOT5, and by means of SPSS, GIS, RS and other spatial analysis tools [46–49].

Materials and Methods

Ethics Statement

No specific permits were required for the described field studies, since the trees chosen in the study are owned and managed by the state including the sites for our sampling are not privately-owned or protected in any way and specific permission for non-profit research, therefore, is not required. The field studies were not involved in endangered or protected species in this area.

Data Acquisition

The study site was located in Beijing (39°26'40"N to 41°03'05"N, 115°25'45"E to 117°30'20"E), The SPOT5 remote

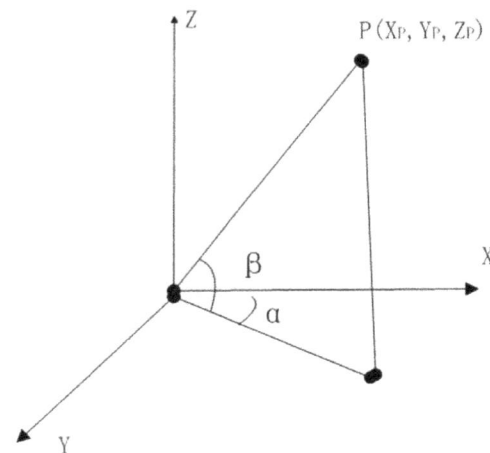

Figure 1. Schematic diagram of scanning point coordinate calculation. X_P——P abscissa values; Y_P——P ordinate values; Z_P——P Elevation Value; α——included angle of P was perpendicular to the YZ plane with X axis; β——included angle of P with XY plane.

sensing image data from four views in summer of 2009 in Beijing were selected in this paper, including resolution of 10 m multi-spectral band and 2.5 m panchromatic band. Besides, there is much supporting information, such as Beijing 1:250,000 administrative map, traffic road map, water maps, maps of forest resources, Worldview remote sensing images of 2008 in Beijing City, the latest Goolge Earth data and so on. The total area of Beijing is 16,800 km^2, of which mountainous areas occupy about 62% and plains take up the rest. Forestry areas is 104,609,637 m, including 65,891,408 m forestation-suitable, 557,631 m open forest, 30,580,843 m shrub, 2,110,388 m young forest and 5,469,367 m other forest. Geographically, Beijing is a transitional zone for southern and northern plants of China. Influenced by warm-temperate continental monsoon climate, its sub-natural flora generally belongs to warm temperate zone deciduous broad-leaved forest and coniferous forest, but due to serious destruction in early years, currently there are only small area forests with sporadically scattered trees. In some higher mountainous planted forest, the Larix principis-rupprechtii forests were originally

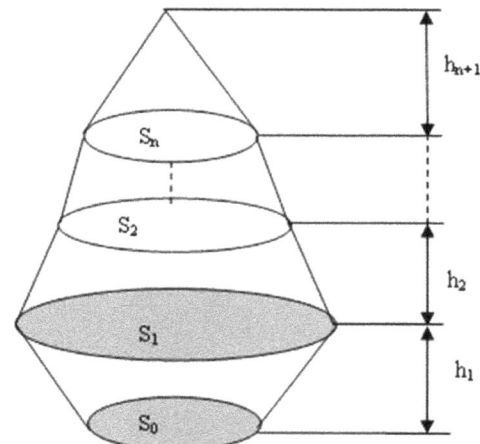

Figure 2. Mimic diagram of three-dimensional laser scanning method for measuring the crown volume.

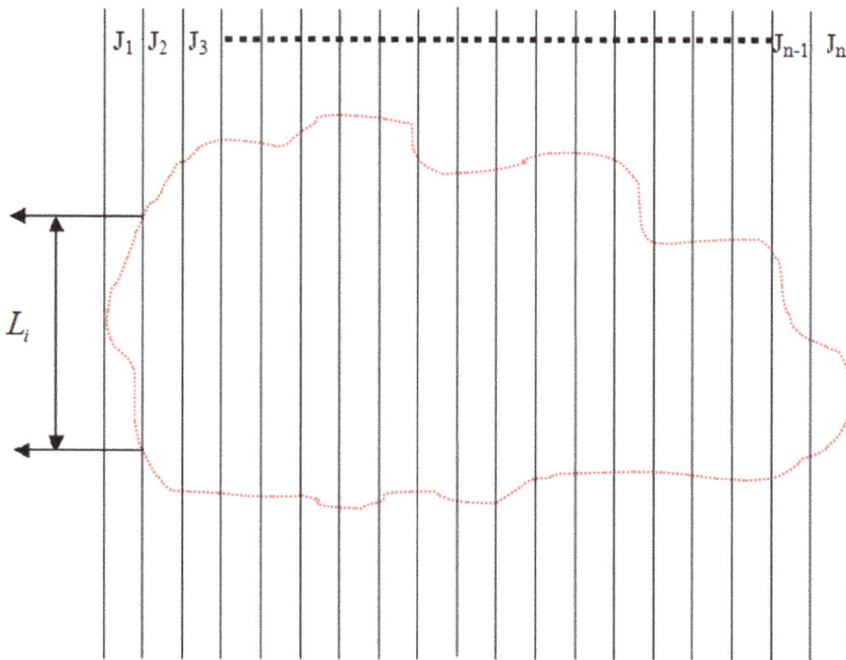

Figure 3. Mimic diagram of measuring crown cross-sectional area.

distributed as the sub-natural compositions, however, shrub and secondary forests are the most widely distributed zonal vegetations of Beijing, such as betula, populus, quercus.

In this study, at first over 1000 trees (30 species and over 30 trees in each species), like pinus bungeana, cedar and pinus tabulaeformis, planting in campuses, parks, roadsides, housing estates and mountain forests in Beijing city, were sampled and scanned systematically and representatively with terrestrial 3D laser scanner FARO LS880, and then their 3D coordinates were measured by means of some instruments like Trimble GPS(Global Positioning System) and Topcon total station. To acquire the point cloud data, we put up a platform, where some parameters of the scanner should be first set. The horizontal direction was 360°, vertical direction was 155° (from −90 to 65) and a resolution of 2 mm in 10 m. Next, three stations should be set up according to

its growth and relative terrain of the tree to be measured, because they can generally constitute an equilateral triangle in theory whose included angle is 120°. Additionally, at least three public spheres should be placed on a non-straight line between two stations and all could be scanned by the above 3 stations without any shelter. The target paper, which only acted as a reference for scanning, were finally posted on the trunk with a height of about 1.3 m above ground northwards (Figure 1). Only when all trees were scanned, could the target ball be removed, otherwise, they had to be re-scanned. To get the volume of the sampling forest, each single tree had to be scanned in about 10 minutes. For field application, scanning for single tree should be done in open forest because too many close planting trees with branches and leaves will lead to serious shadows. Meanwhile, some pictures of the trees

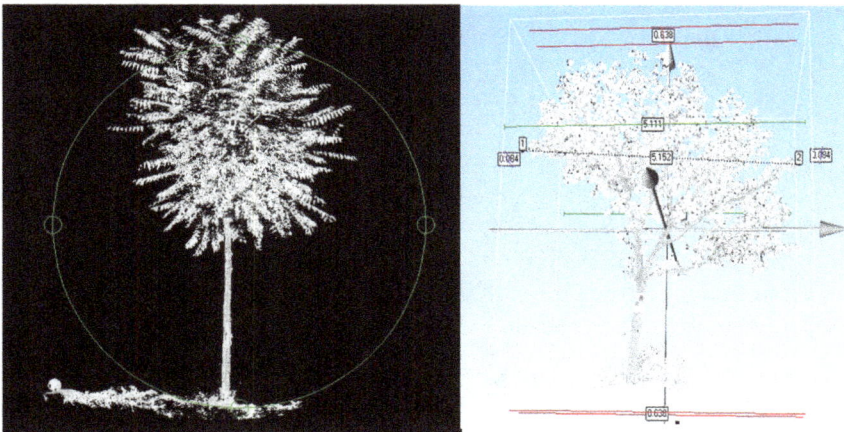

Figure 4. 3D point cloud data.

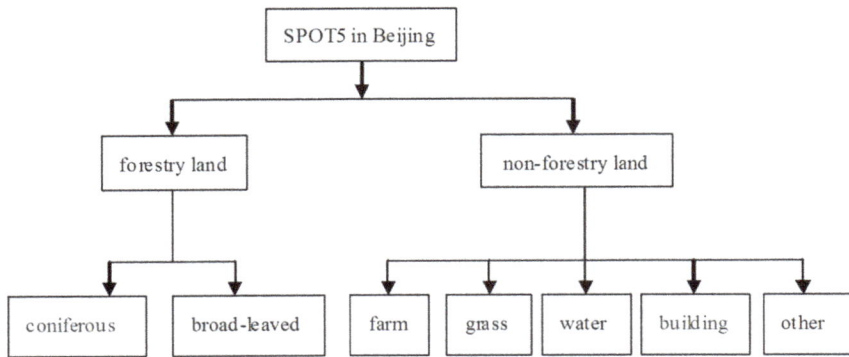

Figure 5. Classification flow diagram of the image information.

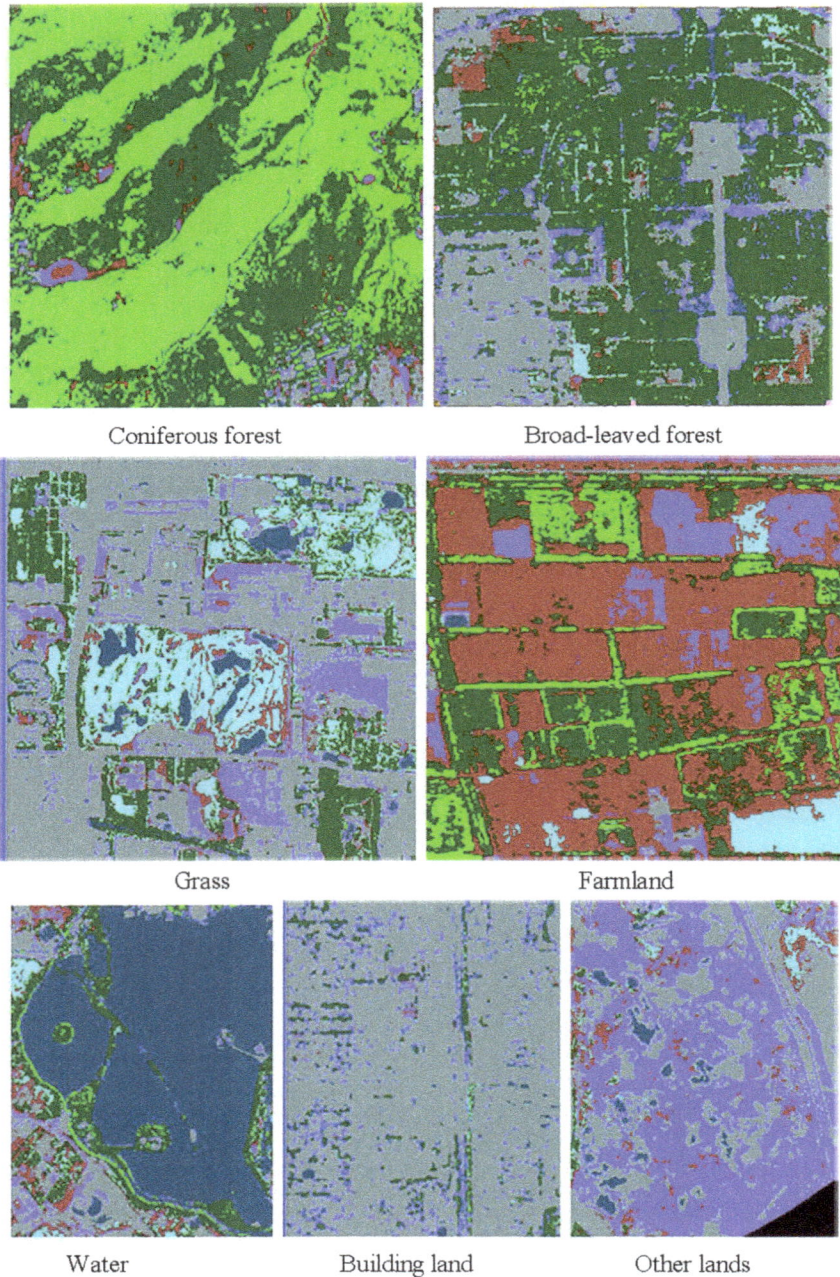

Coniferous forest

Broad-leaved forest

Grass

Farmland

Water

Building land

Other lands

Figure 6. Interpretation signs. Coniferous forest, Broad-leaved forest, Grass, Farmland, Water, Building land and other lands are signed in the figure.

Table 1. Table of accuracy totals of classification.

	Reference Totals	Classified Totals	Number Correct	Producer's Accuracy/%	User's Accuracy/%	Kappa
Coniferous forest	14	7	6	42.86	85.71	0.834
Broadleaf forest	15	15	11	73.33	73.33	0.686
Grass	5	5	4	80.00	80.00	0.790
Farmland	11	15	10	90.91	66.67	0.626
Waters	2	2	2	100	100	1
Construction lands	39	39	37	94.87	94.87	0.916
Other lands	14	17	13	92.86	76.47	0.726
$\sum$	100	100	83			

to be measured as reference for post-processing point cloud trees should be taken, where the scanner was placed.

To achieve the crown 3D green biomass of single tree, we made a calculation on point cloud data of the crown using a CASS software which could computer a volume with digital elevation after a secondary development of CAD (Computer Aided Design). In processing, some pictures with different point clouds of the standing woods were first pieced together at a coordinate system by coordinate match by means of the 3D scanner's software, and then a 3D model of standing wood was developed and saved as .dxf which could be discerned by the digital mapping system after the data were pre-processed and extracted. Next, the point cloud picture captured by the scanner was opened in a digital mapping data acquisition system to set up an independent coordinate system and the elevation could be extracted (Figure 2).The area whose volume should be calculated was outlined with a closed compound line instead of fitting curve in a mesh with 3D triangles, because a fitted curve would be replaced by the broken line so that the precision of the results could be affected. Finally, the point cloud volume was calculated with the help of DTM (Digital Terrain Model) method of the system shown as follow: the points on the crown surface collected at the same height were linked with

the smooth curve to form some contour lines which were then separated into some grids with a regular 2 cm cell size(Certainly, the length can be divided into any other size, but 2 cm here was just for convenience), so that the topmost elevation in each grid could be estimated by linear interpolation and marked at its top right, where the designed elevation was set 0; next, we calculated the volumes with some formulas shown as follows:(1) $V_{cornerpoint} = h*1/4 \, S_{grid}$; (2) $V_{edgepoint} = h*2/4 S_{grid}$;(3) $V_{turningpoint} = h*3/4 \, S_{grid}$;(4) $V_{midpoint} = h*S_{grid}$. (5)$V_{crown} = nV_{grid}$, where h is a canopy height and n is the number of all grids (Figure 2 and 3).

The point cloud data are shown in Figure 4. In remote sensing SPOT5 data, we performed a spatial resolution of 2.5 m panchromatic and 10 m multi-spectral bands using artificial visual interpretation method to extract the vegetation classification information. And we added SPOT5 image gray value, its remote sensing factors and GIS factors into the independent values.

Remote sensing image classification and information extraction

The green biomass retrieval model can be developed by means of various information, methods, monitoring and manual interpretation. In supervised method, at first, some known characteristic parameters are extracted from the spectral features based on samples from training areas, and then other unspecified parameters which can be extracted and classified in sorts from images are analyzed according to prior probability of different kinds of objects. The ground information is demonstrated by pixels in image, while the pixel information is expressed by spectral characteristics of different image bands. Due to the different spatial resolutions and complex grounds, some mixed pixels can appear in images which will result in different objects with the same spectra characteristics or same spectrum with different objects. Therefore, the accuracy will be confined if the urban vegetation is classified only by supervised classification because misclassification and loss classification can arise in some specific classification and extraction. The present vegetation is classified by means of visual interpretation or computer-aided, however, it is of low automation, long hours and low efficiency or likely to be worse because of the unskilled labors or less educated operators (Figure 5 and Figure 6).

In this study, the first information was captured based on the spatial structure feature and spectral brightness of the pixels in different bands of the SPOT5 remote sensing image in Beijing urban regions in 2010. The second was characteristics of landscape, landform and forest resource distribution. The third

Table 2. Factors loading rotation matrix of varimax.

Variables	Remote sensing	principal components	
		1	2
1	B1	−0.716	0.696
2	B2	−0.751	0.658
3	B3	−0.741	0.670
4	SWIR	−0.726	0.686
5	NDVI	0.745	0.667
6	SAVI	0.744	0.668
7	MSAVI	0.742	0.670

Note: B1, B2 are visible bands, where B1 can detect absorption and reflectance of plant green hormone, and B2 belongs to the red light zone capable of distinguishing the color of different types of vegetation from the color difference; Where B3 is near-infrared bands, which can reflect the sensitivity of plants to chlorophyll by the correlation between some acquired strong information and factors like leaf area index and biomass; SWIR (short-wave (length) infrared (band)); NDVI (Normalized Difference Vegetation Index); SAVI (Soil-Adjusted Vegetation Index); MSAVI (Modified Soil-Adjusted Vegetation Index).

Table 3. Correlation coefficient of conifer volume and RS factors.

Variables	Band and combinations	Correlation coefficient with 3D green biomass
1	B1	0.92917
2	B2	0.93602
3	B3	0.93711
4	SWIR	0.93102

was about such maps as current vector, forest distribution, greening investigation data, contour and traffic each year.

Beijing is so large that a lot of random points must appear in the accuracy assessment of classification, thereby they should be chosen at absolute random instead of any mandatory rule in practice and the regional classification maps were assessed under the Accuracy Assessment module of ERDAS IMAGINE software(Table 1).

For classification of remote sensing images in some large areas, its accuracy has been able to meet the needs of the latter analysis and assessment.

Modeling

The Principal Component Analysis(PCA) with varimax rotation was used for factor analysis in this study. As two PCs can be shown in Table 2, the loading values of each independent variable in either the first PC or the second were little changed, indicating that there was only a small amount of correlation between grouping variables. If the eight variables were forcibly added into the model, it could not be guaranteed to high accuracy that some uncorrelated variables to 3D green biomass could be directly regressed by the model. So the factor involved into modeling could be acquired by automatic filtration with the help of stepwise regression modeling.

Based on the results interpreted by the remote sensing images and the green biomass data measured in the plots, the relative remote sensing factors and GIS factors were chosen as the independent values of a model by means of the spatial relationship of RS image rectified by GPS to ground samples, where the remote sensing factors were related to the image gray value in SPOT5 and its linear and nonlinear combination etc, and GIS factors included slope, elevation, aspect etc.(Some factors were not involved in modeling as independent variables like slope and aspect since the subjects in the study were mainly located in Beijing urban areas and topographic relief was not much changed.) The 3D green biomass was repeatedly regressed to model the conifer and broadleaf tree respectively on basis of the correlation between the factors and the green biomass values observed in the plots.

The 3D green biomass model of conifer is shown as follow:

$$V = 1.149 - 0.096\,B1 - 0.1\,B2 + 0.199\,SWIR \qquad (1)$$

The 3D green biomass model of broadleaf tree is shown as follow:

$$V = -40.290 - 0.236\,B1 - 0.188\,B2 + 0.487\,SWIR \qquad (2)$$

Where B1, B2 are visible bands, where B1 can detect absorption and reflectance of plant green hormone, and B2 belongs to the red light zone capable of distinguishing the color of different types of vegetation from the color difference; SWIR band which can well reflect the water feature in the plant leaves is a shortwave infrared zone, by which it is easy to make the identification and classification of vegetation, soil, and water.

Based on validation and accuracy assessment on the model and comparison with the actual measured data in the ground, the data for modeling should be systemized, and then 150 samples were added into modeling and 100 checking samples were chosen to test their accuracy.

It is possible to develop a reliable, scientific and operable model. Analysis on the correlation coefficient between the gray values of four bands in SPOT5 and the 3D green biomass of conifer and broadleaf tree by using EXCEL software, the results were shown in Table 3 and 4.

All tests were performed using version 18.0 of the SPSS software (SPSS Inc., Chicago, IL).

Results and Analysis

Classification results and analysis

All the 3D green biomass of 1015 trees scanned by 3D laser were added in the calibrated remote sensing image so that we could get the gray value at each sampling point, and then by means of remote sensing image Worldview 1 and Google earth with the resolution of 0.5 m, each sampling tree was soon located.

Table 4. Correlation coefficient of broadleaf tree volume and RS factors.

Variables	Band and combinations	Correlation coefficient with 3D green biomass
1	B1	0.86879
2	B2	0.80498
3	B3	0.805022
4	SWIR	0.87112

Table 5. Checking table of model precision.

Species	Correlation Coefficient	Square of Correlation Coefficient	Revised Square	Estimation Error	Statistical Analysis			
					F value	Degree of Freedom1	Degree of Freedom2	Significant Change
Conifer tree	0.951	0.904	0.902	2.21	411.651	3	147	0.0000
Broadleaf tree	0.894	0.799	0.792	7.67	622.611	3	478	0.0000

Based on supervised classification and visual interpretation, the total Beijing urban forest area of conifer and broadleaf tree was 275.08 km^2, of which the area of conifer tree was 54.47 km^2, while that of broadleaf tree was 220.61 km^2. The correlation coefficient between the 3D biomass of conifer tree and RS factors was over 0.9 and that of broadleaf tree was over 0.8, thus revealing that the linear correlation was very close [44–47], thereby a model of 3D green biomass could be developed with the direct remote sensing data by the multiple linear regression.

Model checking

After sampling and drying every organ of the tree, we converted them and got the biomass. Meanwhile, the stem and volume were accurately measured by means of sectional measurement (Table 1), As can be seen from Table 1, the overall classification accuracy reaches more than 80%, and Kappa factor of 0.780, exceeding the requirements of 0.7. After inducing and analyzing the eight variable factors of B1, B2, B3, SWIR(short-wave (length) infrared (band)), NDVI (Normalized Difference Vegetation Index), SAVI (Soil-Adjusted Vegetation Index) and RVI (Ratio Vegetation Index), we got the regression equations of conifer and broadleaf tree shown as Eq.1 and Eq.2. The accuracy of the model was tested with the correlation coefficient and F test to evaluate (Table 5).

It is known that the correlation coefficient of the multiple regression model of Beijing urban 3D green biomass was high(over 0.89) which correlated well with the 3D green biomass, remote sensing factors and GIS factors, and F values also showed that the significant differences existed in the model. The histogram of regression standardized residual in Figure 7 illustrated that it was an ideal bell-shaped normal distribution, and a better diagonal distribution was illustrated in the cumulative probability distribution (Figure 7 and Figure 8).

Accuracy analysis

We performed 241 samples to identify whether the actual measured values and estimated values significant differences existed among the treatments, and when we did, we used the (3D Green Biomass)TGB test to determine which specific combinations of values differed significantly (Table 6)

(1) The residual standard deviation is acquired by formula $S = \sqrt{\dfrac{\sum (TGB_s - TGB_g)^2}{n-2}}$, where S is residual standard deviation, TGB_s is the actual measured value of 3D green biomass, TGB_g is the estimated value of 3D green biomass and n is the number of sampling plots for accuracy test.

(2) The Standard error is acquired by formula $\delta_x = \dfrac{S}{\sqrt{n}}$, where δ_x is Standard error.

(3) The absolute error limit of 95% and 99% is calculated by formula $\Delta = \delta_x t^\alpha_{n-2}$, where Δ is the absolute error limit which acquired by t value distribution table difference.

(4) The relative error limit of 95% and 99% is calculated by formula $E = \dfrac{\Delta}{\bar{X}}$, where E is the relative error limit, and $\bar{X} = \dfrac{\sum TGB_g}{n}$ is the mean value of TGB.

(5) Precision C is calculated by formula $C = 100\% - E$.

The monitoring data in Table 6 demonstrated that the precision of the 3D green biomass of the sample based on the SPOT5 was over 85%, indicating that it could fully meet the requirements.

Dependent Variable: V

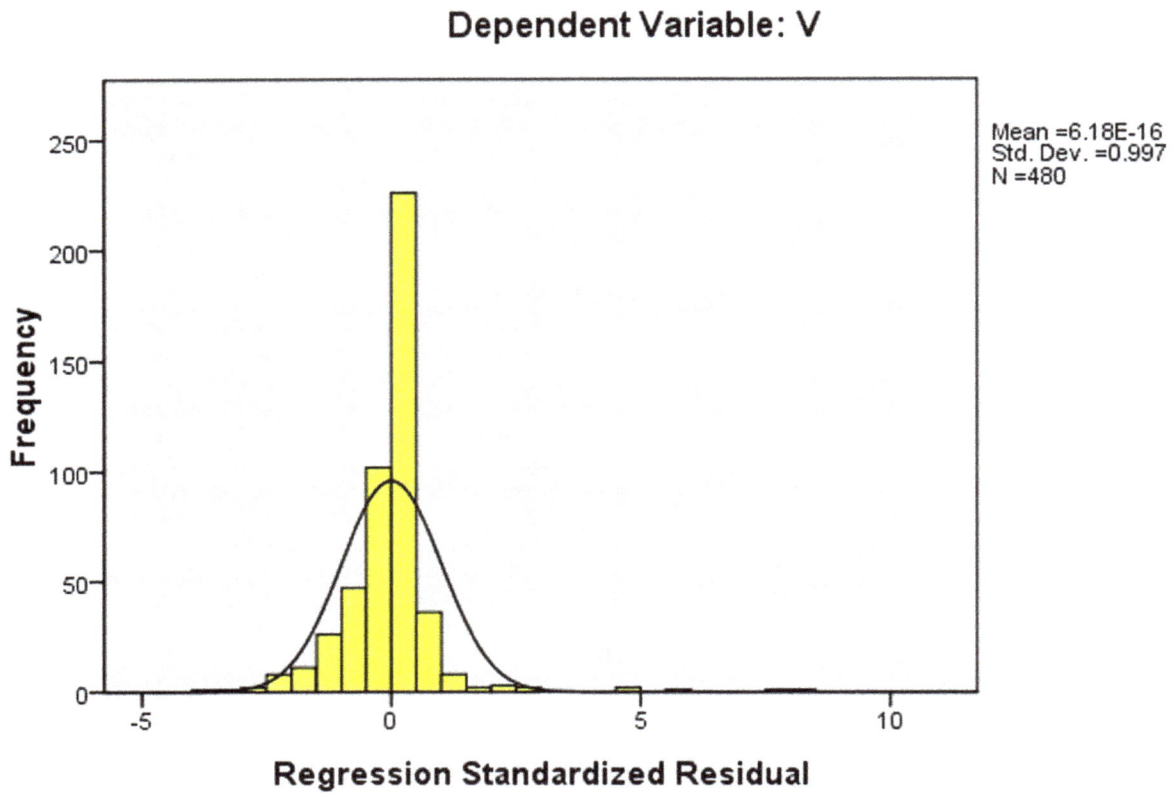

Figure 7. Histogram of regression standardized residual.

Dependent Variable: V

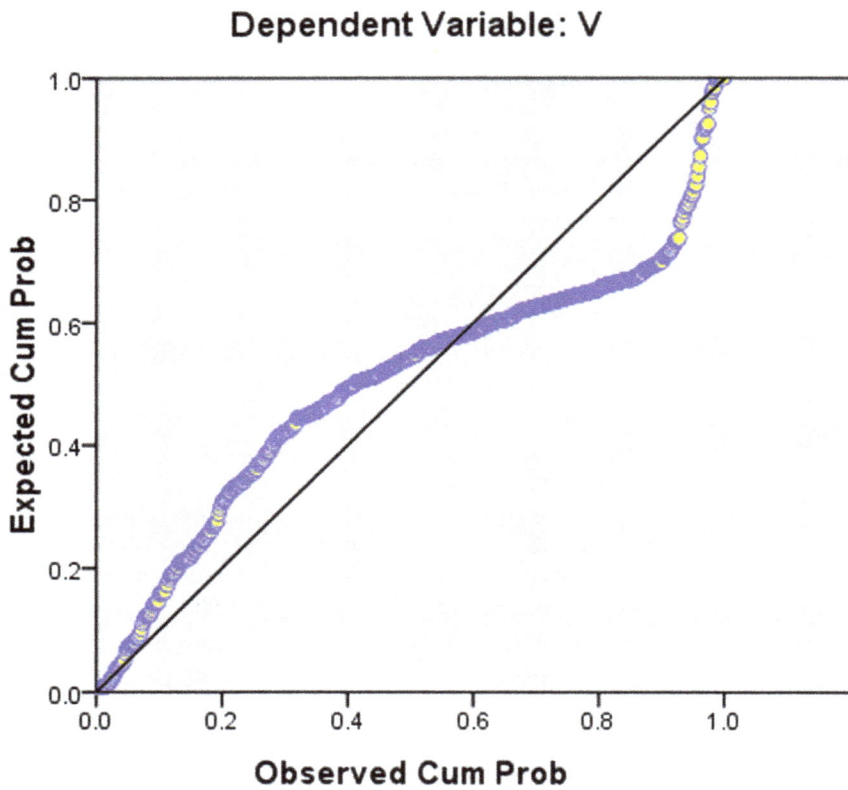

Figure 8. Cumulative probability distribution map.

Table 6. Testing Results of Precision Ratio of TGB Based-on SPOT5 Image.

	Confidence Level	
	a = 0.05	a = 0.01
Sample Number(n)	241	
Total of Actual Measured Values $\sum$	33687.621	
TGB_s Average $\bar{X}$	140.127	
Total of Estimated Values $\sum$	34788.741	
TGB_g Average $\bar{X}$	148.013	
$\sum(TGB_s - TGB_g)$	−1874.453	
$\sum(TGB_s - TGB_g)^2$	3513600.322	
Standard Deviation	122.804	
Standard Error δ_x	8.126	
t_{n-2}^a	1.989	2.620
Absolute Error Δ	15.873	20.958
Relative Error E/%	10.695	14.204
Precision C/%	89.302	85.796

In all, based on the remote sensing image gray values extracted from the model by means of ArcGIS 9.3 and all statistical data calculated on the remote sensing retrieval model of the 3D green biomass, the green biomass in each region of Beijing could be determined. The results shows that the 3D green biomass of Beijing urban forest was 399.1295 million m^3, of which coniferous was 28.7871 million m^3 and broad-leaf was 370.3424 million m^3.

Discussion

As the above statistical data demonstrates, the case study described in this paper confirms that this is possible. Compared with the traditional 2D green indices in forest area, 3D green biomass represents an improvement over the conventional method because 3D index demonstrates that it can both accurately reflect the volume of the vegetation in the region, and scientifically assess the ecological environment of the city, while providing an important basis for urban planning and forest sciences development. The new approach of 3D green biomass illustrates that it is performed more accurately, efficiently, easily, and rapidly than the conventional, because the 3D green biomass not only involves the processing of remote sensing image including identification and classification, but also includes the investigation of forest vegetation on the ground, especially where the same species make great differences in different climatic zones.

The total 3D green biomass of Beijing urban areas can be acquired by the model and its grade distribution of biomass per unit can be also computerized in ArcGIS 9.3 shown in figure 9 [30,31].

Beijing is an important capital urban district of China. In order to fully implement the strategy of "Humanistic Beijing, Scientific Beijing, Green Beijing" and promote the development of urban eco-environment, the study can provide some materials for references in urban green lands.

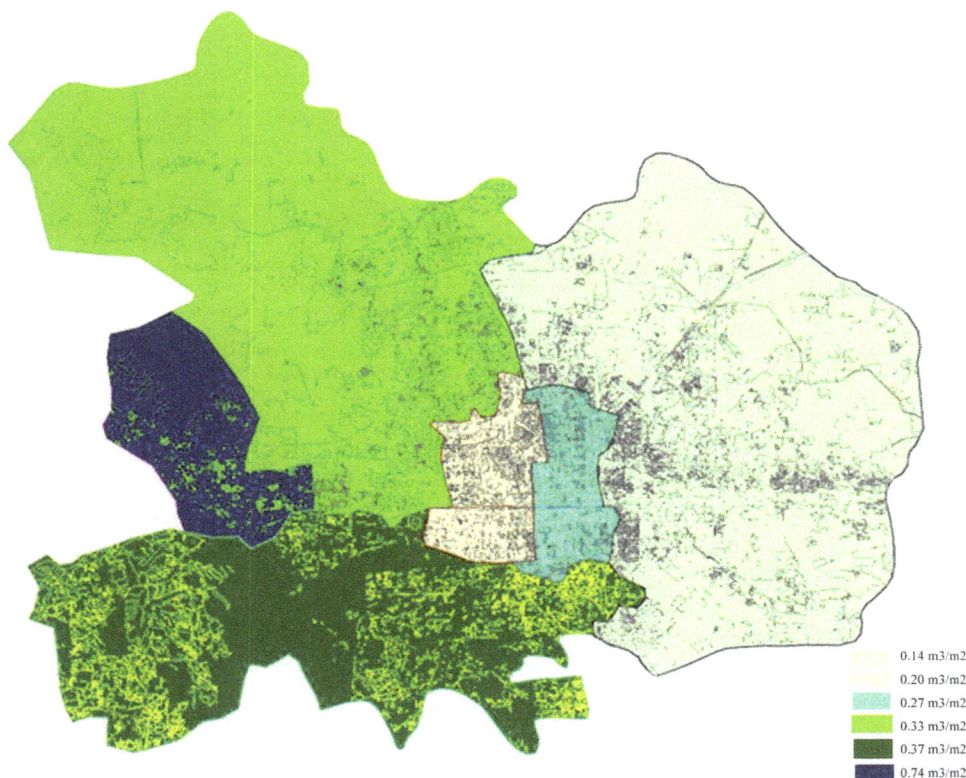

	0.14 m3/m2
	0.20 m3/m2
	0.27 m3/m2
	0.33 m3/m2
	0.37 m3/m2
	0.74 m3/m2

Figure 9. Grade distribution of 3D green biomass per unit area in Beijing.

The crown form is one of non-negligible factors in the calculation of biomass with the help of 3D laser scanner, for example, the shape of the crown will shake in the wind when the scanned point cloud may not fully reflect the true state of the involved crown volume. Therefore, the wood should be scanned at rest to ensure the accuracy of the volume. Meanwhile, only the biomass from the upper half of crowns were involved while the under part, shrubs and herbs were not to be considered. In the future we will focus on the relative research in the field.

The green biomass of 3D in Beijing was estimated by the interpretation and classification of remote sensing data and modeling. In this paper, the urban vegetation was extracted by artificial visual interpretation and computer-aided, or strictly speaking it was still semi-automated and time and labor consuming. The extraction of vegetation is still the hot spot researchers interested in. The resolution in SPOT5 remote sensing image data was not high so that the crown width was greater than 2.5 m like a pixel. However, the gray on the corresponding band can produce the deviation if it is less than a pixel. In all, the biomass acquired in the study work as the exploratory research and reference and more remote sensing images with higher resolution will be used later to study.

Author Contributions

Conceived and designed the experiments: CH SYZ ZKF. Performed the experiments: CH. Analyzed the data: CH ZKF. Contributed reagents/materials/analysis tools: CH SYZ ZKF. Wrote the paper: CH MC.

References

1. Pregitzer KS, Euskirchen ES (2004) Carbon cycling and storage in world forests: biomass patterns related to forest age. Global Change Biology 10: 2052–2077.
2. Running SW (2008) Climate change - Ecosystem disturbance, carbon, and climate. Science 321: 652–653.
3. Yamada T, Zuidema PA, Itoh A, Yamakura T, Ohkubo T, et al. (2007) Strong habitat preference of a tropical rain forest tree does not imply large differences in population dynamics across habitats. Journal of Ecology 95: 332–342.
4. Condit R, Ashton PS, Baker P, Bunyavejchewin S, Gunatilleke S, et al. (2000) Spatial patterns in the distribution of tropical tree species. Science 288: 1414–1418.
5. Pan Y, Birdsey RA, Fang J, Houghton R, Kauppi PE, et al. (2011) A large and persistent carbon sink in the world's forests. Science 333: 998–993.
6. Fang J, Wang G, Liu G, Xu S (1998) Forest biomass of China: An estimate based on the biomass-volume relationship. Ecological Applications 8: 1084–1091.
7. Shi L, Zhao S, Tang Z, Fang J (2011) The changes in China's forests: An analysis using the forest identity. PLoS ONE 6: e20778.
8. Zhang J, Ge Y, Chang J, Jiang B, Jiang H, et al. (2007) Carbon storage by ecological service forests in Zhejiang Province, subtropical China. Forest Ecology and Management 245: 64–75.
9. Keith H, Mackey B, Lindenmayer D (2009) Re-evaluation of forest biomass carbon stocks and lessons from the world's most carbon-dense forests. Proceedings of the National Academy of Sciences of the United States of America 106: 11635–11640.
10. Yang T, Song K, Da L, Li X, Wu J (2010) The biomass and aboveground net primary productivity of Schima superba-Castanopsis carlesii forests in east China. Science in China (Series C: Life Sciences) 53: 811–821.
11. Post WM, Kwon KC (2000) Soil carbon sequestration and land–use change: Processes and potential. Global Change Biology 6: 317–328.
12. Ruiz-Jaen MC, Potvin C (2011) Can we predict carbon stocks in tropical ecosystems from tree diversity? Comparing species and functional diversity in a plantation and a natural forest. New Phytologist 189: 978–987.
13. Kira T (1991) Forest ecosystems of east and southeast-Asia in a global perspective. Ecological Research 6: 185–200.
14. Du YJ, Mi XC, Liu XJ, Ma KP (2012) The effects of ice storm on seed rain and seed limitation in an evergreen broad-leaved forest in east China. Acta Oecologica 39: 87–93.
15. Zhong Z (1987) The typical subtropical evergreen broadleaved forest of China. Journal of Southwest China Normal University 3: 109–121.
16. Wang X, Kent M, Fang X (2007) Evergreen broad-leaved forest in Eastern China: Its ecology and conservation and the importance of resprouting in forest restoration. Forest Ecology and Management 245: 76–87.
17. Piao S, Fang J, Ciais P, Peylin P, Huang Y, et al. (2009) The carbon balance of terrestrial ecosystems in China. Nature 458: 1009–1013.
18. Malhi Y, Wood D, Baker TR, Wright J, Phillips OL, et al. (2006) The regional variation of aboveground live biomass in old-growth Amazonian forests. Global Change Biology 12: 1107–1138.
19. Stegen JC, Swenson NG, Enquist BJ, White EP, Phillips OL, et al. (2011) Variation in aboveground forest biomass across broad climatic gradients. Global Ecology and Biogeography 20: 744–754.
20. Baraloto C, Rabaud S, Molto Q, Blanc L, Fortunel C, et al. (2011) Disentangling stand and environmental correlates of aboveground biomass in Amazonian forests. Global Change Biology 17: 2677–2688.
21. Baker TR, Phillips OL, Malhi Y, Almeida S, Arroyo L, et al. (2004) Variation in wood density determines spatial patterns in Amazonian forest biomass. Global Change Biology 10: 545–562.
22. Lv ME, Pu YX, Huang XY (2000) Application of Remote Sensing on Urban Green, Chinese Landscape Architecture 16(5): 41–43.
23. Zhang LP, Zheng LF, Tong QX (1997) The Estimation of Vegetation Variables Based on High Resolution Spectra, Journal of Remote Sensing 1(2): 111–114.

24. Chen F, Zhou ZX, Wang PC, Li HF, Zhong YF (2006) Green space vegetation quantity in workshop area of Wuhan Iron and Steel Company, Chinese Journal of Applied Ecology 17(4):592–596.
25. Zhou JH, Sun TZ (1995) Study on Remote Sensing Model of Three-Dimensional Green Biomass and the Estimation of Environmental Benefits of Greenery, Remote Sensing of Environment 10(3):162–174.
26. Kra'l K, Jani'k D, Vrska T, Adam D, Hort L, et al. (2010) Local variability of stand structural features in beech dominated natural forests of Central Europe: Implications for sampling. Forest Ecology and Management 260: 2196–2203.
27. Brown S (2002) Measuring carbon in forests: current status and future challenges. Environmental Pollution 116: 363–372.
28. Chen B, Mi X, Fang T, Chen L, Ren H, et al. (2009) Gutianshan forest dynamic plot: Tree species and their distribution patterns. Beijing: China Forestry Publishing House.
29. Legendre P, Mi X, Ren H, Ma K, Yu M, et al. (2009) Partitioning beta diversity in a subtropical broad-leaved forest of China. Ecology 90: 663–674.
30. Yu M, Hu Z, Yu J, Ding B, Fang T (2001) Forest vegetation types in Gutianshan National Natural Reserve in Zhejiang. Journal of Zhejiang University (Agriculture and Life Science) 27: 375–380. (In Chinese).
31. Hu Z, Yu M, Suo F, Wu F, Liu Q (2008) Species diversity characteristics of coniferous broad-leaved forest in Gutian Moutain National Nature Reserve, Zhejiang province. Ecology and Environment 17: 1961–1964. (In Chinese).
32. Du G, Hong L, Yao G (1987) Estimate and analysis the aboveground biomass of a secondary evergreen broad-leaved forest in Northwest of Zhejiang. Journal of Zhejiang Forestry Science and Technology 7: 5–12. (In Chinese).
33. Chen W (2000) Study on the net productivity dynamic changes of the aboveground portion of Alniphyllum fortunei plantation. Journal of Fujian Forestry and Technology 27: 31–34. (In Chinese).
34. Chen Q, Shen Q (1993) Studies on the biomass models of the tree stratum of secondary Cyclobalanopsis glauca forest in Zhejiang. Acta Phytoecologica Sinica 17: 38–47. (In Chinese).
35. Caspersen JP, Pacala SW (2000) Successional diversity and forest ecosystem function. Ecological Research 16: 895–903.
36. Cardinale BJ (2011) Biodiversity improves water quality through niche partitioning. Nature 472: 86–89.
37. Loreau ML, Mouquet N, Gonzalez A (2003) Biodiversity as spatial insurance in hetegrogeneous landscapes. Proceedings of the National Academy of Sciences of the United States of America 100: 12765–12770.
38. Canadell JG, Raupach MR (2008) Managing forests for climate change mitigation. Science 320: 1456–1457.
39. Imai N, Samejima H, Langner A, Ong RC, Kita S, et al. (2009) Co-Benefits of sustainable forest management in biodiversity conservation and carbon sequestration. PLoS ONE 4: e8267.
40. Mascaro J, Asner GP, Muller-Landau HC, van Breugel M, Hall J, et al. (2011) Controls over aboveground forest carbon density on Barro Colorado Island, Panama. Biogeosciences 8: 1615–1629.
41. Laurance WF, Fearnside PM, Laurance SG, Delamonica P, Lovejoy TE, et al. (1999) Relationship between soils and Amazon forest biomass: A landscape-scale study. Forest Ecology and Management 118: 127–138.
42. Tateno R, Takeda H (2003) Forest structure and tree species distribution in relation to topography-mediated heterogeneity of soil nitrogen and light at the forest floor. Ecological Research 18: 559–571.
43. Paoli GD, Curran LM, Slik JWF (2008) Soil nutrients affect spatial patterns of aboveground biomass and emergent tree density in southwestern Borneo. Oecologia 155: 287–299.
44. Ferry B, Morneau F, Bontemps JD, Blanc L, Freycon V (2010) Higher treefall rates on slopes and waterlogged soils result in lower stand biomass and productivity in a tropical rain forest. Journal of Ecology 98: 106–116.
45. McEwan RW, Lin Y, Sun I, Hsieh C, Su S, et al. (2011) Topographic and biotic regulation of aboveground carbon storage in subtropical broad-leaved forests of Taiwan. Forest Ecology and Management 262: 1817–1825.

46. Man X, Mi X, Ma K (2011) Effects of an ice strom on community structure of an evergreen broad-leaved forest in Gutianshan National Natural Reserve, Zhejiang Province. Biodiversity Science 19: 197–205. (In Chinese).

47. Malhi Y, Baker TR, Phillips OL, Almeida S, Alvarez E, et al. (2004) The aboveground coarse wood productivity of 104 Neotropical forest plots. Global Change Biology 10: 563–591.

48. Elser JJ, Bracken MES, Cleland EE, Gruner DS, Harpole WS, et al. (2007) Global analysis of nitrogen and phosphorus limitation of primary producers in freshwater, marine and terrestrial ecosystems. Ecology Letters 10: 1135–1142.

49. Paoli G, Curran L (2007) Soil nutrients limit fine litter production and tree growth in mature lowland forest of southwestern Borneo. Ecosystems 10: 503–518.

50. Schaik CPV (1985) Spatial variation in the structure and litterfall of a Sumatran rain forest. Biotropica 17: 196–205.

51. Zhang L (2010) The effect of spatial heterogeneity of environmental factors on species distribution and community structure. Beijing, China: PhD thesis. Institute of Botany, Chinese Academy of Sciences. (In Chinese).

52. Russo SE, Davies SJ, King DA, Tan S (2005) Soil-related performance variation and distributions of tree species in a Bornean rain forest. Journal of Ecology 93:879–889.

53. Vittoz P, Engler R (2007) Seed dispersal distances: a typology based on dispersal modes and plant traits. Botanica Helvetica 117: 109–124.

14

Non-Destructive Lichen Biomass Estimation in Northwestern Alaska: A Comparison of Methods

Abbey Rosso[1], Peter Neitlich[1]*, Robert J. Smith[2]

1 National Park Service, Winthrop, Washington, United States of America, **2** Department of Botany and Plant Pathology, Oregon State University, Corvallis, Oregon, United States of America

Abstract

Terrestrial lichen biomass is an important indicator of forage availability for caribou in northern regions, and can indicate vegetation shifts due to climate change, air pollution or changes in vascular plant community structure. Techniques for estimating lichen biomass have traditionally required destructive harvesting that is painstaking and impractical, so we developed models to estimate biomass from relatively simple cover and height measurements. We measured cover and height of forage lichens (including single-taxon and multi-taxa "community" samples, $n = 144$) at 73 sites on the Seward Peninsula of northwestern Alaska, and harvested lichen biomass from the same plots. We assessed biomass-to-volume relationships using zero-intercept regressions, and compared differences among two non-destructive cover estimation methods (ocular vs. point count), among four landcover types in two ecoregions, and among single-taxon vs. multi-taxa samples. Additionally, we explored the feasibility of using lichen height (instead of volume) as a predictor of stand-level biomass. Although lichen taxa exhibited unique biomass and bulk density responses that varied significantly by growth form, we found that single-taxon sampling consistently under-estimated true biomass and was constrained by the need for taxonomic experts. We also found that the point count method provided little to no improvement over ocular methods, despite increased effort. Estimated biomass of lichen-dominated communities (mean lichen cover: $84.9 \pm 1.4\%$) using multi-taxa, ocular methods differed only nominally among landcover types within ecoregions (range: 822 to 1418 g m^{-2}). Height alone was a poor predictor of lichen biomass and should always be weighted by cover abundance. We conclude that the multi-taxa (whole-community) approach, when paired with ocular estimates, is the most reasonable and practical method for estimating lichen biomass at landscape scales in northwest Alaska.

Editor: Helge Thorsten Lumbsch, Field Museum of Natural History, United States of America

Funding: This work was funded by the US National Park Service–Western Arctic National Parklands and the US National Park Service–Arctic Network. The corresponding author secured governmental funding for this project via the US National Park Service. The funders had no role in study design, data collection and analysis, decision to publish, or preparation of the manuscript.

Competing Interests: The authors have declared that no competing interests exist.

* Email: peter_neitlich@nps.gov

Introduction

Lichen biomass is an important indicator of grazing impact and the availability of winter forage for caribou, reindeer, muskox, and other animals in northern regions [1], [2]. In northwestern Alaska, these animals rely on forage lichens including many species of *Cladonia* ("reindeer lichen", previously *Cladina*), *Alectoria*, *Bryocaulon*, *Bryoria*, and *Cetraria* [3]. Lichens provide the main winter sustenance for the Western Arctic Caribou Herd, which is one of the largest caribou herds in North America and is important in the subsistence economy of native Alaskans [1]. Aside from utility as wildlife forage, terrestrial lichen biomass can also be used as a vegetation monitoring metric to assess the impact of disturbances such as fire [3]–[5], climate change [6], [7] and air pollution [8], [9]. However, estimating lichen biomass by destructive sampling is very time consuming and does not allow for assessment of the same area over time. Researchers studying epiphytic lichens have previously approached this problem in places such as Norway [10], China [11], British Columbia [12] and the U.S. Pacific Northwest [13] by developing regression equations, yet very few have estimated biomass for ground-dwelling, terrestrial lichens.

Biomass estimation requires accurate volumetric measurements for lichen height and area cover. The method by which cover is measured represents a trade-off between the potential bias of estimates (if estimated subjectively) and the potential to under-represent rare or patchily distributed taxa (if measured quantitatively) [14]. Point-count estimates have the potential to be less biased given a high enough density of points [15], while ocular cover estimates can assess larger areas rapidly and can integrate patchiness. Therefore, ocular and point count methods must be comparatively assessed for accuracy and for their influence on biomass estimates.

Climate and vegetation type are important predictors of lichen biomass [16] and should be accounted for in landscape-level assessments. The landscapes of northwestern Alaska span several climatic ecoregions and a wide variety of vegetation and landcover types including wetlands, spruce woodlands, and tundra dominated by lichens, low shrubs or graminoids [17], [18]. Within these landcover types, there are strong gradients of substrate pH and

Table 1. Number of lichen samples among habitats and ecoregions of northwestern Alaska for all sample types.

Land cover type	Ecoregion	Sample type	Alectoria	Bryocaulon	Cetraria	Cladonia	C. stellaris	Total
L	N	Single-taxon	4	7	1	3	4	19
		Multi-taxa	2	2	1	12	1	18
	S	Single-taxon	0	0	0	0	0	0
		Multi-taxa	0	0	0	0	0	0
M	N	Single-taxon	0	0	5	2	0	7
		Multi-taxa	0	0	0	6	1	7
	S	Single-taxon	0	0	2	3	0	5
		Multi-taxa	0	0	0	2	0	2
P	N	Single-taxon	0	0	7	2	0	9
		Multi-taxa	0	0	0	11	0	11
	S	Single-taxon	3	0	6	3	0	12
		Multi-taxa	0	0	2	14	0	16
S	N	Single-taxon	7	7	0	3	2	19
		Multi-taxa	1	5	0	8	1	15
	S	Single-taxon	0	2	0	0	0	2
		Multi-taxa	1	0	0	1	0	2
Total:			18	23	24	70	9	144

Habitat type codes [17]: L = Dwarf Shrub-Lichen Dominated; M = Mesic/Dry Herbaceous; P = Open Low Shrub Dwarf Birch/Ericaceous; S = Sparse Vegetation. "Cladonia" includes all species except C. stellaris. Ecoregion codes [18]: N = Northern Seward Peninsula/Kotzebue Sound Lowlands (Kotzebue Lowlands) and S = Southern Seward Peninsula/Bering Sea Coast (Bering Coast).

vascular vegetation physiognomy that influence lichen community composition [19]. Because species differ in growth forms and levels of dominance, it could be expected that variability in lichen biomass and bulk density would be associated with distinctive landcover types.

Lichens in northwestern Alaska occur in very diverse, highly mixed assemblages typified by high species richness [3], [19]. Yet, most estimates of terrestrial lichen biomass completed elsewhere have been limited to only a single [5] or a very restricted number of species [2], [20], mainly focusing on the critical forage lichen genus *Cladonia*. Some workers [20] have suggested that in diverse lichen communities, the most rapid and effective way to estimate biomass would be to integrate estimates over entire communities rather than parsing out unique biomass relationships among individual taxa. Whole-community surveys [21] are also the most realistic way to monitor biomass across landscapes that have heterogeneous or patchy distributions of a large number of species, or for landscapes that have large species turnover (beta-diversity) among sites.

This study explored the feasibility of using height and cover measurements to estimate terrestrial lichen biomass in northwestern Alaska. Our inferences focused mainly on forage lichens (including *Alectoria*, *Bryocaulon*, *Cetraria* and *Cladonia* spp.) considered both as individual taxa and as the dominant members of aggregate, multi-taxa lichen communities. Here, we extended the range of species to include multi-taxa "community" samples that integrated all observed species, and we presented regression slope estimates to use as conversion factors applicable to northwestern Alaska. Our primary objectives in this study were: 1) to estimate the slope of zero-intercept linear regressions for biomass vs. volume; 2) to examine model fit between single-taxon vs. multi-taxa sampling techniques; 3) to examine model fit between ocular vs. point count cover estimation techniques; 4) to determine whether observed and predicted biomass in multi-taxa communities differed significantly among landcover types within ecoregions; and 5) to examine whether height or cover by themselves could each be good predictors of lichen biomass. We also place our results in the context of previous studies and make recommendations for best survey practices.

Methods

Ethics Statement

Field sampling was conducted on public lands administered by the US Bureau of Land Management by permission of that agency. We complied with all national and international rules regarding ethics; the research did not involve measurements on humans or animals. Plant material collected for this study was sampled on a very limited scale and therefore had negligible effects on landscapes. We declare no commercial interests or conflicts of interest.

Field Sampling

Field data are available as Dataset S1. We measured lichen height, cover and biomass (both for single taxa and for all taxa aggregated) among two different ecoregions [22] and four landcover types on the Seward Peninsula, northwestern Alaska (Table 1). Although both ecoregions converge on the Seward Peninsula, they differ sharply in rainfall and climate [23]. The first ecoregion, the Northern Seward Peninsula/Kotzebue Sound Lowlands (hereafter, "Kotzebue Lowlands"), has a Chukchi Sea-influenced and more continental climate with mean annual precipitation of 250–500 mm. The second ecoregion, the Southern Seward Peninsula/Bering Sea Coast (hereafter "Bering Sea Coast") is moist with mean annual precipitation ranging from 400–800 mm. Within each ecoregion we sampled four landcover types [17] which formed the basis of long-term lichen monitoring in the Bering Land Bridge National Preserve [24]; these were: Dwarf-Shrub and Lichen Dominated (L), Mesic/Dry Herbaceous (M), Open Low Shrub – Dwarf Birch/Ericaceous (P), and Sparse Vegetation (S) [17]. While there was a newer landcover classification available at the time of sampling, we chose an older data set compatible with NPS's Arctic Network's long-term lichen monitoring studies [24].

In the study area, we assigned plot locations non-randomly. For single-taxon sampling, we targeted sites with continuous coverage of a dominant forage lichen species. For multi-taxa sampling, we targeted sites with continuous coverage of lichens in general, avoiding non-lichen vegetation. Because multi-taxa communities were often dominated by high proportions of forage lichens (*Alectoria* spp., *Bryocaulon divergens*, *Cetraria* spp., or branched-fruticose *Cladonia*), we recorded the dominant lichen taxon within

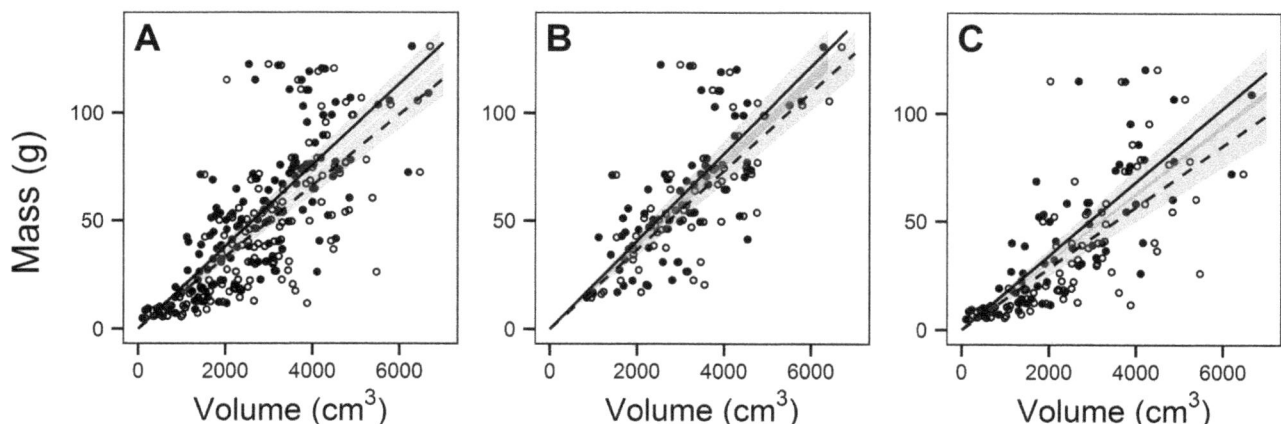

Figure 1. Biomass-volume relations using multiple methods. Shown are the ocular cover method (filled circles and solid lines) and the point count method (open circles and dotted lines), for all sample types combined (Fig. 1A), multi-taxa samples (Fig. 1B), and single-taxon samples (Fig. 1C). Lines indicate fitted model slopes (see Table 2 for estimates), while grey polygons indicate 95% confidence intervals.

Table 2. Estimated parameters and goodness-of-fit statistics for fitted models of forage lichen biomass in northwestern Alaska.

Method	Sample type	Taxon	Slope coefficient (g cm^{-3})			Bayesian Information Criterion (BIC)			Log-likelihood and LLR		
			Ocular	Pt count	Difference	Ocular	Pt count	Difference	Ocular	Pt count	LLR
Combined	All	All	0.0176	-	-	2607.35	-	-	-1298.01	-	-
Either	All	All	0.0189	0.0165	-12.8%	1285.43	1319.60	2.7%	-637.75	-654.83	-34.17
Either	Multi-taxa	All	0.0203	0.0183	-9.8%	641.14	647.12	0.9%	-316.31	-319.30	-5.98
Either	Single-taxa	All	0.0171	0.0143	-16.4%	642.61	666.68	3.7%	-317.01	-329.05	-24.08
Either	Multi-taxa	Alectoria	0.0106	0.0098	-7.2%	35.27	33.63	-4.7%	-16.25	-15.43	1.64
Either	Multi-taxa	Bryocaulon	0.0160	0.0154	-3.8%	54.72	53.66	-1.9%	-25.41	-24.88	1.06
Either	Multi-taxa	Cetraria	0.0177	0.0135	-23.4%	19.87	16.89	-15.0%	-8.84	-7.34	2.98
Either	Multi-taxa	Cladonia	0.0207	0.0186	-10.1%	481.10	486.65	1.2%	-236.56	-239.34	-5.56
Either	Multi-taxa	C. stellaris	0.0299	0.0288	-3.6%	21.56	16.35	-24.2%	-9.68	-7.08	5.21
Either	Single-taxa	Alectoria	0.0083	0.0056	-33.3%	93.63	95.56	2.1%	-44.18	-45.14	-1.93
Either	Single-taxa	Bryocaulon	0.0140	0.0120	-14.6%	128.78	123.45	-4.1%	-61.62	-58.95	5.33
Either	Single-taxa	Cetraria	0.0141	0.0105	-25.5%	157.81	153.10	-3.0%	-75.86	-73.51	4.71
Either	Single-taxa	Cladonia	0.0197	0.0184	-6.6%	155.42	158.15	1.8%	-74.94	-76.30	-2.73
Either	Single-taxa	C. stellaris	0.0221	0.0208	-6.1%	55.66	56.34	1.2%	-26.04	-26.38	-0.68
Height only	Multi-taxa	All	-	11.05 (g cm^{-1})	-	-	653.31	-	-	-322.4	-
Either, cover only	Multi-taxa	All	0.1188 (g cm^{-2})	0.1049 (g cm^{-2})	11.7%	672.70	682.46	1.5%	-332.1	-337.0	-9.76

"Method" column summarizes either ocular or point count methods of estimating volume, both of these, or height only. "Slope coefficient" is in units of g cm^{-3} unless otherwise noted, and is a measure of lichen mat bulk density; uncertainty of each estimate is presented graphically as 95% confidence intervals in Fig. 3. "Difference" between the ocular and point count method is expressed as a proportion of the ocular method. Model goodness-of-fit was determined by likelihood-based methods: better models have values closer to zero for Bayesian Information Criterion (BIC) values and greater log-likelihood values. Log-likelihood ratios (LLR) were calculated holding ocular method as the null hypothesis, therefore, more strongly negative values indicate better fit for models using the ocular method. Each LLR test was significant ($p<0.0001$).

Table 3. Observed and predicted lichen biomass (g m^{-2}) for each sample type and landcover type in northwestern Alaska.

Sample type	Land cover type	Observed Mean	Volume (ocular) Mean	Volume (ocular) Deviation	Volume (pt ct) Mean	Volume (pt ct) Deviation	Height only Mean	Height only Deviation	Cover only (ocular) Mean	Cover only (ocular) Deviation	Cover only (pt ct) Mean	Cover only (pt ct) Deviation
Multi-taxa	L	911.9 (102.1)	943.9 (69.0)	3.5%	916.2 (61.3)	0.5%	907.3 (56.9)	-0.5%	1068.1 (25.0)	17.1%	1021.3 (14.4)	12.0%
Multi-taxa	M	1508.6 (113.1)	1418.5 (104.9)	-6.0%	1379.7 (92.7)	-8.5%	1357.7 (93.6)	-10.0%	1080 (21.9)	-28.4%	1030.7 (10.2)	-31.7%
Multi-taxa	P	789.6 (77.0)	822 (75.8)	4.1%	899.2 (77.5)	13.9%	932.7 (71.4)	18.1%	893.4 (22.3)	13.1%	960.7 (17.3)	21.7%
Multi-taxa	S	1124.1 (118.5)	1021.4 (76.0)	-9.1%	978.6 (59.1)	-12.9%	957.1 (55.6)	-14.9%	1089.7 (28.7)	-3.1%	1034.5 (10.1)	-8.0%
Multi-taxa	All	991.9 (56.8)	976.3 (45.5)	1.6%	983.4 (41.7)	0.9%	986.0 (38.8)	0.6%	1008.3 (16.7)	-1.7%	1002.6 (8.8)	-1.1%
Single taxon	L	671.3 (110.5)	653.7 (86.9)	-2.6%	710.1 (68.9)	5.8%	-	-	-	-	-	-
Single taxon	M	809.9 (192.5)	665.2 (126.7)	-17.9%	651.9 (104.7)	-19.5%	-	-	-	-	-	-
Single taxon	P	435.0 (88.4)	435.0 (89.5)	0.0%	447.0 (76.8)	2.8%	-	-	-	-	-	-
Single taxon	S	676.2 (99.9)	764.5 (74.6)	13.1%	739.3 (62.7)	9.3%	-	-	-	-	-	-
Single-taxon	All	627.5 (58.2)	624.5 (47.0)	0.5%	633.2 (40.0)	-0.9%	-	-	-	-	-	-

Mean lichen biomass, estimated using the "All Taxa" regression slope coefficients from Table 2 and parceled out by landcover type, is reported with standard errors (in parentheses) and mean deviation from observed biomass. Mean deviation is interpreted as measurement error. Landcover type codes: L = Dwarf Shrub-Lichen Dominated; M = Mesic/Dry Herbaceous; P = Open Low Shrub Dwarf Birch/Ericaceous; S = Sparse Vegetation. Caution: these biomass values were based on average quadrat cover >80% and should *not* be considered an inference about areas >1 m^2 (which could have erratic cover).

each community to facilitate later comparison with single-taxa samples, and also recorded the dominant vascular vegetation type for landcover classification according to [17].

On the ground at each sampling location, we placed a quadrat frame made of PVC pipe (internal area 25×25 cm) divided into twenty-five cells (5×5 cm). In the center of each cell, we measured the height of lichens (if present) using a narrow, ruled metal rod (3 mm diameter) lowered to the base of the lichen until firm resistance at the duff/soil layer was encountered, but not pushing into layers below. We calculated lichen height as the average of the twenty-five points within each plot (including "no-hit" zero values) because we wanted to facilitate adaptation to other techniques that might include many lichen absences (e.g., other point count or line intercept surveys). For single-taxon samples, we recorded only the height of the target lichen, while for multi-taxa samples we also recorded the identity of the lichen species and visually estimated its cover to the nearest one percent. Based on ground surface similarity of growth form (i.e., without dislodging the lichen) and bulk density, species found in single-taxon plots were aggregated to the genus level if several were present in the same sample. For example, *Alectoria ochroleuca* was grouped with *A. nigricans*, and *Cetraria laevigata* was grouped with *C. islandica*. Morphologically similar *Cladonia* species (*C. arbuscula* and *C. mitis*; *C. rangiferina* and *C. stygia*) were also grouped, although we measured *C. stellaris* separately because it had apparently different bulk density and a morphology that was readily recognizable in the field.

After recording height and cover, lichens were misted with water, then manually harvested by cutting vertically along the edge of each quadrat and gently separating intact mats from the ground. We removed non-lichen plant parts, non-target species, obviously dead or decaying lichen parts, and any dirt, gravel or extraneous debris, then placed samples in dry paper bags for transport to the lab. Each sample was completely oven-dried at 80°C until reaching a stable, unchanging mass (minimum 8 hours, no more than 12 hours). Note that this is considerably less drying time than the 76 hours used in a similar study [20]. We weighed each sample with a digital scale (precision ± 0.01 g) within just a few minutes of removal from the drying oven to prevent moisture uptake that might bias dry mass estimates. This also differs from [20], in which lichens were allowed to cool to room temperature before weighing.

Calculations

Our goal was to assess the relationship between biomass and lichen volume (calculated as the product of cover×height). We used two methods to calculate volume; the first was based on the ocular (visual) method of estimating cover described above, and the second was based on the point count estimation of cover. For the point count method, we estimated percentage cover by dividing the number of lichen presences by the total number of points (twenty-five) in each plot. For both methods, we multiplied percentage cover by the area of the plot to convert all measurements to cubic centimeters.

We regressed lichen biomass on volume (or height or cover) using ordinary least squares regression with forced zero intercepts implemented in R version 3.0.1 (R Core Development Team 2013) and compared model slope parameters. Although weighted regressions have been used elsewhere [20] to estimate forage lichen biomass, examination of model residuals (not reported here) suggested this was not necessary given our data. Because goodness-of-fit R^2 statistics are artificially inflated and are essentially meaningless for zero-intercept models, we instead report confidence intervals for each slope, and compare models using three metrics: percent difference in slope, Bayesian Information

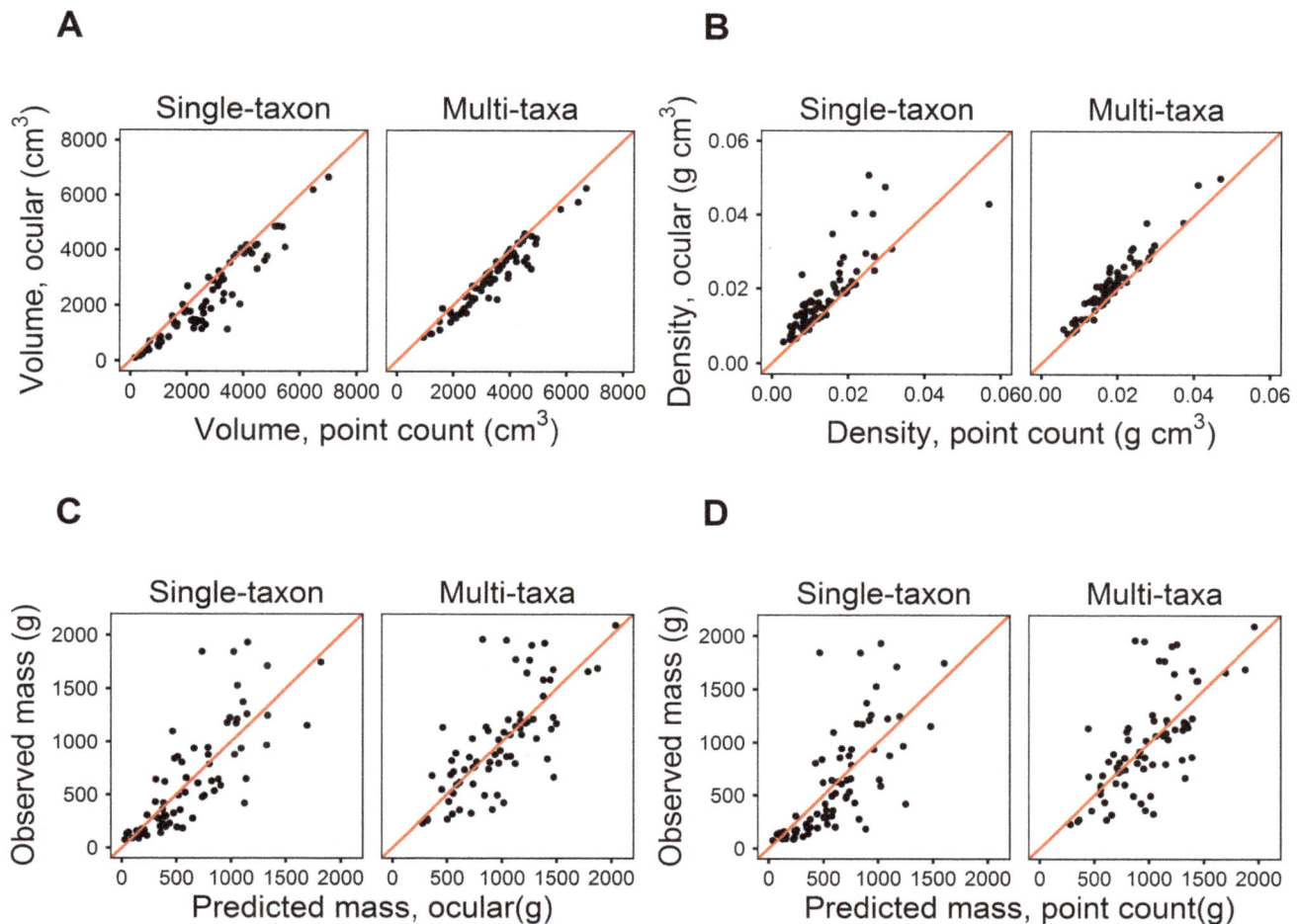

Figure 2. Relative comparisons of bias among two methods. Comparisons are between ocular vs. point count methods for estimates of volume (A) and estimates of bulk density (B); relative comparisons for observed vs. predicted lichen biomass for the ocular method (C) and the point count method (D). Each red line is a hypothetical 1:1 isoline, where deviations from this line indicate magnitude of differences.

Criterion (*BIC*) [25] and log-likelihood ratio (*LLR*) tests [26]. The latter two are likelihood-based methods that do not attempt to assess whether two model parameters differ more than expected by chance (as would a probability-based *F*-test, for example), but instead they provide a relative measure of how much better or worse each model fits, given the data at hand. For our application, this is more useful than *F*-tests because the sign and magnitude of difference between two methods provides more information than simply whether or not any such difference occurs.

To determine whether mean observed and predicted biomass in multi-taxa "communities" differed significantly among landcover types within ecoregions, we used nested analysis-of-variance *F*-tests (landcover type nested within ecoregions), accompanied by Tukey's Honest Significant Difference test for pairwise comparisons; this procedure used only the multi-taxa data to simulate realistic conditions.

Results

Biomass vs. Volume Slope

From the regression of biomass on volume, estimates of the slope ranged from 0.0143 to 0.0203 g cm^{-3} for lichen communities, depending on the method used (Table 2 and Fig. 1A). Individual taxa ranged from 0.0056 g cm^{-3} (*Alectoria*) to

0.0221 g cm^{-3} (*Cladonia stellaris*). All slopes differed significantly from zero ($p<0.05$ from *t*-test of linear slope coefficients). For both observed mass and bulk density, mean values differed among all taxon groups (*F*-tests $p<0.0001$) except the pairwise comparisons among *Bryocaulon* –*Cetraria* and *Cladonia*–*C. stellaris* (for all others, $p<0.05$ from Tukey test).

Single-taxon vs. Multi-taxa Sampling

The single-taxon method consistently underestimated biomass relative to the multi-taxa method (Fig. 1B vs. 1C). Model fit was always better for the multi-taxa method (Table 2), which in all cases also had less variation in biomass than the single-taxon method (Table 3).

Ocular vs. Point Count

The point count method had lower biomass estimates relative to ocular methods (Fig. 1, filled vs. open symbols). This is directly related to the point count method estimating higher volume (Fig. 2A) and lower bulk density (Fig. 2B) relative to the ocular method. Model fit was usually but not always better with the ocular method compared to the point count method (Table 2). Among individual taxa, differences in slope between methods (Fig. 3) were more pronounced for samples dominated by taxa

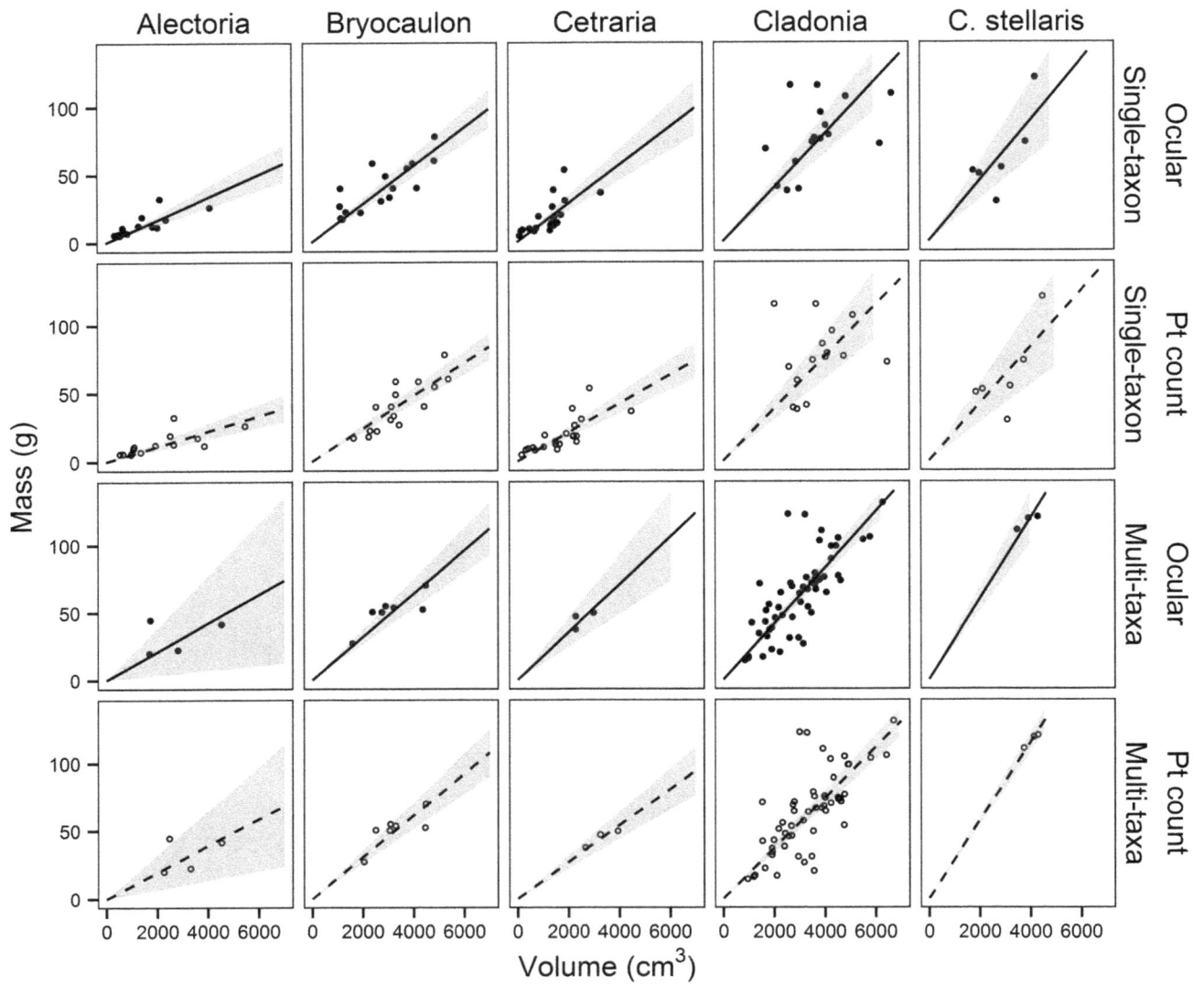

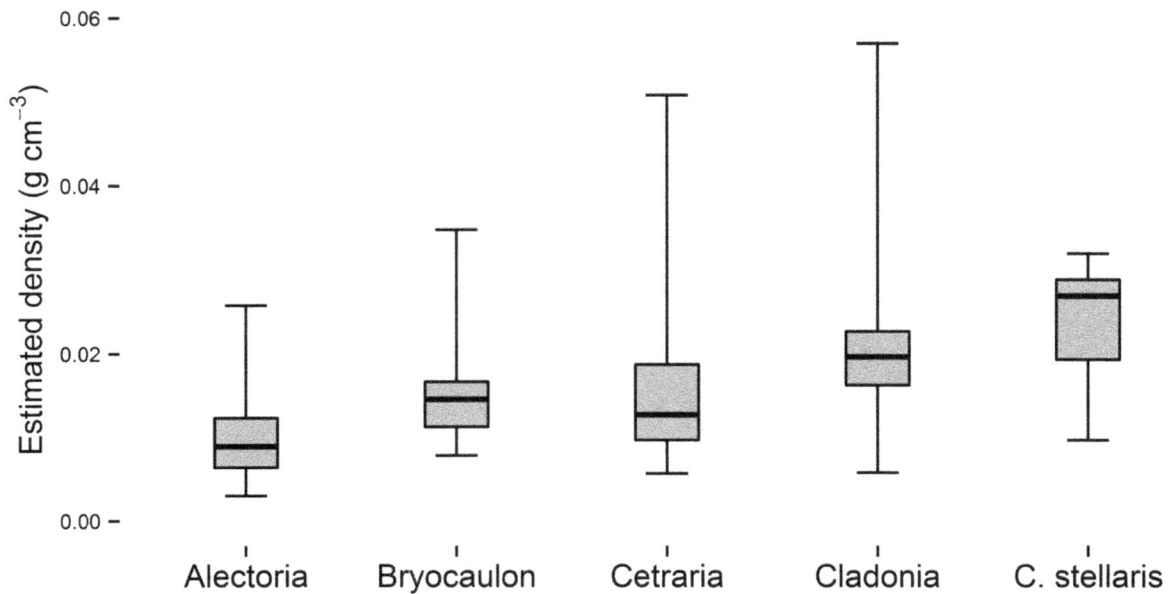

Figure 3. Biomass-volume relations and bulk density distributions for dominant forage lichen taxa in northwestern Alaska. Regressions were fitted using either single-taxon sampling (top two rows) or multi-taxa samples (bottom two rows). Lines indicate fitted model slopes (see Table 2 for estimates), while grey polygons indicate 95% confidence intervals. In the bottom panel, the distribution of estimated bulk density (all methods, grey boxes) for 144 lichen samples from northwestern Alaska is shown, where dark bars in boxes are median values, boxes represent the interquartile range of data values, and whiskers are the maxima/minima within each group. Mean bulk density differs among all species groups (F-test $p < 0.0001$) except the pairwise comparisons among *Bryoria* – *Cetraria* and *Cladonia* – *C. stellaris* (for all others, $p < 0.05$ or less from Tukey HSD test).

with diffuse, filamentous growth forms like *Alectoria*, rather than for denser taxa like *Cladonia* that tended to fill entire sample frames.

Ecoregions and Landcover Types

When slope estimates were used to predict standing biomass for each of the four landcover types, the point count method had generally greater deviation from observed values than the ocular method (Table 3), though this was not true in all cases; magnitudes of deviation ranged from 0.01–19.5%. The biomass values reported here for the four landcover classes studied represents the high end of the lichen biomass distribution for each cover type because sites were selected based on high lichen cover to allow for better biomass-to-volume estimates. Of the seven combinations of landcover types nested within ecoregions (Table 1), there were 21 possible pairwise comparisons. Of the 21 comparisons, Tukey tests revealed very few significant pairwise differences for observed mass (3 significant comparisons), mass predicted using the ocular method (3 comparisons), mass predicted using the point count method (2 comparisons), and mass predicted using only height (0 comparisons). Each of these nominally "significant" pairwise comparisons had p-values that were close to the $p = 0.05$ significance threshold, so we do not report them here.

Biomass vs. Height or Cover

Deviation of estimated values from observed values (Table 3) was similar but of slightly greater magnitude when using height as the sole predictor of biomass in multi-taxa communities than when using volume; magnitudes of deviation ranged from 0.5–18.1% for different landcover types. When using cover as the sole predictor of biomass, the deviation of estimated from observed values was far greater than when using volume (about 3–5 times greater: Table 3); this was true of the ocular method (absolute deviation: 3.1–28.4%) as well as the point count method (absolute deviation: 8.0–31.7%).

Discussion

Biomass as a Function of Volume

When lichen biomass is regressed on volume, the resulting linear slope is equivalent to a mass-to-volume ratio or a measure of the lichen mat's bulk density. Our findings indicate that taxa have different bulk densities as a result of their different growth forms, a trend that is reflected in multi-taxa communities dominated by each growth form. For individual taxa, *Cladonia* was most dense, while *Bryocaulon*, *Cetraria*, and *Alectoria* were successively less dense. There were congruent patterns in multi-taxa communities dominated by each of those genera, with *Cladonia*-dominated communities being densest.

The study most comparable with our methods [20] used volumetric estimation of biomass for four species (three *Cladonia* spp. and *Cetraria islandica*) in Sweden using 50×50 cm plots. Their mean *Cladonia* bulk densities agree closely with our estimates (within 12% of each other for ocular methods), but were slightly lower, perhaps because they manually pulled lichen

mats off the ground (rather than cut them cleanly as we did), which could leave lichen material attached to the ground. By contrast, *Cetraria* spp. bulk densities from our study were nearly 50% less than that of *C. islandica* in Sweden [20]. We assume that unmeasured factors (e.g., nutrient availability, moisture, growing season, fire regime, successional status, herbivory patterns) allow *Cetraria* in Sweden to occur in much denser colonies than on the Seward Peninsula of Alaska.

In Finland, Kumpula et al. [2], reported a much lower bulk density than either our results or those from Sweden [20], possibly resulting from differing definitions of what constitutes a "living" portion of a lichen thallus. Dunford et al. [5] did not report a bulk density because they relied solely on lichen cover (not height or volume) for the single species *Cladonia mitis*. Though their scope of inference extended to 25 sites, Dunford et al. [5] also had a drastically smaller sample size than ours ($n = 8$ vs. 144).

To the extent that we had outlier observations, these were likely due to the unintentional inclusion of non-target vegetation in some plots, especially where lichen thalli were layered over top of unseen moss or shrub tissues beneath. While our volumetric calculations probably included portions of non-target vegetation, we note that this is consistent with all situations in which non-destructive estimates of cover and height are used for applying conversion factors to larger study areas.

Single-taxon vs. Multi-taxa Sampling

Individual genera had generally lower bulk density than the multi-taxa samples. This is probably because lichens on the Seward Peninsula grew in mixed assemblages with intertwined and overlapping layers composed of different growth forms and bulk densities. Thus what superficially appeared to be a continuous cover of one lichen from the surface was in fact usually a tangled mat of several different lichen taxa below. This may be why we found it difficult to locate monotypic mats for every target species, even using a plot size of just 25×25 cm.

Estimating biomass–volume ratios for every species found in northwestern Alaska would be a prohibitive task. Given that multi-taxa surveys require only general taxonomic training and that they capture variation in biomass at levels equivalent to or better than individualistic single-taxon assessments, we conclude that a multi-taxa focus is preferable to single-taxon surveys when lichens across entire landscapes must be faithfully represented.

Ocular vs. Point Count

The use of point-count estimations of cover gave consistently higher cover (and volume) estimates than ocular methods, leading to lower bulk density estimates; this is consistent with other findings [20]. Despite its initial appeal for reducing observer bias, we noted that the point count method often seemed to overestimate cover by "catching" or touching disproportionately small pieces of lichen thallus in areas of the plot that lacked appreciable coverage, especially for taxa that were filamentous (e.g., *Alectoria*) or patchily distributed. Though the ocular method seemed initially less objective, it had the benefit that it allowed observers to visually integrate patchy or diffuse spatial coverage of lichens in a more realistic manner. Point count methods overestimate cover of vascular plants, while visual estimates are more accurate [27], and counting enough points to be useful would be logistically limiting in remote Arctic settings. We note that point count methods may still have utility for plots larger than our 25×25 cm size because ocular cover becomes less precise in larger plots [15]. Point count methods may also be preferable for long-term monitoring where repeatability is paramount [28] and where results may differ among observers [14]. Future work is needed to resolve variation among observers and identify sources of error.

Ecoregions and Landcover Types

The two ecoregions we examined on the Seward Peninsula appeared to have similar bulk densities (g cm^{-3}) of lichens. Bulk densities also differed far less than expected among the four landcover types. We expected that differences in substrate availability, substrate chemistry, moisture and vascular vegetation would be expressed in lichen community biomass because growth forms differ according to those conditions. For example, filamentous groups such as *Bryocaulon* and *Alectoria* dominate the Sparse Vegetation ("S") cover type on the Seward Peninsula, whereas the low-elevation L, P, and M landcover types tend to be dominated by denser *Cladonia* spp. We suspect that ecoregion and landcover would have been more influential had we sampled across a broader latitudinal gradient (see [29] for one example of an Arctic gradient spanning 1800-km north-south). Note that while the biomass values in Table 3 are a starting point for landscape modeling, they portray only one possible scenario (i.e., one with lichen cover averaging ~80% due to our preferential plot assignment). Future landscape estimates must account for patchy or sparse lichen cover, and therefore require a truly random sample of sites from within each landcover type.

Biomass vs. Height or Cover

In some instances, a direct measure of biomass as a function of average height of the lichen can be useful, avoiding the need to estimate cover. For example, line intercept methods can provide copious amounts of height data in a short time frame and may be more applicable than closed-frame methods in the open, unforested landcover types that are commonly found in northwestern Alaska. However, we found a large amount of variation in the estimates that relied on height as the sole predictor, and we advise that height-based estimates must be always weighted by some measure of cover to prevent overestimation. Given that caveat, transect-based line intercept methods could provide biomass estimates if the total area of the study site is known.

If height alone appeared to be an inadequate predictor of lichen biomass, then cover alone was even worse. This makes logical sense – two different lichen patches could cover an equal surface area yet differ by an order of magnitude in height; observers would have no way of knowing their true volume. Because we found that deviations from true values were 3 to 5 times greater when using cover alone as compared to the other methods, we do not recommend using cover as the sole predictor of lichen biomass.

Extending Spatial Coverage

Remote sensing is one way that estimates can be scaled up to landscape-level inferences. Field surveys can be coupled with remote sensing imagery to construct biomass estimates via several algorithms [4]. Other workers [30] have similarly used Landsat imagery to derive leaf area index and biomass for multiple vegetation groups including fruticose, foliose and crustose lichens (in aggregate). Spectral decomposition of remote imagery is a further refinement that can distinguish among taxonomic or functional groups of lichens [31]. Though location-specific correction factors must always be derived and validated on the ground, remote sensing is a promising technology that should allow rapid estimation of biomass for very large, remote or difficult-to-access locations in northern ecosystems.

Management Applications

The biomass functions presented here promise to have utility in several different areas of applied ecology in arctic Alaska. Studies of ungulate biology should be able to use these equations to help scale plot-level vegetation data (where both lichen cover and height measurements are available) to remote sensing imagery, as in [32]. Long-term studies of ungulate grazing exclosures are underway on Alaska's Seward Peninsula [33], [34], and conversion of lichen cover/height data to forage biomass will be an essential component of these studies. The National Park Service has other long-term lichen/vegetation studies in which both lichen cover and height are measured [33]; these may similarly be coupled with remote sensing modeling using our conversion factors for applications including changes in vegetation community structure with climate change effects [35], vegetation response to changing fire regimes [36], and carbon accounting [37]. The National Park Service's Arctic Network has sufficient density of ground-based measurements to model lichen biomass at a landscape scale [24]. Repeat measurements would allow for temporal and spatial depictions of lichen biomass on the landscape. Improved burn severity geospatial layers and historic fire perimeters would increase the resolution of these layers by allowing for the modeling of successional status, and hence lichen height and cover. Lichen winter range for the Western Arctic Caribou Herd is projected to suffer some decline with climate-driven shrub increase [1], [35], more frequent wildfire [36], and increasing inputs of nitrogen and sulfur from regional development [38]. Continued monitoring of lichen biomass will be critical for detecting and addressing ongoing changes.

Conclusions

To summarize, we found that multi-taxa ("bulk") sampling, when coupled with ocular estimation, was the preferred method for biomass estimation because it yielded the most accurate estimates and was logistically most efficient in the field. This method required no more than general taxonomic knowledge in the field and was tractable for statistical purposes. The zero-intercept regression equation (biomass = 0.0203 g cm^{-3}) which we derived by this method is readily applied to lichen communities in northwestern and arctic Alaska, though we caution that this does not necessarily apply to other geographical regions where location-specific equations will be required. Height alone was not the preferred predictor of biomass, and cover alone was a poor predictor; the best estimates should ideally include both height and cover. Because bulk densities varied among the forage lichen species we measured, we suggest that in cases where there is only a single species of interest, investigators may benefit from using separate regression equations for target species. Lichen biomass estimation has a wealth of applications that will help managers estimate wildlife forage, understand successional trends, detect climate and air quality signals, and account for landscape-level carbon. Therefore, continued monitoring will be vital for understanding how ongoing changes in lichen biomass and distribution affect other elements of Arctic ecosystems.

Acknowledgments

We thank Marci Johnson, Gina Hernandez, Matthew Jenkins and Mason London for help with field work and with cleaning samples. We also wish to thank Nichole Andler for providing field crew, and Bruce McCune and Jody Deland for the use of drying ovens.

Author Contributions

Conceived and designed the experiments: AR PN. Performed the experiments: AR PN. Analyzed the data: AR PN RJS. Contributed reagents/materials/analysis tools: AR PN. Contributed to the writing of the manuscript: AR PN RJS.

References

1. Joly K, Chapin FS, Klein DR (2010) Winter habitat selection by caribou in relation to lichen abundance, wildfires, grazing, and landscape characteristics in Northwest Alaska. Ecoscience 17: 321–333. doi:10.2980/17-3-3337.
2. Kumpula J, Colpaert A, Nieminen M (2000) Condition, potential recovery rate, and productivity of lichen (Cladonia spp.) ranges in the Finnish reindeer management area. Arctic 53: 152–160.
3. Holt EA, McCune B, Neitlich P (2008) Grazing and fire impacts on macrolichen communities of the Seward Peninsula, Alaska, USA. Bryologist 111: 68–83. doi:10.1639/0007-2745(2008)111[68:GAFIOM]2.0.CO;2.
4. Arseneault D, Villeneuve N, Boismenu C, Leblanc Y, Deshaye J (1997) Estimating lichen biomass and caribou grazing on the wintering grounds of northern Quebec: an application of fire history and Landsat data. J Appl Ecol 34: 65–78.
5. Dunford JS, McLoughlin PD, Dalerum F, Boutin S (2006) Lichen abundance in the peatlands of northern Alberta: implications for boreal caribou. Ecoscience 13: 469–474. doi:10.2980/1195-6860.
6. Heggberget TM, Gaare E, Ball JP (2002) Reindeer (Rangifer tarandus) and climate change: importance of winter forage. Rangifer 22: 13–31. doi:10.7557/2.22.1.388.
7. Lang SI, Cornelissen JHC, Shaver GR, Ahrens M, Callaghan TV, et al. (2012) Arctic warming on two continents has consistent negative effects on lichen diversity and mixed effects on bryophyte diversity. Glob Chang Biol 18: 1096–1107. doi:10.1111
8. Geiser LH, Neitlich PN (2007) Air pollution and climate gradients in western Oregon and Washington indicated by epiphytic macrolichens. Environ Poll 145: 203–218. doi:10.1016/j.envpol.2006.03.024.
9. Pardo LH, Fenn ME, Goodale CL, Geiser LH, Driscoll CT, et al. (2011) Effects of nitrogen deposition and empirical nitrogen critical loads for ecoregions of the United States. Ecol Appl 21: 3049–3082. doi:10.1890/10-2341.1.
10. Gauslaa Y, Lie M, Ohlson M (2008) Epiphytic lichen biomass in a boreal Norway spruce forest. Lichenologist 40: 257–266.
11. Li S, Liu W, Wang L, Ma W, Song L (2011) Biomass, diversity and composition of epiphytic macrolichens in primary and secondary forests in the subtropical Ailao Mountains, SW China. For Ecol Manage 261: 1760–1770. doi:10.1016/j.foreco.2011.01.037.
12. Price K, Hochachka G (2001) Epiphytic lichen abundance: effects of stand age and composition in coastal British Columbia. Ecol Appl 11: 904–913. doi:10.1890/1051-0761(2001)011[0904:ELAEOS]2.0.CO;2.
13. Rhoades FM (1981) Biomass of epiphytic lichens and bryophytes on Abies lasiocarpa on a Mt. Baker lava flow, Washington. Bryologist 84: 39–47. doi:10.2307/3242976.
14. McCune B, Lesica P (1992) The trade-off between species capture and quantitative accuracy in ecological inventory of lichens and bryophytes in forests in Montana. Bryologist 95: 296–304.
15. Godínez-Alvarez H, Herrick JE, Mattocks M, Toledo D, Van Zee J (2009) Comparison of three vegetation monitoring methods: Their relative utility for ecological assessment and monitoring. Ecol Indic 9: 1001–1008. doi:10.1016/j.ecolind.2008.11.011.
16. Joly K, Jandt RR, Klein DR (2009) Decrease of lichens in Arctic ecosystems: the role of wildfire, caribou, reindeer, competition and climate in north-western Alaska. Polar Res 28: 433–442.
17. Markon CJ, Wesser S (1997) The Bering Land Bridge National Preserve Land Cover Map and Its Comparability with 1995 Field Conditions. USGS Open File Report No 97-103. Reston, Virginia: US Department of the Interior, Geological Survey.
18. Jorgenson MT, Roth JE, Miller PE, Macander MJ, Duffy MS, Wells AF, Frost GV, Pullman ER (2009) An Ecological Land Survey and Landcover Map of the Arctic Network. Natural Resource Technical Report NPS/ARCN/NRTR—2009/270. Fort Collins, Colorado: US Department of the Interior, National Park Service.
19. Holt EA, McCune B, Neitlich P (2009) Macrolichen communities in relation to soils and vegetation in the Noatak National Preserve, Alaska. Botany 87: 241–252. doi:10.1139/B08-142.
20. Moen J, Danell Ö, Holt R (2007) Non-destructive estimation of lichen biomass. Rangifer 27: 41–46.
21. Bergerud AT (1971) Abundance of forage on the winter range of Newfoundland caribou. Can Field Nat 85: 39–52.
22. Jorgenson MT, Roth JE, Emers M, Davis WA, Schlentner SF, et al. (2004) Landcover Mapping for Bering Land Bridge National Preserve and Cape Krusenstern National Monument, Northwestern Alaska. Fort Collins, Colorado: US Department of the Interior, National Park Service.
23. Daly C (2009) Annual Mean Average Temperature for Alaska 1971–2000. PRISM Climate Group at Oregon State University, Corvallis, Oregon, USA. Available: http://www.prism.oregonstate.edu. Accessed 11 January 2011.
24. Holt EA, Neitlich PN (2010) Lichen Inventory Synthesis: Western Arctic National Parklands and Arctic Network, Alaska. Natural Resource Technical Report NPS/AKR/ARCN/NRTR—2010/385. Fort Collins, Colorado: US Department of the Interior, National Park Service.
25. Schwarz G (1978) Estimating the dimension of a model. Ann Stat 6: 461–464. doi:10.1214/aos/1176344136.
26. Engle RF (1984) Wald, likelihood ratio, and Lagrange multiplier tests in econometrics. In: Griliches Z, Intriligator MD, editors. Handbook of Econometrics. Elsevier, Vol. 2. pp. 775–826.
27. Bråkenhielm S, Qinghong L (1995) Comparison of field methods in vegetation monitoring. Water Air Soil Pollut 79: 75–87. doi:10.1007/BF01100431.
28. Fancy SG, Gross JE, Carter SL (2009) Monitoring the condition of natural resources in US national parks. Environ Monit Assess 151: 161–174. doi:10.1007/s10661-008-0257-y.
29. Epstein HE, Walker DA, Raynolds MK, Jia GJ, Kelley AM (2008) Phytomass patterns across a temperature gradient of the North American arctic tundra. J Geophys Res Biogeosci 113: 1–11. doi:10.1029/2007JG000555.
30. Chen W, Li J, Zhang Y, Zhou F, Koehler K, et al. (2009) Relating biomass and leaf area index to non-destructive measurements in order to monitor changes in Arctic vegetation. Arctic 62: 281–294. doi:10.14430/arctic148.
31. Nelson PR, Roland C, Macander MJ, McCune B (2013) Detecting continuous lichen abundance for mapping winter caribou forage at landscape spatial scales. Remote Sens Environ 137: 43–54. doi:10.1016/j.rse.2013.05.026.
32. Swanson DK, Neitlich PN (2014) Terrestrial Vegetation Monitoring Protocol for the Arctic Alaska Network: Establishment, Sampling, and Analysis of Permanent Monitoring Plots. Natural Resource Technical Report NPS/AKR/ARCN/NRTR. Fort Collins, Colorado: US Department of the Interior, National Park Service. In Preparation.
33. Moore KE (2013) Collaborative Monitoring of Seward Peninsula, Alaska Reindeer (Rangifer tarandus tarandus) Grazing Lands: A Monitoring Protocol. M.S. thesis. Fairbanks, AK: University of Alaska, Fairbanks. 55 p.
34. Swanson DK (2013) Three Decades of Landscape Change in Alaska's Arctic National Parks: Analysis of Aerial Photographs, c. 1980–2010. Natural Resource Technical Report NPS/ARCN/NRTR—2013/668. Fort Collins, Colorado: US Department of the Interior, National Park Service.
35. Tape K, Sturm M, Racine C (2006) The evidence for shrub expansion in Northern Alaska and the Pan-Arctic. Glob Chang Biol 12: 686–702.
36. Higuera PE, Chipman ML, Barnes JL, Urban MA, Hu FS (2011) Variability of tundra fire regimes in Arctic Alaska: millennial scale patterns and ecological implications. Ecol Appl 21: 3211–3226. doi:10.1890/11-0387.1.
37. Porada P, Weber B, Elbert W, Pöschl U, Kleidon A (2013) Estimating global carbon uptake by lichens and bryophytes with a process-based model. Biogeosciences 10: 6989–7033. doi:10.5194/bg-10-6989-2013.
38. Linder G, Brumbaugh W, Neitlich P, Little E (2013) Atmospheric deposition and critical loads for nitrogen and metals in Arctic Alaska: review and current status. Open J Air Poll 2(4): 76–99. doi: 10.4236/ojap.2013.24010.

15

A DNA-Based Semantic Fusion Model for Remote Sensing Data

Heng Sun[1]*, Jian Weng[1], Guangchuang Yu[2], Richard H. Massawe[3]

1 Department of Computer Science, College of Information Science and Technology, Jinan University, Guangzhou, People's Republic of China, **2** Key Laboratory of Functional Protein Research of Guangdong Higher Education Institutes, Institute of Life and Health Engineering, College of Life Science and Technology, Jinan University, Guangzhou, People's Republic of China, **3** International School, Jinan University, Guangzhou, People's Republic of China

Abstract

Semantic technology plays a key role in various domains, from conversation understanding to algorithm analysis. As the most efficient semantic tool, ontology can represent, process and manage the widespread knowledge. Nowadays, many researchers use ontology to collect and organize data's semantic information in order to maximize research productivity. In this paper, we firstly describe our work on the development of a remote sensing data ontology, with a primary focus on semantic fusion-driven research for big data. Our ontology is made up of 1,264 concepts and 2,030 semantic relationships. However, the growth of big data is straining the capacities of current semantic fusion and reasoning practices. Considering the massive parallelism of DNA strands, we propose a novel DNA-based semantic fusion model. In this model, a parallel strategy is developed to encode the semantic information in DNA for a large volume of remote sensing data. The semantic information is read in a parallel and bit-wise manner and an individual bit is converted to a base. By doing so, a considerable amount of conversion time can be saved, i.e., the cluster-based multi-processes program can reduce the conversion time from 81,536 seconds to 4,937 seconds for 4.34 GB source data files. Moreover, the size of result file recording DNA sequences is 54.51 GB for parallel C program compared with 57.89 GB for sequential Perl. This shows that our parallel method can also reduce the DNA synthesis cost. In addition, data types are encoded in our model, which is a basis for building type system in our future DNA computer. Finally, we describe theoretically an algorithm for DNA-based semantic fusion. This algorithm enables the process of integration of the knowledge from disparate remote sensing data sources into a consistent, accurate, and complete representation. This process depends solely on ligation reaction and screening operations instead of the ontology.

Editor: Guy J-P. Schumann, NASA Jet Propulsion Laboratory, United States of America

Funding: This study was supported by the Natural Science Foundation of China (Grant No. 61272413, No. 60903178, and No. 61272073) (URL: http://www.nsfc.gov.cn/). The funders had no role in study design, data collection and analysis, decision to publish, or preparation of the manuscript.

Competing Interests: The authors have declared that no competing interests exist.

* E-mail: tsunheng@jnu.edu.cn

Introduction

As the hereditary basis of every living organism, DNA has an ability to store and process information. This information is determined by the sequence of four distinct bases (A, C, G, T). An oligonucleotide is a short, single-stranded DNA molecule, and the complementary base pairing enables hybridization into a double-stranded polymer. These features of DNA have inspired the idea of DNA computing [1–3]. DNA computing, known also under the name of molecular computing, has great advantages of in vivo computing and in vitro computing, such as massive parallelism, extraordinary information density and exceptional energy efficiency. In contrast to traditional silicon-based technology, DNA computing has the natural potential of semantic fusion and reasoning for big data.

Nowadays, ontology has gained more and more acceptance as one of semantic technologies to solve the problem of heterogeneous knowledge sharing [4]. Many research efforts have been devoted to ontology modeling over the past decade [5–9], and quite a few running systems based on manual ontologies have been developed [10–12]. However, data is accumulating at an astounding rate with increasing computing power. Many activities,

for instance encoding an organism's DNA [13], collecting satellite data [14], and conducting scientific experiments at the Large Hadron Collider [15], can create a staggering amount of data. The growth of these big data outstrips the capacities of current ontology engineering practices and tools. In bioinformatics, the semantic integration of big data has been identified as a new frontier [16]. The same trend can also be observed in other scientific domains. For example, with a vast amount of geographical data becoming available from satellites, especially the recent opening of the Landsat archive [17], there comes an increasing demand for automatic semantic processing of remote sensing images (RSIs) in a reasonable amount of time. Up to now, reasoning from big data is challenging. As the winner of the Semantic Web Challenge, Williams provided the experimental results showing that reasoning over the Billion Triple Dataset required 3712 processors from IBM LS21 blade servers and the computation time was 1314 seconds per processor [18]. Although this dataset contains 898,966,813 triples and the size of combined dataset is around 17 GB, the amount of data obtained from satellite devices and open sources on the Internet per day is much higher and beyond the capabilities of analyst to process the data with the help of ontology [19]. Novel tools and approaches are needed to address this problem that has arisen during the current period of rapid data and knowledge growth.

Now DNA computing has become an active research area [20–24]. DNA-based parallel computing takes advantage of many different DNA molecules to solve the NP-complete problems in polynomial or even linear time, while exponentially increasing time is required in silicon-based computer. In this paper, a DNA model is introduced for semantic fusion of the RSIs. It utilizes DNA computing and ontology technologies to enable the complete representation of the RSI's knowledge in linear time regardless of the amount of data obtained.

There is few published work in the literature about the application of DNA-based approach to semantic fusion. Tsuboi proposed a pattern matching algorithm based on stickiness of DNA molecules [25]. Semantic network technology is used to solve information recognition problem. However, the fusion of semantic relationship is not involved. This restricts the analysis and reasoning capacity of the processing system. Moreover, the encoding scheme in this algorithm is not suitable for arbitrary digital information and the different data objects have to be encoded by different oligonucleotides. However, an exhaustive representation is considered unrealistic. Church proposed a novel strategy to store digit information in DNA [20]. In Church's work, all data blocks can be programmed into a bitstream and then encoded onto thousands of oligonucleotides. But the sequential conversion code (Perl) faces the challenge from big data. Xu provided a new DNA computing model for graph vertex coloring problem [26], which can effectively reduce the solution space by seminested polymerase chain reaction. All these approaches described above lack support for semantic reasoning and little attention has been given to big data, which have become the key problems of knowledge sharing and semantic representation in the web environment.

In an attempt to overcome these difficulties, we propose here a novel DNA-based semantic fusion model as an extension of our previous research for distributed data application in remote sensing field [27]. In previous work, we have implemented a semantic fusion and reasoning system for the RSIs' retrieval. At present, the use of DNA computing in semantic fusion presents numerous opportunities for our future DNA reasoner. The inherent massive parallelism of DNA strands allows for big data storage and reasoning. The main efforts in this paper are to 1) develop a remote sensing data ontology with 1,264 concepts and 2,030 semantic relationships to annotate the RSIs; 2) encode arbitrary semantic properties, property values, semantic relationships and data types in DNA, and organize the semantic information into directed acyclic graph; 3) evaluate the performance of our parallel conversion method against the sequential approach with the Rest dataset [28]; 4) create an algorithm that takes advantage of the biochemical reaction to fuse the semantic information.

Results and Discussion

Remote sensing data ontology

Ontology, as a formal representation of both implicit and explicit domain knowledge, can help to deal with heterogeneous representations of data and their interrelationships. There exist several forms of ontology with different semantic richness. As a specification developed by World Wide Web Consortium, the Resource Description Framework (RDF) [29] can present semantic information of web resources. RDF Schema [30] provides a type system for RDF and defines classes and properties that may be used to describe classes, properties and other data

resources. It can also be used to build a lightweight ontology by describing RDF vocabularies.

Figure 1 illustrates the remote sensing data ontology by using RDF Schema language. The computer code of the ontology is provided in File S1. All terms in the ontology vocabulary are divided into five groups (namely, Identification Information, Data Quality Information, Spatial Data Organization Information, Instrument Information, and Location Information) to represent the content, quality, condition, and other characteristics of data. To enable the extensibility of the ontology, we evaluated the suitability of several existing geospatial metadata standards, including the Content Standard for Digital Geospatial Metadata: *Extension* for Remote Sensing Metadata [31], ISO 19115 [32] and ISO/TS 19319 [33]. The *Extension* defines the metadata elements published by the U.S. Federal Geographic Data Committee and documents digital remote sensing datasets in the US. While ISO 19115 does only provide a structure for describing digital geographic data and many elements in ISO 19115 are from the *Extension* standard. ISO/TS 19139 defines an XML schema implementation derived from ISO 19115. These two ISO standards are very simple but not suitable for ontology modeling. Considering the fact that the conceptual model in the *Extension* does not provide enough semantic description of geographic data, we construct a hierarchical structure of the ontology. The relationships among specific classes are encoded into the ontology structure. The RDF Schema properties *rdfs:range* and *rdfs:domain* describe the relationships between specific properties and classes, and a lot of image data relationships have been described using the domain properties from the *Extension* standard.

The real RSIs must be first preprocessed with semantic annotation technique, where semantic tags defined in the ontology are assigned to the phrases in the descriptive metadata of the RSIs. This facilitates the fusion and reasoning based on image semantics. RDF instance of an RSI is shown in Figure 2, where the metadata of RSI *103001001E1EB700* are annotated with the properties such as *imagequal* (image quality), *Cloud_Cover* and *spatresv* (spatial resolution value), etc. The property values are numerous "intermediate" anonymous resources to represent constant values (called literals) such as *Excellent*, *0*, *1.85*, or aggregate concepts such as RSI's structured *Nominal_Spatial_Resolution* values. Anonymous resources cannot be referred to from outside their defining RDF instance, and hence do not require meaningful names.

Semantic property and data type

In order to convert the classes and properties representing data semantics into the sequence of nucleotides, we propose the property representation and type design suited for DNA implementation. For example, this paper annotates three RSIs *E1EB7*, *D87C9* and *B8EF1* with three properties: *city* (*ct*), *imagequal* (*qa*) and *Cloud_Cover* (*cc*). The first image's property values are *Guang Zhou* (*GZ*), *Excellent* (*E*), and *0*, respectively. The other two's values are *Hong Kong* (*HK*), *Good* (*G*), *0*, and *HK, G, 16*. Considering the linear structure of DNA strands, we arrange these properties and their values in sequence as shown in Figure 3. The label of a vertex is denoted as two-tuples (property name, property value). The edge denotes the connection between the vertices in the directed graph. To simplify the graphic structure, two new vertices labeled as "*Start*" and "*End*" are added to the directed graph and the vertices are integrated into one if they have the same property and property values. As shown in Figure 4, there are directed paths representing the annotation results of the RSIs between initial and terminal vertex in property network.

Everything would be simple if the type of property to be recorded was obviously in the form of the simple character string

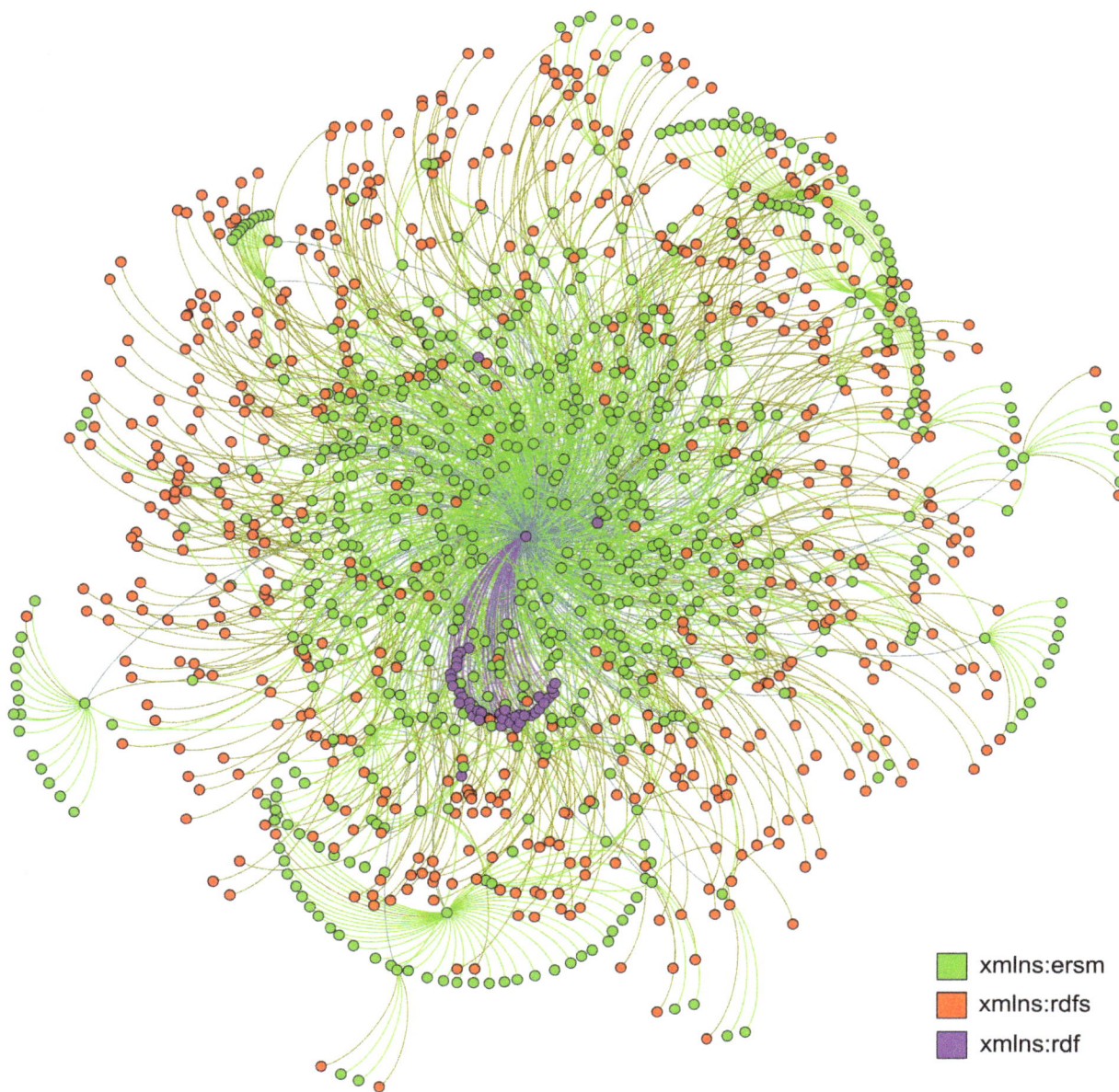

Figure 1. RDF graph of the remote sensing data ontology. This figure contains 1,264 nodes and 2,030 edges. Nodes are a set of classes and concepts in the remote sensing domain, such as *Worldwide_Reference_System*, *Multiple_Image_Alignment*, and *Spatial_Domain*, etc. Edges are a set of specific properties that characterize these classes. Classes, properties, and domains are all considered as ontology elements. All the elements are partitioned according to their namespaces. The namespaces in ontology vocabulary show the Uniform Resource Identifier References (URIrefs) as the URLs of web resources that provide further information about this vocabulary. The xmlns:ersm (http://cs.jnu.edu.cn/sun/ontology/ersm), xmlns:rdfs (http://www.w3.org/2000/01/rdf-schema), and xmlns:rdf (http://www.w3.org/1999/02/22-rdf-syntax-ns) are used mainly in our remote sensing data ontology. (For interpretation of the references to color in this figure, the reader is referred to the web version of this paper.)

literal (plain literal) illustrated so far. However, most RSIs data involve structures that are more complex than that. Many constant values that serve as property values in the RSIs are numbers (e.g. the value of a *Nominal_Spatial_Resolution* property) or some other kinds of more specialized values. For example, Figure 4 illustrates a network diagram recording information about three RSIs, where the values of RSIs' *Cloud_Cover* property are literals "0%" and "16%". However, there is no explicit indication that "0%" or "16%" should be interpreted as a number. The common practice in computer programming or database systems is to provide additional information about how to interpret a literal by

associating a data type, such as integer, boolean, or string, with this literal. In our new DNA model, 4-nt oligonucleotides are used to provide this kind of information. Since DNA strand has no built-in data type system of its own, our model simply provides a way to explicitly indicate, for a given data type, what oligonucleotide should be associated with it. Table 1 shows the common data types. The data types in this model refer to the XML Schema Datatypes defined in [34]. An advantage of this approach is that it gives our model the flexibility to directly represent information obtained from various RSIs or web sources. It is worth noting that

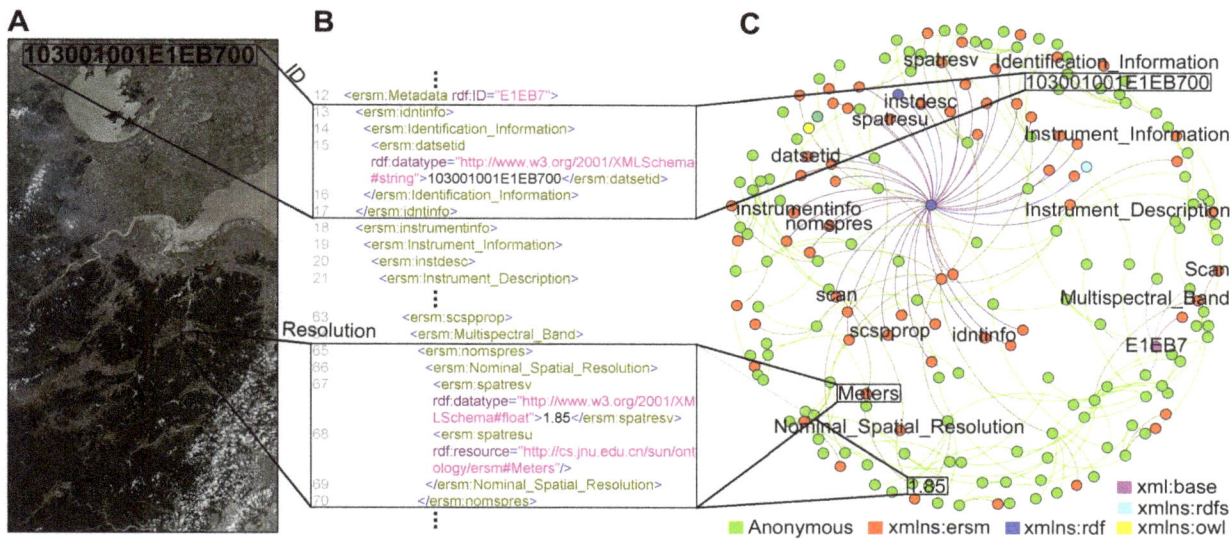

Figure 2. RDF instance description and visualization of an RSI. This figure includes three interactive parts: an RSI in A, an RDF annotation of the RSI in B, and data instance visualization in C. (A) One example RSI's ID is 103001001E1EB700 and its resolution is 1.85 meter. (B) The RDF identifies the data instance using the URIref and the image data can be described by making statements. A statement, such as "An RSI *103001001E1EB700* has a *nomspres* (Nominal Spatial Resolution) whose value is *1.85 meter*", is represented by these two RDF/XML statement blocks. File S2 provides the complete RDF code of catalog ID *103001001E1EB700* imagery. (C) The 193 classes and concepts are partitioned into six colors according to their namespaces. Most of them (120 green nodes) represent blank nodes. They provide a way to more accurately make statements about data because constant values and most aggregate concepts may not have URIs. The other namespaces include xml:base (http://cs.jnu.edu.cn/sun/ontology/103001001E1EB700), xmlns:rdfs (http://www.w3.org/2000/01/rdf-schema), xmlns:ersm (http://cs.jnu.edu.cn/sun/ontology/ersm), xmlns:rdf (http://www.w3.org/1999/02/22-rdf-syntax-ns), and xmlns:owl (http://www.w3.org/2002/07/owl). (For interpretation of the references to color in this figure, the reader is referred to the web version of this paper.)

type conversions may still be required when moving data between systems having different sets of data types.

Moreover, a property value may sometimes appear to be simple, but may actually be more complex. For example, the unit information of the spatial resolution for satellite imagery is meter, but in some cases such information is not explicitly given and omitted in contexts where it can be assumed that anyone accessing the property value will understand the unit information being used. However, this assumption is generally unsafe in the wider context of the imagery. One might give a resolution value in kilometer or degree, whilst others might assume that is in meter. In general, a comprehensive consideration should be given to the explicit representation of unit information.

Encoding the semantic information

Before the semantic information is converted into DNA, an encoding model is required. Although diverse coding strategies for DNA sequences have been developed and some have been demonstrated [20,35,36], no standard model exists. Church GM [20] first proposed a simple, universal strategy. In Church's work, arbitrary digital information can be converted into bitstreams by utilizing the ASCII code. These bits are then encoded onto the oligonucleotide library. Unlike conventional approaches, Church encodes one bit per base in order to meet the appropriate GC-content and introduces a 19-nt oligonucleotide to represent the data's address space.

However, the common type system is not considered in Church's encoding method. Thus, we propose a novel data encoding approach for semantic information. Firstly, the vertices and edges in Figure 4 are converted into DNA sequences in order to efficiently represent the semantic properties. Every vertex is associated with a 48-nt oligonucleotide which is denoted *V*. The full description about the mapping from the vertex property to the

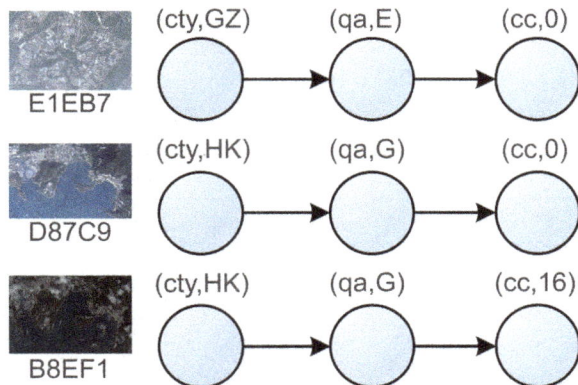

Figure 3. The linear model of semantic properties in three RSIs.

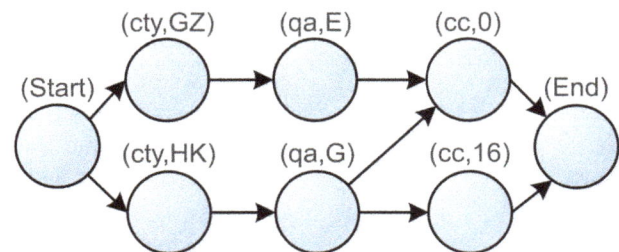

Figure 4. Network diagram of semantic property set.

Table 1. Mapping from the data types to the oligonucleotides.

Data types	Oligonucleotides
string	TCGA
boolean	CTGA
float	GTCA
dateTime	AGTC
duration	TAGC
URI	ACGT
RName	GCTA
integer	CATG
undefined	TGCA

A

B

Figure 5. Conversion performance on the test dataset. The result dataset contain DNA sequence information corresponding to the test data. (A) The conversion time is about 4,937 seconds, 31,426 seconds and 81,536 seconds for three programming languages. Error bars depict Standard Error of the mean. (B) The sizes of the datasets are both 54.51 GB for the sequential C and the parallel C. The size is 57.89 GB for the Perl program because the code uses different data block size.

DNA sequence is provided in the Materials and Methods section. Now each V, except the *start* and *end* vertices, is decomposed into four oligonucleotides whose lengths are 24, 4, 4, 16: $V = NTUA$. N, T, U, and A represent the property name, data type, unit (or comment), and property value respectively. The unit value U depends on N and T. For example, the property name cc and property value 0 in the vertex $(cc,0)$ are represented by the first and last parts of $V_{(cc,0)}$ respectively, where $N_{(cc,0)}$ = aaCgaagagC-TaagCCgCCgaaTC and $A_{(cc,0)}$ = gaCTgagaggTTggag. The oligonucleotide GCAT in $V_{(cc,0)}$ represent the unit %, as shown in Table 2.

Since the volume of electronic data expands rapidly, it is important to choose the optimal computer architecture for converting big data set. Conversion solutions range from cluster-based computing [37] to cloud-based computing [38]. Considering the cost-effective way to achieve a supercomputer performance, we use the cluster computing. All the conversion experiments in this paper were carried out in the HPC-JNU cluster system. The description about the HPC-JNU is provided in the Materials and Methods section. The sequential and parallel codes in C language are provided in File S3 and File S4 respectively. To evaluate the performance of these conversion programs, our semantic data are partly from the Rest dataset in BTC2012 dataset (http://km.aifb. kit.edu/projects/btc-2012/rest/). This dataset is encoded in NQuads format [39] and includes three data files that range in size from 409.99 MB to 2.69 GB. Figure 5 shows the conversion results of 4.34 GB source dataset in the HPC-JNU cluster system. As an explanatory scripting language, the Perl language has poor IO disk performance. The result of the parallel method shows the

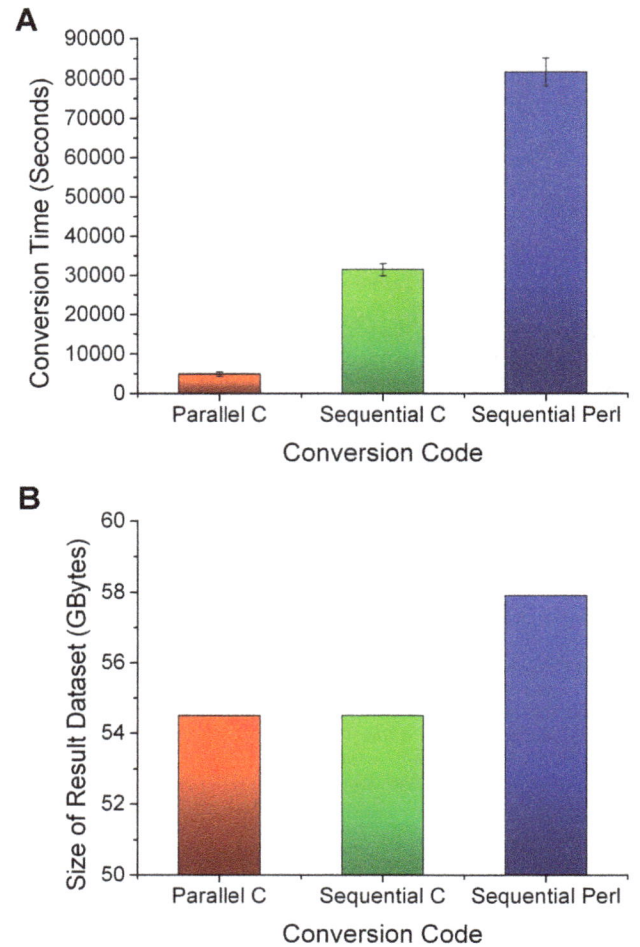

Table 2. The oligonucleotides representing the vertex properties.

Vertex	Oligonucleotides	Denotation
start	5′-ggTaagagaTTCgaCCaCTCaCgagCCaaggTgTCTaaCagTCTgCag-3′	V_{start}
(cty,GZ)	5′-aCCggaTTgTCCgCaggCCTTggCTCGATGCAaTagaCCTaCgTTaCa-3′	$V_{(cty,GZ)}$
(qa,E)	5′-gaTaagaaaTTCaagTgTTggagTTCGATGCAaaCggagagTgagTaT-3′	$V_{(qa,E)}$
(qa,null)	5′-gaTaagaaaTTCaagTgTTggagTTGCATGCAaaCggagagaCagaag-3′	$V_{(qa,null)}$
(cc,0)	5′-aaCgaagagCTaagCCgCCgaaTCCATGGCATgaCTgagaggTTggag-3′	$V_{(cc,0)}$
(cc,null)	5′-aaCgaagagCTaagCCgCCgaaTCTGCATGCAgaCagagaggTaggag-3′	$V_{(cc,null)}$
end	5′-ggTaaggaggTaggagagTaaggagCCggTgCgCCaCCTggTTggTaa-3′	V_{end}

best performance although the user of the cluster system has a maximum limit of 80 cores.

DNA's storage density

At present, remote sensing data are dramatically increasing in volume. For example, the U.S. National Climatic Data Center holds the world's largest archive of weather data and has archived 3 PB (petabyte) satellite imagery [40]. The extreme compactness of DNA is incredible. Because the mean molecular weight of a nucleotide is 330 g/mol [41] and a 200 bp encodes 128 bits in our encoding method, one gram of DNA can store 5.84×10^{20} bits. We approximate DNA's density to water's density (10^{-3} g/mm^3), then the volume of all DNA sequences encoding 3 PB data is 4.63×10^{-2} mm^3. We compare favorably contemporaneous storage technologies in Table 3 [42–50]. DNA storage has obviously the potential of storing data 100 times more compactly than other technologies.

Semantic fusion based on DNA

Semantic fusion is the key operation that ontology technology supports. It can automatically implement the union of the properties and semantic relationships. A resource, such as an RSI, and its replicas may be widely distributed over several image replicas databases. The owners of the resource may select different kinds of feature properties to annotate this RSI. We must merge these properties and relationships in order to improve the efficiency and accuracy of the knowledge. As shown in Figure 6, the semantic fusion enables image's semantic information from disparate data sources to be merged. The initial properties dissolve in the new properties and do not preserve their duplicate internal structures. However, the performances of ontology fusion and reasoning degrade rapidly as data grows. Therefore, we build a semantic fusion model based on DNA.

Table 2 shows a set of oligonucleotides representing the possible properties labeling the vertices in Figure 6A. As regards orientation, all of the oligonucleotides are written 5′ to 3′. Now each V in Figure 6A is divided into two oligonucleotides, each of length 24: $V = V'V''$. V' and V'' are the first and second half of V. An edge from the vertex i to the vertex j is encoded as a 48-nt oligonucleotide, obtainable as the Watson-Crick complement of the second and the first halves of the oligonucleotides encoding the vertices i and j touching the edge. For example, the encoding of an edge from the vertex (cty,GZ) to the vertex (qa,E) is given: $e_{(cty,GZ) \to (qa,E)}$ = AGCTACGTTaTCTggaTgCaaTgTCTaTTC TTTaagTTCaCaaCCTCa. For every vertex and every edge

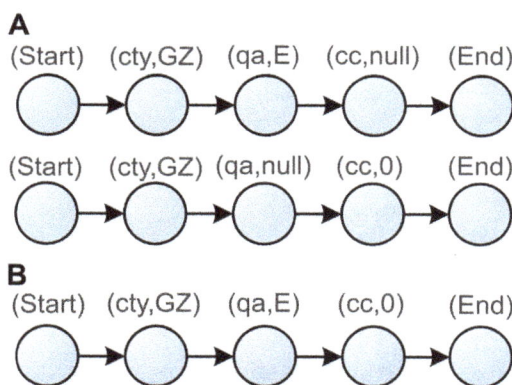

Figure 6. Semantic fusion pattern of an RSI. (A) Two owners of the RSI *E1EB7* select different properties to annotate it. One of them selects the properties *cty* and *qa*. The other selects the properties *cty* and *cc*. The property value *null* means the unannotated property. Certainly, both its data type and its unit are *undefined*. (B) The result property string after semantic fusion represents the complete semantic information of this RSI.

in Figure 6A, large quantities of V_i and e$_{ij}$ are mixed together in the hybridization and ligation reaction as shown in Figure 7. The oligonucleotides V_i served as splints to bring oligonucleotides associated with compatible edges together for ligation. Consequently, many DNA molecules encoding the property string are created. The remaining steps, as well as the conclusion in the output, are filtering and screening procedures. We use the Adleman style [1,51] algorithm for obtaining the result property string:

Input: DNA molecules generated randomly in large quantities.

Step 1: Reject all DNA molecules that do not begin with V_{start} and end in V_{end}.

Step 2: Reject all DNA molecules encoding property strings that do not involve exactly 5 vertices.

Step 3: Reject all DNA molecules that contain the oligonucleotide TGCATGCA encoding the *null* value.

Output: Read out the property strings (if any).

As shown in Figure 8, we can obtain the result property string by using the semantic fusion method based DNA. It is consistent with the semantic properties in Figure 6B.

Table 3. Storage volume calculations for 3 PB data.

Medium type	Year	Volume (mm³)	Notes
CD-ROM [42]	1982	6.24×10^{10}	1.2 mm thickness, 120 mm diameter, 700 MB
DVD-R (single layer) [43]	1996	9.08×10^{9}	1.2 mm thickness, 120 mm diameter, 4.7 GB
Blu-ray (single layer) [44]	2002	1.71×10^{9}	1.2 mm thickness, 120 mm diameter, 25 GB
Flash memory [45]	2013	1.25×10^{8}	72 mm×26.94 mm×21 mm, 1 TB
Magnetic tape (LTO-6) [46]	2012	8.02×10^{7}	6.1 μm thickness, 846 m length, 12.65 mm width, 2.5 TB
Hard disk [47]	2013	1.98×10^{5}	10 TB/inch², platter 1 mm thickness
Quantum storage [48–50]	2012	5.16	5×7 bit/10×10 nm² on the Cu(111) surface, the average height of Cu(111) terrace 65 nm, bilayer cobalt nano-islands 0.8 nm, two additional capping layer 1 nm
This paper	2013	4.63×10^{-2}	

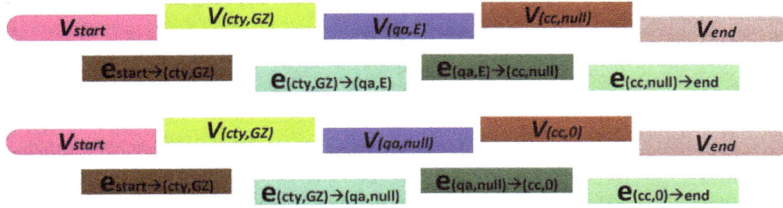

Figure 7. The oligonucleotides in the hybridization and ligation reaction. For each property *i* including the labels *start* and *end*, a 48-nt oligonucleotide V_i is generated. For each edge ij, an oligonucleotide e_{ij} is derived from the 3′ 24-nt of V_i and the 5′ 24-nt of V_j.

Abstract representation of semantic fusion

The above algorithm can be formally described by an abstract model. This abstract model is based on the data structure of the tubes. A tube is a multi-set of finite strings over the alphabet {A, C, G, T}, namely the DNA alphabet. Given a tube, one can perform the following operations:

1. pre-separate(*T*, *s*)/post-separate(*T*, *s*)/sub-separate(*T*, *s*). Given a tube *T* and a string *s* over the alphabet {A, C, G, T}, this operation creates a tube containing all strands in *T* that have the string *s* as a prefix/postfix/substring.

2. length-separate(*T*, *n*). Given a tube *T* and integer *n*, this operation creates a tube containing all strands in *T* with length less than or equal to *n*.

3. detect(*T*). Given a tube *T*, this operation outputs true if *T* contains at least one DNA molecule, otherwise outputs false.

In our model, each of the oligonucleotides in *T* is of length 48. Thus,

SemanticFusion(*T*):

1. input(*T*)
2. *T* ← pre-separate(*T*, V_{start})
3. *T* ← post-separate(*T*, V_{end})
4. *T* ← length-separate(*T*, 240)
5. *T* ← sub-separate(*T*, TGCATGCA)
6. detect(*T*).

This model starts with the input tube *T*, containing the result of the ligation reaction. All separate operations select the oligonucleotides and thus require the amplification of the resulting tubes by the PCR (polymerase chain reaction).

Indeed, semantic fusion problem have been shown to be an NP-complete problem [52,53], which means that it is unlikely to find an algorithm working in polynomial time. The semantic fusion on image properties of modest size requires an altogether impractical amount of time on conventional electronic computer [54,55]. However, we use a finite sequence of ligation reaction and screening operations described above to solve the semantic fusion problem. A fusion starts with an initial tube and ends with one final tube. The fusion time depends solely on the total time of ligation reaction and five screening steps instead of the number of semantic properties and ontology complexity. Then the massive parallelism of DNA renders exponential time complexity in semantic fusion to linear time.

Conclusions

Semantic fusion is a process that is ubiquitous in nature. In this paper, a novel DNA-based semantic fusion model is proposed. The model combines organically parallel strategy with DNA encoding, which makes semantic conversion more efficient and storage density higher. Furthermore, we describe the abstract representation of semantic fusion and thus show that the fusion time of semantic properties in remote sensing images depends solely on the biochemical reactions and operations instead of the ontology. However, there are still many issues to be considered. Foremost issue is error. DNA molecules are fragile and they break easily. The errors of separate operations with DNA strands can make a really dramatic difference. Thus, steps towards coping with errors should be taken in. In future work, we also implement the ligation reaction and screening procedures based on biochemical techniques and clarify details in another paper.

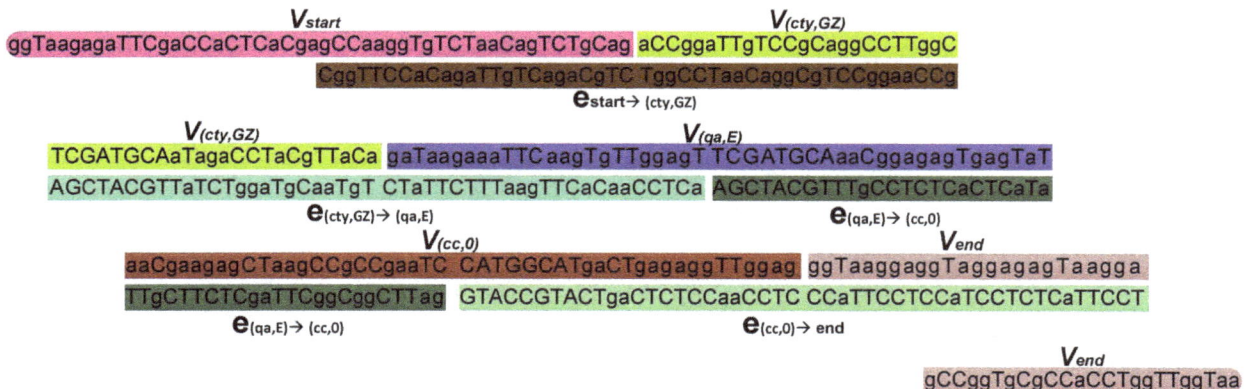

Figure 8. DNA sequence representing the complete semantic information.

Table 4. Specifications of the HPC-JNU cluster system.

Hardware			Software	
	Computational node	**Storage node**		
CPU	AMD Opteron 2.4 GHz	Intel Xeon 2.13 GHz	OS	CentOS 6.2
Number of nodes	20	1	MPI	Open MPI 1.6
Number of CPU cores/node	24	4	File System	NFS 4.1
Number of CPU cores	480	4	Queue Scheduler	Torque 3.3
Memory/node	48 GB	8 GB		
Disk	300 GB	26TB RAID5 Array		
Interconnection network	40 G QDR InfiniBand	40 G QDR InfiniBand		

Materials and Methods

Mapping from semantic information to an oligonucleotide

All properties and property values are converted to binary strings based on ASCII encoding. Each character corresponds to an 8-bit binary code. For example, the property *cty* has the binary code *01100011011101000111001*. Conversion code in File S4 can then convert these bits to a or g for 0 and T or C for 1. Bases are chosen randomly according to the result of function *rand()*. Considering the big dataset, we add a 32-bit address starting from *00000000000000000000000000000000*. For example, the properties and property values of an RSI *E1EB7* in Figure 3 is represented by the string *startctyGZ qa E cc00 end*, where the symbol represents a whitespace character, *start* and *end* are the labels of the new vertices added in Figure 4. This property string has an ASCII code *001000000111001101110100011000010111001001110100 011000110111010001110010100011101011010001000001110 001011000010010000001000101001000000110001101100011100 110000000110000001000000010000000100000011001010110111 001100100*. It is then encoded to two 200 nt oligonucleotides by the conversion code given in File S4. Each encodes a 128-bit data block (128 nt). Before synthesized, the sequence is augmented to include the bases representing data type and data unit. For example, an oligonucleotide *aCCggaTTgTCCgCaggCCTTggCaTa-gaCCTaCgTTaCa* is the result of encoding the property *ctyGZ* in the vertex (cty,GZ). Considering the data type is *string* and data unit is *undefined*, we add TCGA and TGCA to the original oligonucleotide according to Table 1. Thus, the final oligonucleotide of the vertex (cty,GZ) is *aCCggaTTgTCCgCaggCCTTggCTCGATGCAaTa-gaCCTaCgTTaCa*, as shown in Table 2.

Specification of the cluster system

The HPC-JNU cluster system (http://hpc.jnu.edu.cn/) has 20 computational nodes. Each node is connected via the InfiniBand network. Table 4 shows the specifications of the HPC-JNU cluster system. Figure S1 and Figure S2 show the photographs of the computational nodes and the storage node.

Supporting Information

Figure S1   Photograph of the computational nodes. (JPG).

Figure S2   Photograph of the storage node. (JPG).

File S1   Code for remote sensing data ontology (see also http://cs.jnu.edu.cn/sun/ontology). Computer code in the RDF Schema language is used to generate the remote sensing data ontology in **Figure 1**. The RDF/OWL API is required. (RDFS).

File S2   Code for ID 103001001E1EB700 instance (see also http://cs.jnu.edu.cn/sun/ontology). Computer code in the RDF language is ontology annotation file of remote sensing data (catalog ID 103001001E1EB700) instance in **Figure 2**. The RDF/OWL API is required. (RDF).

File S3   The sequential conversion code in C language. The code accesses and converts the data stored contiguously on disk. Despite the cache provided by the operating system, an application that performs a large number of reads, conversions and writes usually faces the performance challenge. GCC compiler is required. (C).

File S4   The parallel conversion code in C language. To support the run-time allocation of conversion tasks, a manager/worker-style parallel C program has been built. The multiple processes of this parallel program can simultaneously access and convert big data by utilizing the MPI-IO. The MPI API is required. (C).

Acknowledgments

The authors would like to gratefully acknowledge the useful comments of Dr. Guy J-P. Schumann (Academic Editor) and the anonymous reviewer. We would also like to acknowledge input and advice from Dr. Shubin Cai, Dr. Shun Long, Hao Jiang, Shuanghuan Lv and Puihang Wong.

Author Contributions

Conceived and designed the experiments: HS JW GY. Performed the experiments: HS JW RHM. Analyzed the data: HS JW GY RHM. Contributed reagents/materials/analysis tools: HS JW GY. Wrote the paper: HS GY RHM.

References

1. Adleman LM (1994) Molecular computation of solutions to combinatorial problems. Science 266: 1021–1024.

2. Lipton R (1995) DNA solution of hard computational problems. Science 268: 542–545.

3. Bancroft C, Bowler T, Bloom B, Clelland CT (2001) Long-term storage of information in DNA. Science 293: 1763–1765.

4. Renear A, Palmer C (2009) Strategic reading, ontologies, and the future of scientific publishing. Science 325: 828–832.

5. Yoder MJ, Miko I, Seltmann KC, Bertone MA, Deans AR (2010) A gross anatomy ontology for hymenoptera. PLoS ONE 5(12): e15991.

6. Janowicz K (2012) Observation-driven geo-ontology engineering. Trans GIS 16: 351–374.

7. Alterovitz G, Xiang M, Hill D, Lomax J, Liu J, et al. (2010) Ontology engineering. Nat Biotechnol 28: 128–130.

8. Iribarne L, Padilla N, Asensio JA, Criado J, Ayala R, et al. (2011) Open-environmental ontology modeling. IEEE Trans Syst Man Cybern A Syst Hum 41: 730–745.

9. Hastings J, Chepelev L, Willighagen E, Adams N, Steinbeck C, et al. (2011) The chemical information ontology: Provenance and disambiguation for chemical data on the biological semantic web. PLoS ONE 6(10): e25513.

10. Ashburner M, Ball C, Blake J, Botstein D, Butler H, et al. (2000) Gene ontology: Tool for the unification of biology. Nat Genet 25: 25–29.

11. Hey T, Trefethen A (2005) Cyberinfrastructure for e-Science. Science 308: 817–821.

12. Ma XG, Carranza EJ, Wu CL, Meer FD (2012) Ontology-aided annotation, visualization, and generalization of geographic time-scale information from online geographic map services. Comput Geosci 40: 107–119.

13. Gerstein M (2012) Genomics: ENCODE leads the way on big data. Nature 489: 208.

14. Mervis J (2012) Agencies rally to tackle big data. Science 336: 22.

15. Lynch C (2008) Big data: How do your data grow. Nature 455: 28–29.

16. Jones M, Schildhauer M, Reichman O, Bowers S (2006) The new bioinformatics: Integrating ecological data from the gene to the biosphere. Annu Rev Ecol Evol Syst 37: 519–544.

17. Woodcock CE, Allen R, Anderson M, Belward A, Bindschadler R, et al. (2008) Free access to Landsat imagery. Science 320: 1011.

18. Williams G, Weaver J, Atre M, Hendler J (2010) Scalable reduction of large datasets to interesting subsets. J Web Semant 8: 365–373.

19. Schneider T, Hashemi A, Bennett M, Brady M, Casanave C, et al. (2012) Ontology for big systems: The ontology summit 2012 communique. Appl Ontol 7: 357–371.

20. Church GM, Gao Y, Kosuri S (2012) Next-generation digital information storage in DNA. Science 337: 1628.

21. Ke YG, Ong LL, Shih WM, Yin P (2012) Three-dimensional structures self-assembled from DNA bricks. Science 338: 1177–1183.

22. Halvorsen K, Wong WP (2012) Binary DNA nanostructures for data encryption. PLoS ONE 7(9): e44212.

23. Borresen J, Lynch S (2012) Oscillatory threshold logic. PLoS ONE 7(11): e48498.

24. Bryant B (2012) Chromatin computation. PLoS ONE 7(5): e35703.

25. Tsuboi Y, Ibrahim Z, Ono O (2005) DNA-based semantic memory with linear strands. Int J Innov Comput I 1: 755–766.

26. Xu J, Qiang XL, Yang Y, Wang BJ, Yang DL, et al. (2011) An unenumerative DNA computing model for vertex coloring problem. IEEE T Nanobiosci 10: 94–98.

27. Sun H, Li SX, Li WJ, Ming Z, Cai SB (2005) Semantic-based retrieval of remote sensing images in a grid environment. IEEE Geosci Remote Sens Lett 2(4): 440–444.

28. Konrath M, Gottron T, Staab S, Scherp A (2012) SchemEX: Efficient construction of a data catalogue by stream-based indexing of linked data. J Web Semant 16: 52–58.

29. Hendler J (2003) Science and the semantic web. Science 299: 520–521.

30. Wang XS, Gorlitsky R, Almeida JS (2005) From XML to RDF: How semantic web technologies will change the design of omic standards. Nat Biotechnol 23: 1099–1103.

31. Tsou M (2004) Integrating web-based GIS and image processing tools for environmental monitoring and natural resource management. J Geogr Syst 6: 155–174.

32. Wei YX, Di LP, Zhao BH, Liao GX, Chen AJ (2007) Transformation of HDF-EOS metadata from the ECS model to ISO 19115-based XML. Comput Geosci 33: 238–247.

33. Batchellera J, Reitsma F (2010) Implementing feature level semantics for spatial data discovery: Supporting the reuse of legacy data using open source components. Comput Environ Urban Syst 34: 333–344.

34. Geneves, Pierre G, Nabil L, Vincent Q (2011) Impact of XML schema evolution. ACM Trans Internet Technol 11: 1–27.

35. Gibson DG, Glass JI, Lartigue C, Noskov VN, Chuang RY, et al. (2010) Creation of a bacterial cell controlled by a chemically synthesized genome. Science 329: 52–56.

36. Clelland CT, Risca V, Bancroft C (1999) Hiding messages in DNA microdots. Nature 399: 533.

37. Afek Y, Alon N, Barad O, Hornstein E, Barkai N, et al. (2011) A biological solution to a fundamental distributed computing problem. Science 331: 183–185.

38. Fox A (2011) Cloud computing-what's in it for me as a scientist. Science 331: 406–407.

39. Cyganiak R, Harth A, Hogan A (2012) N-quads: Extending n-triples with context. Available: http://sw.deri.org/2008/07/n-quads/. Accessed 2012 Nov 29.

40. Lattanzio A, Schulz J, Matthews J, Okuyama A, Theodore B, et al. (2013) Land surface albedo from geostationary satellites. B Am Meteorol Soc 94: 205–214.

41. Kneuer C, Sameti M, Bakowsky U, Schiestel T, Shirra H, et al. (2000) A nonviral DNA delivery system based on surface modified silica-nanoparticles can efficiently transfect cells in vitro. Bioconjug Chem 11: 926–932.

42. Imai H (1982) Sony CDP-101 co player. Stereo Review 12: 63.

43. Mimura H (1997) DVD-video format. Proceedings of IEEE COMPCON 97. San Jose, California, , United States: IEEE. 291–294.

44. Blu-ray Disc Association (2010) White paper blu-ray disc format. Available: http://www.blu-raydisc.com/en/Technical/TechnicalWhitePapers/General. aspx. Accessed 2013 Jul 11.

45. Kingston (2013) Kingston digital ships its fastest, world's largest-capacity USB 3.0 flash drive. Available: http://www.kingston.com/us/company/ press?article = 6487. Accessed 2013 Jul 11.

46. Rivera R, Vargas G, Vazquez M (2012) IBM system storage LTO ultrium 6 tape drive performance white paper. Available: http://public.dhe.ibm.com/ common/ssi/ecm/en/tsw03182usen/TSW03182USEN.PDF. Accessed 2013 Jul 11.

47. Hussain S, Kundu S, Bhatia CS, Yang H, Danner AJ (2013) Heat assisted magnetic recording (HAMR) with nano-aperture VCSELs for 10 Tb/in^2 magnetic storage density. Proceedings of SPIE 8639, Vertical-Cavity Surface-Emitting Lasers XVII, 863909. San Francisco, California, , United States.

48. Brovko OO, Stepanyuk VS (2012) Quantum spin holography with surface state electrons. Appl Phys Lett 100: 163112.

49. Oka H, Ignatiev PA, Wedekind S, Rodary G, Niebergall L, et al. (2010) Spin-dependent quantum interference within a single magnetic nanostructure. Science 327: 843–846.

50. Figuera JDL, Prieto JE, Ocal C, Miranda R (1993) Scanning-tunneling-microscopy study of the growth of cobalt on Cu(111). Phys Rev B 47: 13043–13046.

51. Paun G, Rozenberg G, Salomaa A (1998) DNA computing: New computing paradigms. Berlin: Springer-Verlag. 43–50 p.

52. Glimm B, Horrocks I, Lutz C, Sattler U (2008) Conjunctive query answering for the description logic SHIQ. J Artif Intell Res 31: 157–204.

53. Calvanese D, Giacomo GD, Lembo D, Lenzerini M, Rosati R (2005) DL-lite: Tractable description logics for ontologies. Proceedings of the 20th National Conference on Artificial Intelligence and the 17th Innovative Applications of Artificial Intelligence Conference, AAAI-05/IAAI-05. Pittsburgh, PA, United states. 602–607.

54. Leida M, Gusmini A, Davies J (2012) Semantics-aware data integration for heterogeneous data sources. J Ambient Intell Humaniz Comput. Available: http://link.springer.com/content/pdf/10.1007%2Fs12652-012-0165-4.pdf. Accessed 2013 Jul 24.

55. Lewis JJ, Callaghan RJO, Nikolov SG, Bull DR, Canagarajah N (2007) Pixel- and region-based image fusion with complex wavelets. Inf Fusion 8: 119–130.

16

Effective Key Parameter Determination for an Automatic Approach to Land Cover Classification Based on Multispectral Remote Sensing Imagery

Yong Wang[1], Dong Jiang[1]*, Dafang Zhuang[1], Yaohuan Huang[1], Wei Wang[2], Xinfang Yu[1]

1 State Key Laboratory of Resources and Environmental Information System, Institute of Geographical Sciences and Natural Resources Research, Chinese Academy of Sciences, Beijing, China, **2** School of Computer and Information Engineering, Beijing Technology and Business University, Beijing, China

Abstract

The classification of land cover based on satellite data is important for many areas of scientific research. Unfortunately, some traditional land cover classification methods (e.g. known as supervised classification) are very labor-intensive and subjective because of the required human involvement. Jiang et al. proposed a simple but robust method for land cover classification using a prior classification map and a current multispectral remote sensing image. This new method has proven to be a suitable classification method; however, its drawback is that it is a semi-automatic method because the key parameters cannot be selected automatically. In this study, we propose an approach in which the two key parameters are chosen automatically. The proposed method consists primarily of the following three interdependent parts: the selection procedure for the pure-pixel training-sample dataset, the method to determine the key parameters, and the optimal combination model. In this study, the proposed approach employs both overall accuracy and their Kappa Coefficients (KC), and Time-Consumings (TC, unit: second) in order to select the two key parameters automatically instead of using a test-decision, which avoids subjective bias. A case study of Weichang District of Hebei Province, China, using Landsat-5/TM data of 2010 with 30 m spatial resolution and prior classification map of 2005 recognised as relatively precise data, was conducted to test the performance of this method. The experimental results show that the methodology determining the key parameters uses the portfolio optimisation model and increases the degree of automation of Jiang et al.'s classification method, which may have a wide scope of scientific application.

Editor: Guy J-P. Schumann, NASA Jet Propulsion Laboratory, United States of America

Funding: This study was partially funded by the China Postdoctoral Science Foundation (20060400496), the Chinese Academy of Sciences (KZZD-EW-08), the National Natural Science Foundation of China (41001279), and the National Scientific and Technological Support Projects of China (2011BAJ07B01). The funders had no role in study design, data collection and analysis, decision to publish, or preparation of the manuscript.

Competing Interests: The authors have declared that no competing interests exist.

* E-mail: jiangd@igsnrr.ac.cn

Introduction

As research on global change has grown in depth and scope, Land Use and Land Cover Change (LUCC) has increasingly become a core part of global environmental change research [1,2,3]. Multispectral satellite imagery is an important data source for LUCC research [4,5,6,7,8,9]. One of the most common uses of satellite images is the mapping of LUCC via image classification. Various methods or algorithms have been successfully employed in LUCC classification and change detection, including visual interpretation classification [10], unsupervised classification [11], supervised classification [12,13] (e.g. artificial neural network algorithms [14,15], support vector machine algorithms [16,17]), object oriented classification [18,19], and decision tree algorithms [20,21]. Different methods have their own scope, advantages and disadvantages [22,23,24,25,26,27]. Some new methods of land classification imagery that lack historical and coincidental ground information to either calibrate data, validate data or assess identification accuracy have been proposed [28,29,30], which can increase classification accuracy. However, some important classification steps including invariant feature identification, training samples establishment, classification accuracy assessment and so on all require human participation, which made the classification procedure hard to be carried out automatically.

To overcome the problems mentioned above, a promising solution in land cover classification is to better utilise a prior, high-precision classification map instead of independently classifying the remote sensing images. Jiang et al. proposed a simple but robust method for land cover classification using a prior classification map [1]. In that study, the prior high-precision classification map and the multispectral remote sensing image were first employed to obtain pure pixels and constitute a semi-automatic classification dataset of training samples. Principal component analysis (PCA) was then performed on the data in all spectral bands of each land cover class extracted from the region of interest. The satellite images in that study were automatically classified using only the prior land cover map, thus requiring less human interaction or interpretation. Jiang et al.'s classification results showed that the classification method is appropriate for different environmental condition land cover classification. Although Jiang et al.'s method was capable of producing a reasonably accurate land cover classification map in a cost-effective way, the method was only a semi-automatic approach, not an automatic one, because the key parameters used (P_a and

P_{buffer}) could not be selected automatically. Two questions about these parameters arise to which clear answers are not available in the literature:

1. How should the parameters P_a (P_a is the accumulation area threshold of a certain class of land cover) and P_{buffer} (P_{buffer} is the area threshold for buffer analysis) be determined?

2. How should the optimal combination of (P_a, P_{buffer}) be determined?

We address these two questions in this study, as they are the key and most important parts of the study of the semi-automatic approach to land cover classification based on multispectral remote sensing imagery [1]. The purpose of this article is to present the procedures of these analyses.

In spite of its limitations, an approach using "pure pixels" representative of the major land use classes as training samples to classify images is promising. Therefore, we developed a new approach for automatically selecting the key parameters using the efficient computer technique and the test algorithm of image classification accuracy based on the existing method, and we improved Jiang et al.'s strategy through automatically adjusting and choosing the vital two parameters. The proposed approach aims to achieve two objectives: (1) to select the optimal parameters P_a and P_{buffer} and (2) to determine the portfolio optimisation model (P_a, P_{buffer}). In the proposed approach, the overall classification accuracy and their derived Kappa Coefficient (KC), which is widely used to assess the accuracy of classification results in application studies, is employed to select the optimal parameters P_a and P_{buffer}. TC (Time-Consuming), which signifies the classification efficiency, is employed to determine the best combination of the two parameters. The greatest challenge in our method is to properly select the training sample. We propose an iterated procedure to automatically select a different percentage of pure pixels as training samples based on a prior classification map to ensure that the method is completely automatic. This approach was evaluated by using it to generate a land cover classification of Weichang County in Hebei Province, China, applying a Landsat TM (Thematic Mapper) image and a prior land cover classification map. Our method is expected to be more practicable for automatic land cover classification than traditional classification methods or algorithms such as the Maximum Likelihood approach [13].

Methods

The proposed approach includes three main, interdependent components: the selection procedure for the pure-pixel training sample dataset, the method for determining the key parameters, and the optimal combination model. The general flowchart is shown in Fig. 1.

Automatic Selection of Pure-pixel Training Samples

Fundamentally, training samples provide descriptive statistical information for each class in multispectral remote sensing imagery that may be used to classify an image. The key step in each classification approach is the proper selection of the training samples. Traditionally, accurate training samples are selected manually depending entirely on the knowledge of the analyst or on field investigations, which reduce the automation of land cover classification [1]. Jiang et al. introduced a novel idea of extracting sterling pixels of land cover semi-automatically using an accurate, existing land cover dataset as prior knowledge. Similar to supervised classification or/and object-oriented classification,

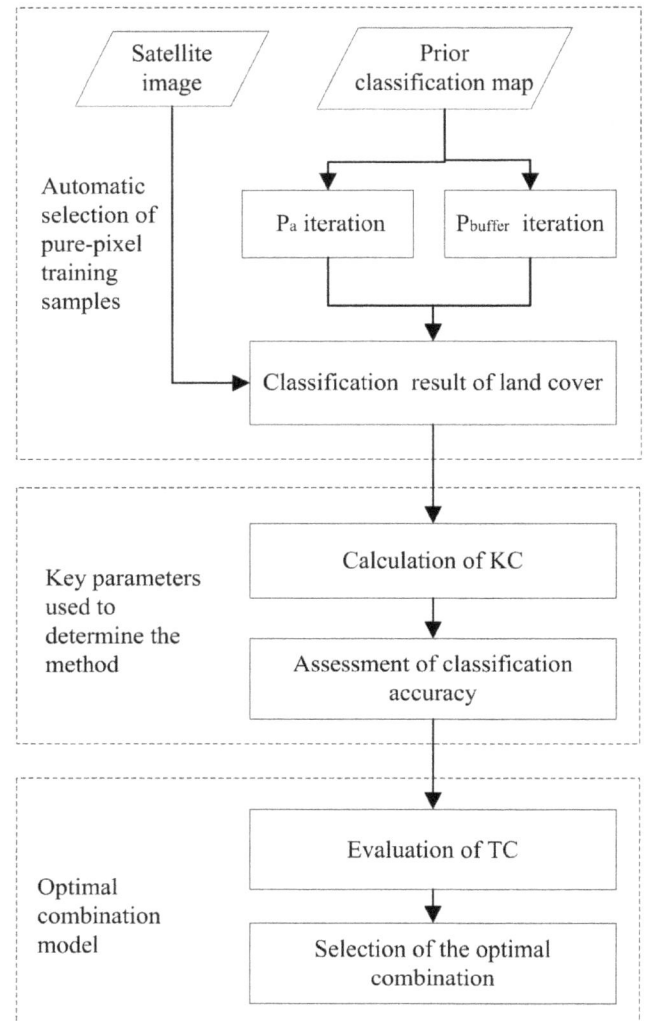

Figure 1. General flowchart of proposed approach (KC: Kappa Coefficient; TC: Time-Consuming).

these selected pixels as training samples are used to characterise the classes and ultimately convey the information to three-dimensional feature space. Samples of different types of land cover selected have an accumulation area threshold (P_a) value as follows:

$$P_a \leq \frac{\sum_{j=0}^{x} Ac_{ij}}{As_i} \qquad (1)$$

where P_a is the accumulation area threshold of the ith class of land cover, Ac_{ij} is the area of the patches of the ith class of land cover sorted in descending order, x is the number of the ith class of land cover, and As_i is the total area of the ith class of land cover. According to ecological theory, a joint region with different types of land cover is discarded during spatial buffer analysis, and the buffer analysis distance is variable. The buffer analysis distance is defined as follows:

$$P_{buffer} = \frac{Ab_d}{A}, d < 0 \qquad (2)$$

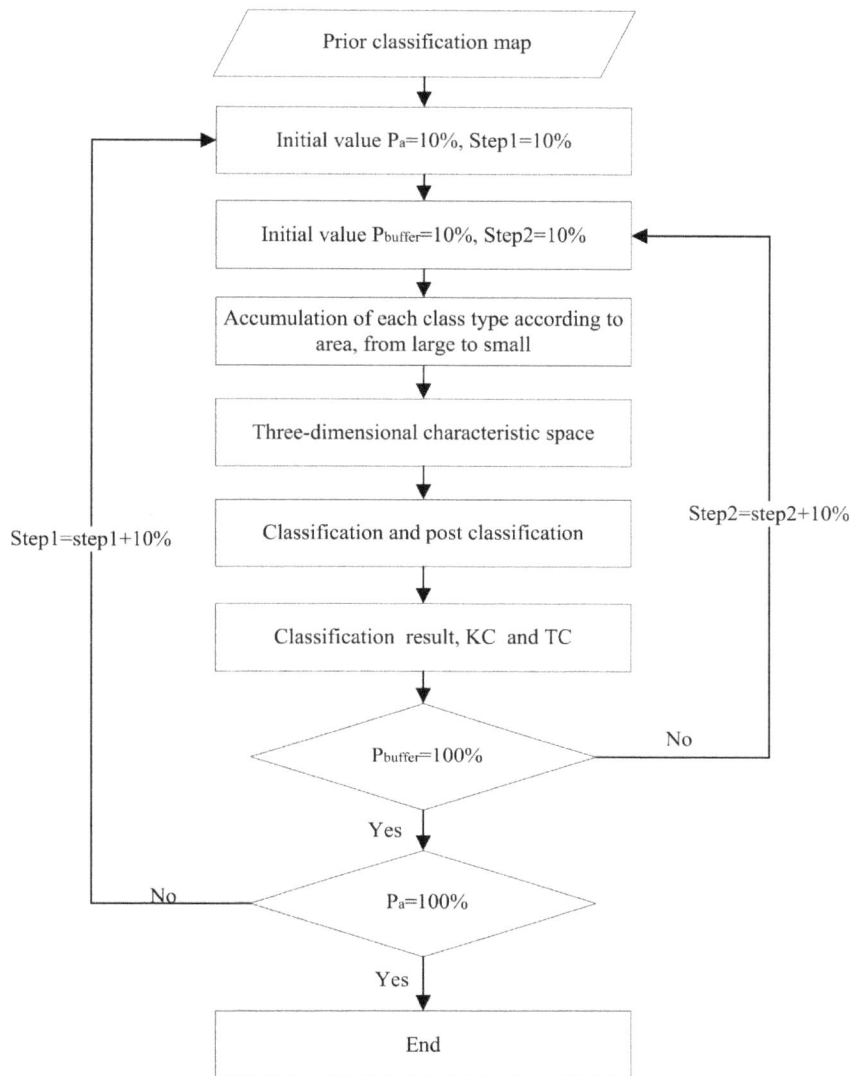

Figure 2. Flowchart of the iterated procedure used to determine the key parameters (P_a and P_{buffer}) (KC: Kappa Coefficient; TC: Time-Consuming).

where P_{buffer} is area threshold for buffer analysis, Ab_d is the buffer area of the patch with a distance of d. The variable d is negative, indicating that the representative area was reduced. A is the area of the patch. The pure pixels within buffer region are chosen as training samples, different buffer regions constitute the diversely automatic training samples collection, and the accuracy of the collection depends on the key parameters P_a and P_{buffer}. The combination of (P_a, P_{buffer}) is used to determine which samples are selected optimally for land cover classification.

The drawback of Jiang et al.'s method was not an automatic one because the two critical parameters P_a and P_{buffer} were chosen to be 60% and 50%, respectively, by a series of experiments using data in 4 different test areas (the determination procedure we called test-decision), instead of by automatic calculation. These experiments used an exploratory rather than an automatic method because satisfactory classification results require the proper calibration of various model parameters. In order to choose P_a and P_{buffer} automatically, we propose an iterated procedure reliably based on computer technologies to ensure that the method

is completely automatic. Additionally, an approach based on this iterated algorithm is employed to reduce the "salt-and-pepper" error that usually occurs in pixel-based classification methods. This iterated procedure refining more pure, sterling pixels within the changed/unchanged area as the training samples is expected to improve classification accuracy. After iteration, the final changed/unchanged sampling results are obtained and used for the images' classification. The iterated procedure used to determine the key parameters is shown as follows:

As described in Fig. 2, the key parameters are run through the classification process and determine the classification results by verifying the collection of the pure pixels that constitute the training samples dataset. This determination process is composed of a dual circulation. The purpose of the inner loop is to increase P_{buffer}, which has an initial value of 10%, incrementally with 10% steps until it reaches 100%. In the outer loop, P_a increases incrementally from 10% to 100% with 10% steps. In each re-circulation, the PCA, the establishment of three-dimensional feature space, classification, and post-classification is carried out in

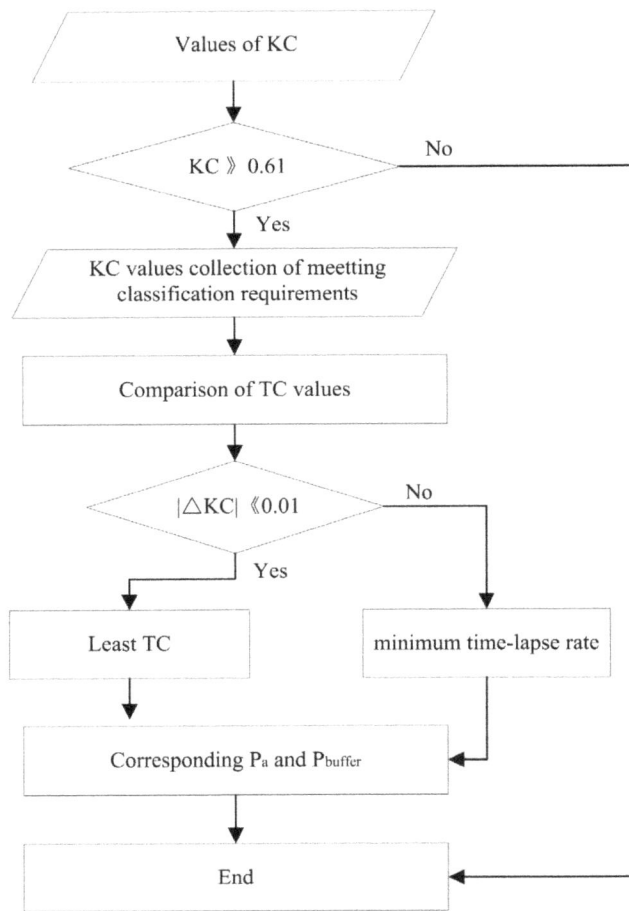

Figure 3. Selection process of portfolio optimisation model (KC: Kappa Coefficient; TC: Time-Consuming).

turn. Ultimately, there are many different levels of training sample datasets (are comprise of different buffer regions), and different levels of land cover classification results and different values of TC are generated, which help determine the key parameters. By evaluating the entire process, we adopt an enumeration method to determination the key parameters in the following part.

Determining the Key Parameters

Different values of P_a and P_{buffer} determine different classification results. Accurate classification results correspond to optimal classification parameters. In our strategy, we determined the key parameters through evaluating the classification results. The procedure of accuracy assessment of the classification results is carried out by comparing these results to a known, accurate classification map, which has been visually interpreted and classified using high spatial resolution images and validated by intensive field surveys. In LUCC research, KC is a very important index value to the accuracy assessment of land cover classification [31,32,33,37] because KC value provides both a better overall measure of accuracy and incorporates information about the errors of omission and commission [34,35].

Furthermore, the KC measures the association between the two inputs (the known classification map and the TM image) and helps to evaluate the output images (different classification images) [36,37,38]. KC value denotes the agreement degree between the

two comparative maps/images. Blackman and Landis assigned a scale for Kappa values between 0 and 1 for the analysis of map agreement degree, and this scale has become the standard measure of agreement between maps in classification applications [39,40]. According to the iterated procedure observed in Fig. 2, different accuracy levels of land cover classification results derived from variables P_a and P_{buffer} using the proposed approach are generated. Different group KC values are automatically calculated and a new matrix of KC values related to different combinations of P_a and P_{buffer} is also generated. For the Kappa value equal to or greater than 0.61 is considered to be in good agreement [39,40]. Thus, we select the KCs whose values are equal to or greater than 0.61, and the key parameters P_a and P_{buffer} corresponding to those KC values are chosen for the alternative combination, implying that the KCs are in substantial agreement or perfect agreement.

Optimal Combination Model

In a general respect, the computer TC value is proportional to the image complexity or to the number of vector plaques of a map. A greater number of vector plaques or a more complex image results in a greater computer TC value. The manual multispectral remote sensing image classification by visual interpretation usually takes considerable time. In this sense, the computer TC value reflects the complexity of the vector graphics or remote sensing image as well as the artificial process.

We construct our portfolio optimisation model based on two principles:

1. If the requirements are met accurately and a small difference exists between the classification results (for example, the absolute value of difference of KC is no less than 0.01), we chose a less TC combination of P_a and P_{buffer} to form the optimal combination model.

2. If the requirements are met accurately and the increased rate of accuracy was significantly less than the time-lapse rate, we chose a combination of P_a and P_{buffer} with a time-lapse rate minimum to form the optimal combination model.

The selection process of portfolio optimisation model is shown in Fig. 3

We choose the optimal combination model and determine the key parameters. Applying these two parameters to the semi-automatic classification proposed by Jiang et al., the fully automatic classification algorithm is formed. The problem of determination of the key parameters by test-decision is completely resolved.

Case Study

Study area and data sources

Study area. A case study of Weichang County of Hebei Province, China (41°35′–42°40′N, 116°32′–118°14′E) is conducted to confirm the effectiveness of the proposed approach. This study area covers 9219 km^2 and encompasses over 25% of Zhangjiakou District, Hebei Province. Weichang is also a Manchu and Mongolian Autonomous County, the largest county in Hebei Province, and the most northern junction to the Inner Mongolia autonomous region (Fig. 4).

There are six types of land cover in the study area: cropland, forestland, grassland, water, residential/construction land, and bare land. Forestland and grassland is dominant, next is cropland, and residential/construction land, water and bare land are relatively fewer. The study area is located in the transition zone of the Inner Mongolia Plateau and the northern Hebei Mountains,

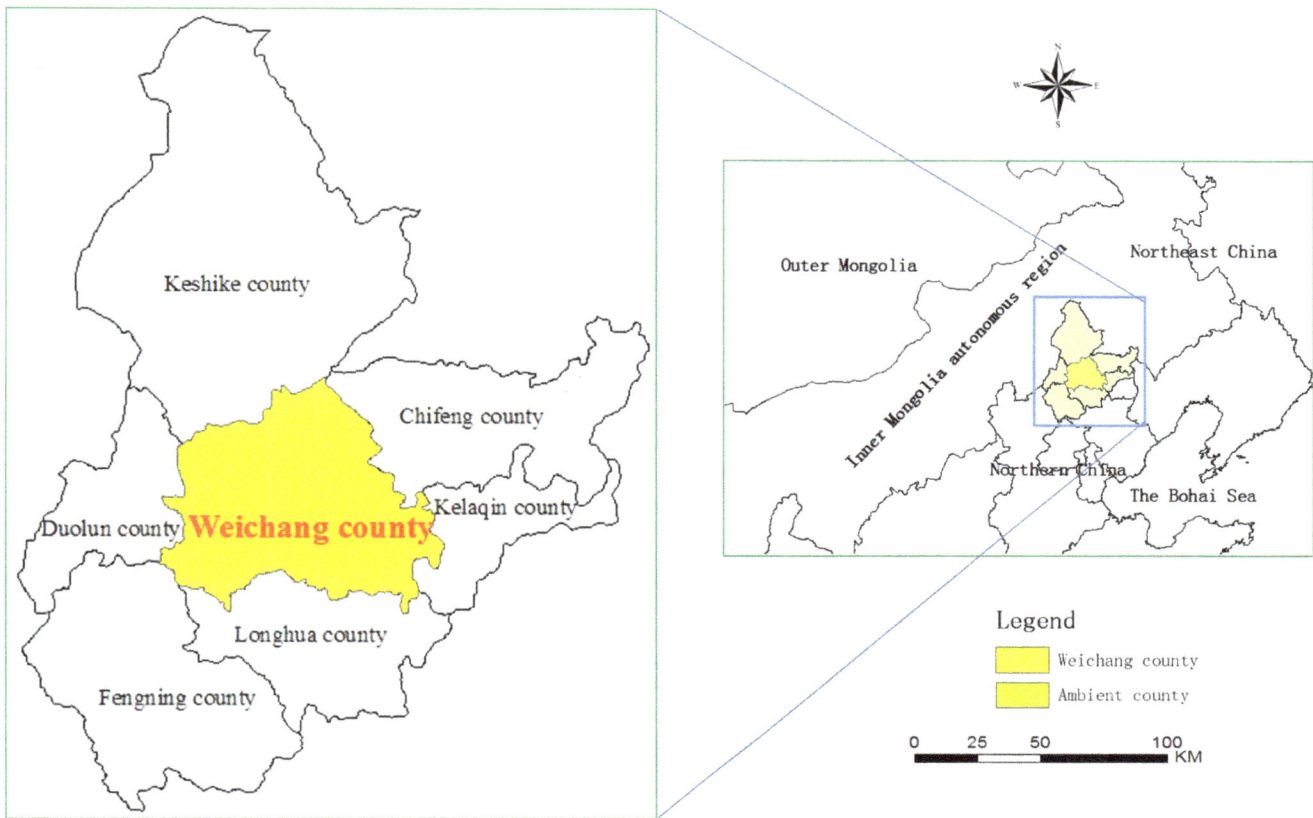

Figure 4. Location of the study area: Weichang County, Hebei Province, China.

with an elevation gradient ranging from 750 meters to 2067 meters above sea level [41]. There are three types of area distinguished based on changing degrees of land cover: dramatic change area, moderate changes area, and little changes area. In the central and southern regions of Weichang County, similar to other cities in China, around the county town of Weichang expanded rapidly in the last decade, and rapid economic growth in the areas with residential/construction land extended to over 100 km^2 from 1995 to 2010. This growth caused the dominant land change of the area to be a loss of cropland, grassland and

forestland. These areas belong to dramatic changes area. In the north and northeast of Weichang County, there are lots of natural forest conservation regions growing with a large number of deciduous, pine, etc, with little land cover change, belonging to little change area. In the east and west of Weichang County, some types of land cover have changed, but the change is not very significant. For example, some cropland was restored to forestland since the implementation of the "Returning crops to forest" policy in 2000. Due to the three different types of land cover changing

Table 1. The results of KC (Kappa Coefficient).

$P_{buffer} \backslash P_a$	10%	20%	30%	40%	50%	60%	70%	80%	90%	100%
10%	0.011	0.013	0.015	0.017	0.018	0.021	0.019	0.018	0.016	0.015
20%	0.099	0.149	0.207	0.312	0.353	0.381	0.331	0.283	0.197	0.091
30%	0.132	0.282	0.323	0.401	0.437	0.476	0.441	0.317	0.204	0.169
40%	0.201	0.278	0.391	0.452	0.528	0.539	0.516	0.476	0.361	0.292
50%	0.276	0.324	0.441	0.524	0.581	0.692	0.628	0.528	0.456	0.308
60%	0.332	0.398	0.492	0.568	0.621	0.762	0.760	0.612	0.489	0.362
70%	0.328	0.364	0.486	0.557	0.579	0.623	0.619	0.573	0.477	0.213
80%	0.294	0.347	0.424	0.502	0.544	0.592	0.585	0.486	0.392	0.271
90%	0.192	0.297	0.316	0.392	0.473	0.492	0.468	0.395	0.226	0.107
100%	0.019	0.222	0.248	0.258	0.287	0.268	0.257	0.238	0.201	0.014

Table 2. TC (Time-Consuming, unit: second) of the proposed methodology with different parameters.

$P_{buffer}\backslash P_a$	10	20	30	40	50	60	70	80	90	100
10	25	26	31	39	42	51	80	103	111	149
20	32	37	45	57	68	87	122	170	205	242
30	47	57	71	89	93	119	181	251	285	346
40	57	69	86	109	119	149	221	309	336	415
50	65	83	102	111	138	168	273	337	403	476
60	78	99	136	161	185	236	370	432	479	573
70	88	115	148	168	199	247	375	445	492	582
80	100	133	159	178	217	258	389	468	525	601
90	157	183	212	239	273	318	467	547	621	698
100	134	182	235	272	298	336	515	569	657	721

degrees of land use across the area, we select Weichang County as an ideal case study to evaluate the effective automatic approach.

Data sources. TheLandsat-5/TM data of 2010 (WRS-2 123/31 for 2010/8/24) and 1:100,000 land cover maps of two dates are used for this experiment. The known classification maps were produced by the Chinese Academy of Sciences (CAS) with consistent classification schemes that have an overall accuracy of 95% for all land use classes validated by intensive field surveys [1,42,43]. Here, we acquired the land cover maps of 2005 and 2010 from CAS, which were visually interpreted and classified using high spatial resolution images (QuickBird) and field surveys, respectively. The map of land cover of 2005 was used as prior knowledge for choosing the pure-pixel training samples, while the land cover map of 2010 was used as a reference map for assessing classification accuracy. The multispectral TM image was radio-metrically corrected by CAS. The resulted image covers the whole area of the Weichang County that was used for land cover classification.

Determination P_a and P_{buffer}

As a critical component of the proposed methodology, the effect of the training sample automatic selection is determined by the two parameters P_a and P_{buffer}. These key parameters are determined using the iterative procedure described in the previous section. We assess classification accuracy by calculating the KCs. As observed in Figure 2. The P_a and P_{buffer} are interval of (10%, 100%)with 10% step increments independently to generate totals of 100 land cover classification maps. Using equation of Kappa Coefficient [31,32], 100 KC results are calculated (Table 1).

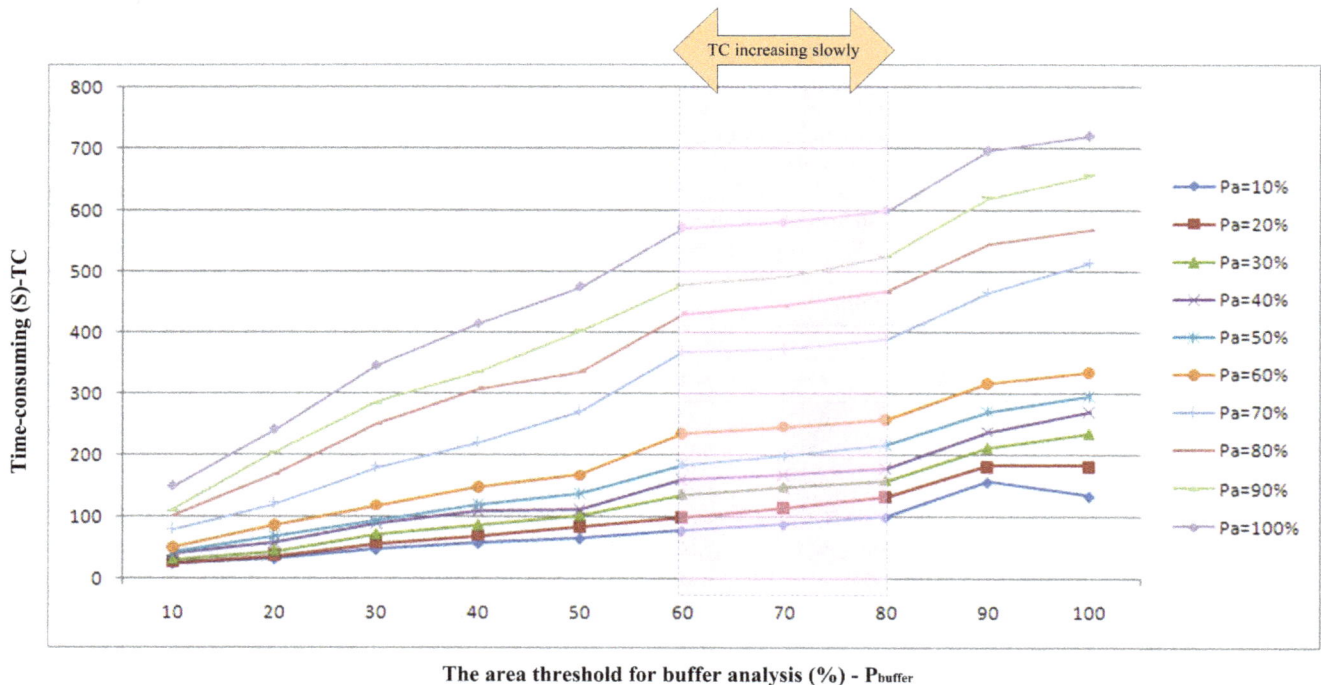

Figure 5. Relationship among P_a, P_{buffer} and TC (Time-Consuming, unit: second).

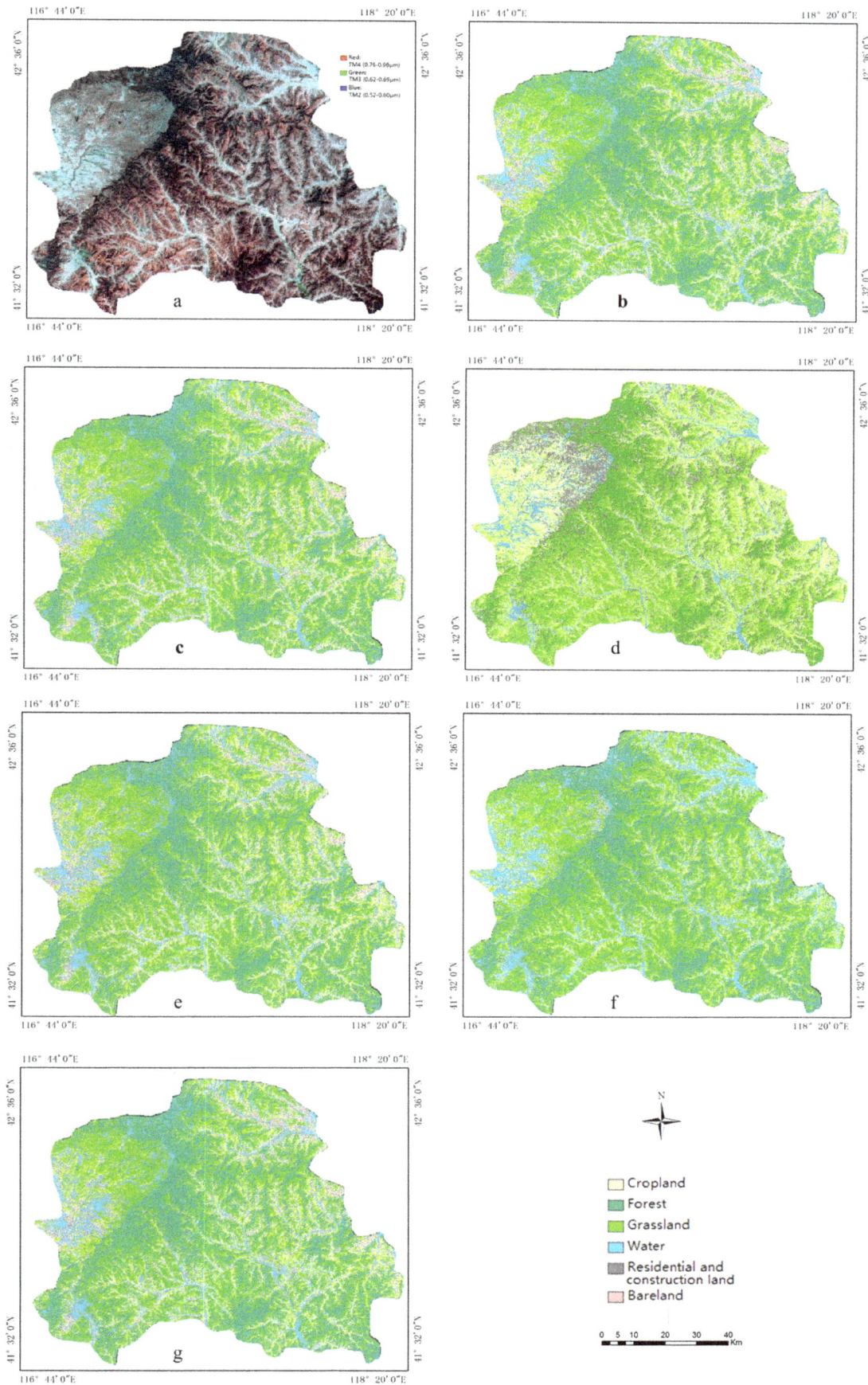

Cropland
Forest
Grassland
Water
Residential and construction land
Bareland

0 5 10   20   30   40
                          Km

Figure 6. Comparison of land cover classification in Weichang.

Table 1 shows the calculated results for the KCs of different combinations of P_a and P_{buffer}. Clearly, the KC increases with P_a and P_{buffer} until it reaches its maximum, and it then decreases with increasing P_a and P_{buffer} values. As observed in Table 1, the values 0.692, 0.628, 0.621, 0.762, 0.760, 0.612, 0.623 and 0.619 are considered to represent land cover classifications in good agreement, and the values 0.762 and 0.760 show the best agreement among the results. We choose the P_a and P_{buffer} values, $P_a = 60\%$, $P_{buffer} = 60\%$ and $P_a = 70\%$, $P_{buffer} = 60\%$, with the chosen combinations of P_a and P_{buffer} being (60%, 60%) and (70%, 60%). Because the difference between 0.762 and 0.760 is very small, the optimal combination is not immediately clear. According to the principles of the portfolio optimisation model, we recorded the TC value of each parameter changed as shown in Table 2 and mapped the relationship among P_{buffer} P_a and TC value, as shown in Figure 5.

Table 2 clearly shows the TC of the proposed methodology with different values for P_a and P_{buffer}. Fig. 5 shows the relationship among P_a, P_{buffer} and TC. The TC value of land cover classification increases as the two parameters' incremental change increases, in spite of the ratio of TC difference. As shown in Fig. 5, if the threshold for buffer analysis is below 60% or over 80%, the TC is greatly increased. If P_{buffer} is in the interval of 60% to 80% (as marked with carmine colour in Fig. 4), land cover classification does not significantly increase the computational cost, which also proves that the combinations (60%, 60%) and (70%, 60%) are the optimal combinations. For P_a, the accuracies of the two combinations are similar, while the value of TC based on $P_a = 60\%$ performs notably better. The tendencies of the two lines are also similar. In detail, the classification accuracy based on $P_a = 70\%$ is 0.762, while the accuracy based on $P_a = 60\%$ is 0.760. Both of the combinations maintain high accuracy with slight differences. However, the TC of the two lines $P_a = 60\%$ and $P_a = 70\%$ have prominent changes of 236 s (second) and 370 s, respectively. According to the prerequisite of classification accuracy, the TC of $P_a = 60\%$ is less than that of $P_a = 70\%$, so the combination (60%, 60%) is chosen using the portfolio optimisation model.

Results

In light of the aforementioned results, we selected the pure-pixel samples based on the portfolio optimisation model (60%, 60%) as training samples for automatic classification. The land cover classification results are shown in Figure 6. Five types of land cover maps were compared to evaluate the result of the final classification: (1) the visual interpretation of land cover classes of 2010, recognised as relatively precise data(Fig. 6-b); (2) the classification result using Maximum Likelihood(ML) approach (Fig. 6-d); (3) the automatic classification of land cover of 2010 based on the TM image of the same area using the portfolio optimisation model (60%, 60%)(Fig. 6-e); (4) the automatic classification result using the combination model (20%, 20%)(Fig. 6-f); and (5) the automatic classification result using the combination model (80%, 80%)(Fig. 6-g).. Fig. 6 exhibits the data sources including the Landsat-5/TM image of 2010 (Fig. 6-a), the visual interpretation results of 2010 as a standard classified map (Fig. 6-b)and visual interpretation results of 2005 as a prior, exact classification map (Fig. 6-c). Figs. 6-e, 6-f and 6-g show the classification results based on different combination models using our proposed approach. 6-d displays the classification results using ML approach. The classification results (Figs. 6-e, 6-f and 6-g) show that forests and grasslands in Weichang are more predominant than the other four types of land cover, which is consistent with the known, accurate classification map (Fig. 6-b).

For better quantitative assessment, absolute values (pixel number) were converted to percentage values in each error matrix.As shown in Tables 3,4, 5 and 6, each table uses a different combination model/classification approach. For overall classification accuracy evaluation, the overall accuracy are 83.4% (using the portfolio optimisation model (60%, 60%)), 37.7% (using the combination model (20%, 20%)), 62.2% (using the combination model (80%, 80%)) and 58.4% (using the ML approach), respectively. Apparently, using the portfolio optimisation model can improve the overall classification accuracy significantly. Similarly, the commission errors and the omission errors using the portfolio optimisation model are reduced significantly than using other combination models or using ML classification

Table 3. Error matrix of the combination model (80%, 80%).

	Cropland[2]	Forest[2]	Grassland[2]	Water[2]	Residential and construction land[2]	Bareland[2]	Sum	Omission error
Cropland[1]	16.08	3.54	4.63	0.82	0.19	1.15	26.40	39.1
Forest[1]	4.84	22.85	8.41	0.95	0.02	0.70	37.77	39.5
Grassland[1]	3.30	2.97	16.35	1.00	0.11	1.12	24.85	34.2
Water[1]	0.52	1.19	0.15	3.01	0.03	0.18	5.09	40.7
Residential and construction land[1]	0.06	0.02	0.06	0.10	0.43	0.08	0.75	43.3
Bareland[1]	0.29	0.46	0.41	0.49	0.03	3.46	5.14	32.7
Sum	25.09	31.04	30.01	6.38	0.80	6.69	100.00	
Commission error	35.9	26.4	45.5	52.7	46.8	48.2		

Note: 1) Land cover types with number 1 (i.e. Cropland[1], Forest[1], Grassland[1], Water[1], Residential and construction land[1], and Bareland[1]) stand for land cover results of the visual interpretation; Land cover types with number 2 stand for land cover results of Automatic classification. 2) For better quantitative assessment, absolute values (pixel number) were converted to percentage values in each error matrix. 3) For automatic classification result, overall accuracy = 62.2%, KC = 0.486, sample size = 8,658,588.

Table 4. Error matrix of the combination model (20%, 20%).

	Cropland[2]	Forest[2]	Grassland[2]	Water[2]	Residential and construction land[2]	Bareland[2]	Sum	Omission error
Cropland[1]	7.85	6.88	5.06	5.06	0.12	0.66	25.65	69.4
Forest[1]	5.73	19.98	11.14	3.33	0.21	1.26	41.66	52.0
Grassland[1]	4.64	7.15	7.58	4.25	0.49	1.09	25.20	69.9
Water[1]	0.24	1.48	0.14	1.81	0.11	0.05	3.82	52.6
Residential and construction land[1]	0.08	0.09	0.06	0.04	0.19	0.02	0.48	61.1
Bareland[1]	0.43	0.59	0.56	1.18	0.12	0.32	3.20	89.9
Sum	18.96	36.18	24.54	15.67	1.24	3.41	100.00	
Commission error	58.6	44.8	69.1	88.4	85.0	90.5		

Note: 1) Land cover types with number 1 (i.e. Cropland[1], Forest[1], Grassland[1], Water[1], Residential and construction land[1], and Bareland[1]) stand for land cover results of the visual interpretation; Land cover types with number 2 stand for land cover results of Automatic classification. 2) For better quantitative assessment, absolute values (pixel number) were converted to percentage values in each error matrix. 3) For automatic classification result, overall accuracy = 37.7%, KC = 0.149, sample size = 8,658,588.

approach. In order to test whether the KC values are statistical significance, a Z-test on the portfolio optimisation model, other combination models and using ML classification approach were performed respectively, as shown in Table 7. The calculated results showed P values in bold were statistically significant (p<0.0001). The values of asymptotic standard error (ASE) were all less than 0.0002. Statistical comparisons against percentage of pixels reveal a significant difference between the portfolio optimisation model and other combination models and using ML classification approach with confidence intervals (CI) values ranging interval difference.

As shown in the tables, the accuracies of the three combinations and ML approach are significantly different; the result of the portfolio optimisation model is much more accurate than that of the other combination models or ML approach. The structures of the four error matrices are also different. Water showed the highest individual classification accuracy due to its lower reflectance values, whereas there was much misclassification between forest and cropland, forest and grassland, and residential and construction land and bare land because of the similar reflectance values of these land cover types. Residential and construction land and bare land occupied small proportions of the entire study area, which increased the inaccurate effect on the map

agreement maybe one of the misclassification reasons. Another main reason for misclassification was the land cover classification system, each type of land cover includes many subcategories, i.e. cropland includes two subcategories of paddy field and dry farming field, residential and construction land includes three subcategories of urban land, rural residential and other construction land, etc.

In Tables 3 and 4, whether by omission or commission error, much misclassification is apparent using the combination model (20%, 20%) and the combination model (80%, 80%). The fundamental reason for these errors in classification is related to the pure-pixel training samples. When the combination model (20%, 20%) is used, there are not enough pure-pixel training samples, which causes classification error. If the combination model (20%, 20%) is chosen, some pixel training samples that are not pure (we refer to these as "Noise") are used as training samples, leading to classification error. Fig. 7 describes this problem as follows:

As shown in the Fig. 7, the primary type of land cover was forest in 2005 (the entire range of the blue line in Fig. 7). In 2010, more land cover changed to grassland and bare land, due to human activities and natural environmental changes, respectively, (adjacent to the outer blue line, within the red line in Fig. 7). The area

Table 5. Error matrix of the combination model (60%, 60%).

	Cropland[2]	Forest[2]	Grassland[2]	Water[2]	Residential and construction land[2]	Bareland[2]	Sum	Omission error
Cropland[1]	10.06	1.04	0.91	0.01	0.02	0.19	12.23	17.7
Forest[1]	1.03	32.45	4.83	0.02	0.07	0.40	38.80	16.4
Grassland[1]	0.91	4.84	29.62	0.08	0.11	0.43	35.99	17.7
Water[1]	0.01	0.02	0.07	1.13	0.01	0.02	1.25	9.7
Residential and construction land[1]	0.02	0.09	0.11	0.01	1.28	0.12	1.64	22.1
Bareland[1]	0.19	0.42	0.50	0.02	0.11	8.85	10.09	12.3
Sum	12.23	38.86	36.04	1.26	1.61	10.01	100.00	
Commission error	17.7	16.5	17.8	10.4	20.6	11.6		

Note: 1) Land cover types with number 1 (i.e. Cropland[1], Forest[1], Grassland[1], Water[1], Residential and construction land[1], and Bareland[1]) stand for land cover results of the visual interpretation; Land cover types with number 2 stand for land cover results of Automatic classification. 2) For better quantitative assessment, absolute values (pixel number) were converted to percentage values in each error matrix. 3) For automatic classification result, overall accuracy = 83.4%, KC = 0.760, sample size = 8,658,588.

Table 6. Error matrix of common classification approach (Maximum Likelihood Approach).

	Cropland[2]	Forest[2]	Grassland[2]	Water[2]	Residential and construction land[2]	Bareland[2]	Sum	Omission error
Cropland[1]	13.34	4.92	3.22	1.04	1.32	0.12	23.96	44.3
Forest[1]	1.28	20.69	9.14	0.59	3.74	0.25	35.69	42.0
Grassland[1]	2.86	7.01	13.18	0.81	0.27	0.10	24.24	45.6
Water[1]	0.41	0.84	0.05	3.93	0.08	0.08	5.38	27.0
Residential and construction land[1]	0.64	0.98	0.04	0.20	5.67	0.08	7.61	25.5
Bareland[1]	0.56	0.50	0.09	0.35	0.06	1.56	3.11	50.0
Sum	19.08	34.94	25.73	6.92	11.14	2.18	100.00	
Commission error	30.1	40.8	48.8	43.2	49.1	28.7		

Note: 1) Land cover types with number 1 (i.e. Cropland[1], Forest[1], Grassland[1], Water[1], Residential and construction land[1], and Bareland[1]) stand for land cover results of the visual interpretation; Land cover types with number 2 stand for land cover results of Maximum Likelihood approach classification. 2) For better quantitative assessment, absolute values (pixel number) were converted to percentage values in each error matrix. 3) For Maximum Likelihood approach classification, overall accuracy = 58.4%, KC = 0.4482, sample size = 8,658,588.

of pure-pixel training samples was Ab_{d1} (the area within the yellow line in Fig. 7) or Ab_{d2} (the area within the green line in Fig. 7) after being buffered inward by the distances d_1 or d2, respectively. Different P_{buffer} values were calculated by applying equation 2. We found that if d_1 is too small, the pure-pixel training samples include much "Noise" (forest samples include many grassland or bare land samples), which causes misclassification. In a similar manner, if d_2 is too big, the number of the pure-pixel training samples are too low, which also causes classification error (grassland or bareland was classified forestland). Therefore, the buffer distance (transformed into P_{buffer} which is understood easily) is a key parameter in the selection of the pure-pixel training samples. By the same token, P_a is also a key determinant in the selection of these training samples. The selection of these two key parameters, ultimately constituting the optimal combination model (P_a, P_{buffer}), which is closely related to the classification accuracy, is satisfactorily accomplished by automatic selection.

In order to test the applicability of our approach, we also applied our approach in other different regions chosen based on their changing degrees of land cover. The four regions are: 1) Anshan city in Liaoning Province, China, with moderate changes; 2) Neijiang county in Sichuang Province, China, with moderate changes; 3) Shuangtaihe natural conservation area in Liaoning Province, China, with little changes; 4) Qingpu District in Shanghai city, China, with dramatic changes. The testing results show that our method performs well in the regions with normal-to-high rates of land cover change, especially in rapid changing area.

Despite the accuracy is slightly lower in regions with little land cover change, but is acceptable. The mainly changed rules were described in literature 1 [1]. The land cover with little change was natural affected without rules which increased error of ultimate result of land cover classification. Some following work, such as improving upon our algorithm our strategies (i.e. consider double-kernel combination method, narrow the double-loop step), may be useful in improving the performance of the method in the future.

Conclusions and Discussion

This study improved Jiang et al.'s strategy and developed a new approach for automatically selecting the key parameters used in land cover automatic classification. Three main, interdependent parts consist of our approach: the selection procedure of the pure-pixel training-sample dataset, the method to determine the key parameters, and the optimal combination model. The main achievements in this study include: (1) Selection the two key parameters automatically instead of using a test-decision, which avoids subjective bias and (2) Determination the portfolio optimisation model (P_a, P_{buffer}) used to select the pure-pixel samples as training samples for automatic classification.

The study area experimental results showed that the methodology determining the key parameters can automatically select the portfolio optimisation model and the classification results based on different combination models using our proposed approach and using the ML approach demonstrate that: (1) The portfolio

Table 7. Comparison of the Z values in each model/approach.

Model/Approach	Kappa	Asymptotic standard error (ASE)	95% confidence lower limit	95% confidence upper limit
combination model (80%, 80%)	0.4856	0.0002	0.4852	0.4860
combination model (20%, 20%)	0.1490	0.0002	0.1485	0.1494
combination model (60%, 60%)	0.7607	0.0002	0.7603	0.7611
Maximum Likelihood Approach	0.4482	0.0002	0.4478	0.4487

Note. Sample size was 8,658,588.
The 95% confidence intervals (2.5% each side) all were less than 0.0001.
P values in bold were statistically significant (p<0.0001).

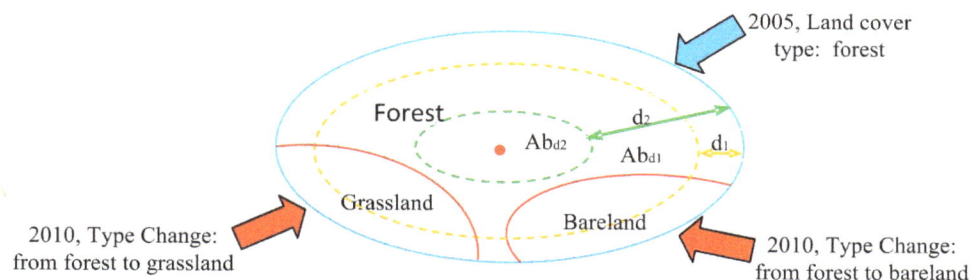

Figure 7. Sketch map of automatic dataset of pure-pixel training samples.

optimisation model produces more precise results, higher overall classification accuracy and lower omission errors/commission errors; (2) The portfolio optimisation model performs well in the region with normal-to-high degree changes of land cover and may have a wide scope of scientific application; (3) The proposed iterated training-sample selection process can refine the training samples and improve classification accuracy while does not significantly increase the computational cost.

Of course, we recognise that this new methodology has possible limitations. First, this methodology is subject to the limitations of the Jiang et al.'s classification model. If the original assumption is flawed, the selection of the two key parameters is undoubtedly affected. For example, Jiang et al. assume that the classification system of the prior, exact classification map and the subsequent image is the same. A new type of land cover will not be accurately classified if it is not available on the prior, exact classification map but appears on the subsequent image. In this case, we must first improve upon our original method (i.e. consider double-kernel combination classification instead of PCA and three-dimensional feature analysis). One promising option is to use a one-class classifier [44] to identify a new type of land cover because pure-pixel training samples of the new class are needed. Another limitation is that we are using a method that may be limited in scope to obtain the P_a and P_{buffer} to form the optimal combination model. The intervals between P_a and P_{buffer} values determine the accuracy of the new methodology. In this study, we set the two steps both equal to 10%. In setting these steps to 10%, perhaps we have obtained the approximate optimal values, rather than the global optimums. In the future, we must improve our algorithm to a narrower step (for example, 5% or 1%) to obtain an accurate global optimum.

Supporting Information

Figure S1 Comparison of land cover classification of region with much land cover change (Anshan): (a) TM image of 2010;(b) land cover of 2005;(c) land cover of 2010 from visual interpretation; (d) land cover classified by our method.

Figure S2 Comparison of land cover classification of region with normal land cover change degree (Neijiang): (a) TM image of 2010;(b) land cover of 2005;(c) land cover of 2010 from visual interpretation; (d) land cover classified by the proposed method.

Figure S3 Comparison of land cover classification around Shuangtaihe natural conservation region: (a) TM image of 2010;(b) land cover of 2005;(c) land cover of
2010 from visual interpretation; (d) land cover classified our method.**

Figure S4 Comparison of land cover classification of region with much land cover change (Qinpu): (a) TM image of 2009;(b) land cover of 2005;(c) land cover of 2009 from visual interpretation; (d) land cover classified by our method.

Table S1 Statistics of six land cover classes of the three classification results in region with much land cover change (Anshan).

Table S2 Confusion matrix of two classification algorithms of Anshan, 2010.

Table S3 Statistics of four land cover classes of the three classification results in region with normal land cover change (Neijiang).

Table S4 Confusion matrix of two classification algorithms of Neijiang, 2010.

Table S5 Statistics of six land cover classes of the three classification results in natural conservation region with little land cover change (Shuangtaihe).

Table S6 Confusion matrix of two classification algorithms of Shuangtaihe natural conservation region, 2010.

Table S7 Statistics of five land cover classes of the three classification results in natural conservation region with dramatic land cover change (Qinpu district).

Table S8 Confusion matrix of two classification algorithms of QinPu district, 2009.

Acknowledgments

We thank Dr. Xiaopeng Qi in National Center for Public Health Surveillance and Information Services, Chinese Center for Disease Control and Prevention for her help in Z-test statistical analyses. We like to thank the editors and anonymous reviewers for their helpful remarks.

Author Contributions

Conceived and designed the experiments: YW DJ YH DZ. Performed the experiments: YW DJ YH. Analyzed the data: YW DJ WW. Contributed reagents/materials/analysis tools: YH DZ XY. Wrote the paper: YW DJ.

References

1. Jiang D, Huang Y, Zhuang D, Zhu Y, Xu X, et al. (2012) A Simple Semi-Automatic Approach for Land Cover Classification form Multispectral Remote Sensing Imagery. PLoS ONE 7(9): e45889, doi: 10.1371/journal.pone.0045889.

2. Liu JY, Deng XZ (2010) Progress of the research methodologies on the temporal and spatial process of LUCC. Chinese Sci Bull 55: 1354–1362.

3. Denis R, Antoine T, Christian P, Claudine D (2011) Comparison of methods for LUCC monitoring over 50 years from aerial photographs and satellite images in a Sahelian catchment. International Journal of Remote Sensing 32(6): 1747–1777.

4. Daniel EO, Bethany AB, Jeff A, John FM, Steven PH (2011) How much is built? Quantifying and interpreting patterns of built space from different data sources. International Journal of Remote Sensing 32(9): 2621–2644.

5. Gutman G, Janetos AC, Justice CO, Moran EF, Mustard J, et al. (2004) Land change science: Observing, monitoring and understanding trajectories of change on the earth's surface. New York: Kluwer Academic Publishers. pp. 108–118.

6. Abdullah SA, Nakagoshi N (2006) Changes in landscape spatial pattern in the highly developing state of Selangor, peninsular Malaysia. Landscape Urban Planning 77: 263–275.

7. Veldkamp A, Lambin EF (2001) Predicting land-use change. Agriculture. Ecosystems and Environment 85:1–6.

8. Robert F, Ian O, Mélanie C, Alice D, Darren P (2012) A method for trend-based change analysis in Arctic tundra using the 25-year Landsat archive. Polar Record 48(1): 83–93.

9. Berberoğlu S, Akin A, Atkinson PM, Curran PJ (2010) Utilizing image texture to detect land-cover change in Mediterranean coastal wetlands. International Journal of Remote Sensing 31(11): 2793–2815.

10. Liu J, Liu M, Tian H, Zhuang D, Zhang Z, et al. (2005) Spatial and temporal patterns of China's cropland during 1990–2000: An analysis based on Landsat TM data. Remote Sensing of Environment 98: 442–456.

11. Loveland T, Merchant J, Ohlen D, Brown J (1991) Development of a land-cover characteristics database for the conterminous U.S.. Photogrammetric Engineering and Remote Sensing 57: 1453–1463.

12. Zhao Y (2003) The Application Principle and Method of Remote Sensing. Beijing: Science Press. pp. 57–73.

13. Richards JA, Jia X (1999) Remote Sensing Digital Imaging Analysis: an Introduction, third ed. Springer, Berlin. pp. 102–115.

14. Foody GM, Lucas RM, Curran PJ, Honzak M (1997) Nonlinear mixture modelling without end-members using an artificial neural network. International Journal of Remote Sensing 18(4): 937–953.

15. Wang YC, Feng CC (2011) Patterns and trends in land use/land cover change research explored using self-organizing map. International Journal of Remote Sensing 32(13): 3765–3790.

16. Robin P, Antoine C (2013) Spatial location and ecological content of support vectors in an SVM classification of tropical vegetation. Remote Sensing Letters 4(7): 686–695.

17. Jaime PG, Jean FM, Gerard M, Jordi C, Martí OM, et al. (2013) Enhanced land use/cover classification of heterogeneous tropical landscapes using support vector machines and textural homogeneity. International Journal of Applied Earth Observation and Geoinformation 23: 372–383.

18. Geneletti D, Gorte BGH (2003) A method for object-oriented land cover classification combining Landsat TM data and aerial photographs. International Journal of Remote Sensing 24(6): 1273–1286.

19. Francisco FDS, John MK, Patrick L (2013) An object-oriented classification method for mapping mangroves in Guinea, West Africa, using multipolarized ALOS PALSAR L-band data. International Journal of Remote Sensing 34(2): 563–586.

20. Schneider A, Friedl MA, Potere D (2010) Mapping global urban areas using MODIS 500 - m data: New methods and datasets based on 'urban ecoregions'. Remote Sensing of Environment 114: 1733–1746.

21. Hansen M, Dubayah R, Defries R (1996) Classification trees: an alternative to traditional land cover classifiers. International Journal of Remote Sensing 17:1075–1081.

22. Weng Q (2011) Advances in Environmental Remote Sensing: Sensors, Algorithms and Applications. CRC Press/Taylor and Francis, Boca Raton, FL, USA. pp. 63–78.

23. Aitkenhead MJ, Aalders IH (2011) Automating land cover mapping of Scotland using expert system and knowledge integration methods. Remote Sensing of Environment 115 (5): 1285–1295.

24. Chen J, Gong P, He C, Pu R, Shi P (2003) Land-use/land-cover change detection using improved change-vector analysis. Photogrammetric Engineering & Remote Sensing 69 (4): 369–379.

25. Gong P, Mahler S, Biging G, Newburn D (2003) Vineyard identification in an oak woodland landscape with airborne digital camera imagery. International Journal of Remote Sensing 24: 1303–1315.

26. Canty MJ (2006) Image Analysis, Classification and Change Detection in remote Sensing with Algorithms for ENVI/IDL. Taylor & Francis, CRC Press. pp. 77–82.

27. Lambin EF, Geist H (2006) Land-use and land-cover change: local processes with global impacts. New York: Springer. pp. 38–49.

28. Xie Y, Sha Z, Bai Y (2010) Classifying historical remotely sensed imagery using a tempo-spatial feature evolution (T-SFE) model. ISPRS Journal of Photogrammetry and Remote Sensing 65(2): 182–190.

29. Fortier J, Rogan J, Woodcock C, Runfola DM (2011) Utilizing temporally invariant calibration sites to classify multiple dates of satellite imagery. Photogrammetric Engineering & Remote Sensing 77(2): 181–189.

30. Xian G, Collin H, Fry J (2009) Updating the 2001 National Land Cover Database land cover classification to 2006 by using Landsat imagery change detection methods. Remote Sensing of Environment 113 (6): 1133–1147.

31. Cohen J (1960) A coefficient of agreement for nominal scales. Educational and Psychological Measurement 20: 37–46.

32. Yvonne MMB, Stephen EF, Paul WH, Richard JL, Frederick M (1975) Discrete Multivariate Analysis: Theory and Practice. MIT Press, Cambridge, MA. pp. 58–67.

33. Jensen JR (2005) Introductory Digital Image Processing: A Remote Sensing Perspective. New York: Pearson Prentice Hall, 3rd ed., 495–515.

34. Allouche O, Tsoar A, Kadmon R (2006) Assessing the accuracy of species distribution models: Prevalence, kappa and the true skill statistic (TSS). Journal of Applied Ecology 43: 1223–1232.

35. Foody G (2007) Map comparison in GIS. Progress in Physical Geography 31:439–445.

36. Congalton RG (1991) A review of assessing the accuracy of classifications of remotely sensed data. Remote Sensing of Environment 37: 35–46.

37. Congalton R, Green K (1999) Assessing the accuracy of remotely sensed data: principles and practices, Boca Raton: CRR/Lewis Press. 137p.

38. Gupta M, Srivastava PK (2010) Integrating GIS and remote sensing for identification of groundwater potential zones in the hilly terrain of Pavagarh, Gujarat, India. Water International 35: 233–245.

39. Landis J, Koch G (1977) The measurement of observer agreement for categorical data. Biometrics 33: 159–174.

40. Blackman N, Koval J (2000) Interval estimation for Cohen's Kappa as a measure of agreement. Statistics in Medicine 19: 723–741.

41. Baidu. Introduction of Mongolia Weichang Manchu Autonomous County, China. Available: http://baike.baidu.com/view/1214694.htm.

42. Liu J, Liu M, Deng X, Zhuang D, Zhang Z, et al. (2002) The land use and land cover change database and its relative studies in China. Journal of Geographical Sciences 12: 275–282.

43. Liu J, Liu M, Zhuang D, Zhang Z, Deng X (2003) Study on spatial pattern of land-use change in China during 1995–2000. Science in China, Series D 46:373–384.

44. Chen J, Chen X, Cui X, Chen J (2011) Change vector analysis in posterior probability space: a new method for land cover change detection. IEEE Geoscience and Remote Sensing Letters 8(2): 317–321.

17

Fuzzy Nonlinear Proximal Support Vector Machine for Land Extraction Based on Remote Sensing Image

Xiaomei Zhong[1], **Jianping Li**[2,3]*, **Huacheng Dou**[2,3], **Shijun Deng**[2,3], **Guofei Wang**[2,3], **Yu Jiang**[2,3], **Yongjie Wang**[2,3], **Zebing Zhou**[2,3], **Li Wang**[2,3], **Fei Yan**[4]

1 Tianjin Chengjian University, Tianjin, China, **2** Tianjin Institute of Geotechnical Investigation and Surveying, Tianjin, China, **3** Tianjin StarGIS Information Engineering Company Limited, Tianjin, China, **4** Beijing Forestry University, Beijing, China

Abstract

Currently, remote sensing technologies were widely employed in the dynamic monitoring of the land. This paper presented an algorithm named fuzzy nonlinear proximal support vector machine (FNPSVM) by basing on ETM$^+$ remote sensing image. This algorithm is applied to extract various types of lands of the city Da'an in northern China. Two multi-category strategies, namely "one-against-one" and "one-against-rest" for this algorithm were described in detail and then compared. A fuzzy membership function was presented to reduce the effects of noises or outliers on the data samples. The approaches of feature extraction, feature selection, and several key parameter settings were also given. Numerous experiments were carried out to evaluate its performances including various accuracies (overall accuracies and kappa coefficient), stability, training speed, and classification speed. The FNPSVM classifier was compared to the other three classifiers including the maximum likelihood classifier (MLC), back propagation neural network (BPN), and the proximal support vector machine (PSVM) under different training conditions. The impacts of the selection of training samples, testing samples and features on the four classifiers were also evaluated in these experiments.

Editor: Guy J.-P. Schumann, NASA Jet Propulsion Laboratory, United States of America

Funding: Financial support for this study was provided by Xiaomei Zhong, Jianping Li, Hucheng Dou, and they had a key role in algorithm study, data collection and analysis, accuracy assessment.

Competing Interests: The authors declare that they have no conflict of interest, they have no financial and personal relationships with other people or organizations that can inappropriately influence their work, there is no professional or other personal interest of any nature or kind in any product, service and/or company that could be construed as influencing the position presented in, or the review of, the manuscript entitled, "Fuzzy Nonlinear Proximal Support Vector Machine for Land Extraction Based on Remote Sensing Image". Of the authors, Li Jianping, Dou Huacheng, Deng Shijun, Wang Guofei, Jiang Yu, Wang Yongjie, Zhou Zebing, and Wang Li, are currently employed by Tianjin StarGIS Information Engineering CO,.LTD. And the authors hereby declare that this affiliation does not cause any competing interests.

* E-mail: ljpzl@126.com

Introduction

Remote sensing (RS) plays a key role in the dynamic monitoring of lands[1–3]. Approaches of land extraction that are based on remote sensing image basically include manual visual interpretation and computerized auto-classification. Due to the large number of drawbacks in manual visual interpretation, numerous classification algorithms for computerized auto-classification have been developed; among the most popular are the maximum likelihood classifier, neural network classifiers and decision tree classifiers [4]. The maximum likelihood classifier is a popular classifier on the basis of the assumption that classes in the input data follow a Gaussian distribution. However, there will be errors in the results if the sample data size is not sufficient, where the input data set does not follow the Gaussian distribution and/or the classes have much overlap in their distribution, and therefore resulting in poor separability. The back propagation neural network model is widely applied because of its simplicity and its power to extract useful information from samples [5,6]. It is a hierarchical design consisting of fully interconnected layers or rows of processing units (with each unit comprising several individual processing elements, which will be explained below). Back propagation belongs to the class of mapping neural network architectures and therefore the information processing function that it carries out is the approximation of a bounded mapping [7]. Furthermore, the approach can effectively avoid some of the problems associated with MLC by simulating the processing patterns of the human brain, although it also has some disadvantages including a slow learning convergent velocity and being easily converging to local minimum [8]. Lastly, the basic idea of decision tree classifier is to break down a complex decision-making process into a collection of simpler decisions, thus providing a solution which is often easier to interpret.

Support vector machine (SVM) is based on statistical learning theory, and aims to determine the location of decision boundaries that produce the optimal separation of classes [9]. This approach, a new classification technique in the field of remote sensing as compared to the above three methods, has quickly gained ground in the past ten years. The SVM classifier can achieve higher accuracies than both the ML (Maximum Likelihood) and ANN (Artificial Neural Network) classifiers [10] can, thus recently it has been applied to classify remote sensing images [11]. Although perfect performance and high classification accuracy can be achieved by basing on the SVM approach, there still are some shortcomings. One of such shortcomings is that the SVM mainly aims at the classification of a small number of training samples, and the cost of calculation increases rapidly with larger data size, especially so for remote sensing data. In order to resolve such issue of high calculation cost, Fung and Mangasarian [12]proposed proximal support vector machine (PSVM), which can also be interpreted as regularized least squares and considered in the much more general context of regularized networks, wherein classifies points are assigned to the closet of two parallel planes that are pushed apart as far as possible. In addition, the method is much more efficient than traditional SVM in terms of running speed because it merely requires the solution of a single system of linear equations. Accuracy and speed of classification are deemed significant in the classification that's based on remote sensing images. A variety of factors would affect the accuracy and speed of classification: training data size, selection of feature, algorithm parameter setting, just to name a few. Often, real data sets contain noises and the noisy samples might not be representative of a class, as if there is an uncertainty with regard to the class to which they belong. The noises tend to corrupt the data samples, and the optimal hyperplane obtained by the PSVM may be sensitive to noises or outliers in the training sets. As a result, a classifier might not be able to correctly classify some of the data samples having noisy data, so the fuzzy support vector machines [13,14] and fuzzy linear proximal support machines [15,16] were proposed to address the problem.

Normally however, real data set is not linearly separable. In this paper, we proposed the fuzzy nonlinear proximal support vector machine (FNPSVM) to extract different types of lands, and this technique is actually a fuzzy non-linear extension of the existing PSVM methods. In addition, we defined a fuzzy membership function that assigned a fuzzy membership to each data point, such that different data points could have different effects in the learning of the separating hyperplane. Additionally, for the purpose of improving algorithm performance, we presented the approaches of some key parameters of this algorithm, as well as the approaches of feature extraction and feature selection. And lastly, we compared our algorithm with the other three classifiers (MLC, BPN, and PSVM).

The paper is organized as follows:

Section 2 discusses in detail the architectures of PSVM and FNPSVM.

Training algorithm of FNPSVM is shown in section 3.

Experimental results of the algorithm and discussion are presented in section 4.

Section 5 contains the concluding remarks.

Architectures of PSVM and FNPSVM

Architecture of PSVM

To deduce our FNPSVM algorithm, we briefly introduce the binary category proximal support vector machine first. Let the data set consisting of m points in the n-dimensional real space R^n be represented by the $m \times n$ matrix, and let each point be represented by an n-dimensional row eigenvector $A_i(i=1,2,\cdots,m)$. In the case of binary classification, each data point A_i in the class of $A+$ or $A-$ is specified by a given $m \times m$ diagonal matrix D, with +1 or -1 elements along its diagonal. The target is separating the m data points into $A+$ and $A-$, as depicted in Figure 1. For the problem, the proximal support vector machine with a linear kernel [12] is given by the following quadratic program with parameter $c>0$ (which controls the tradeoff between the margin and the error) and linear equality constraint:

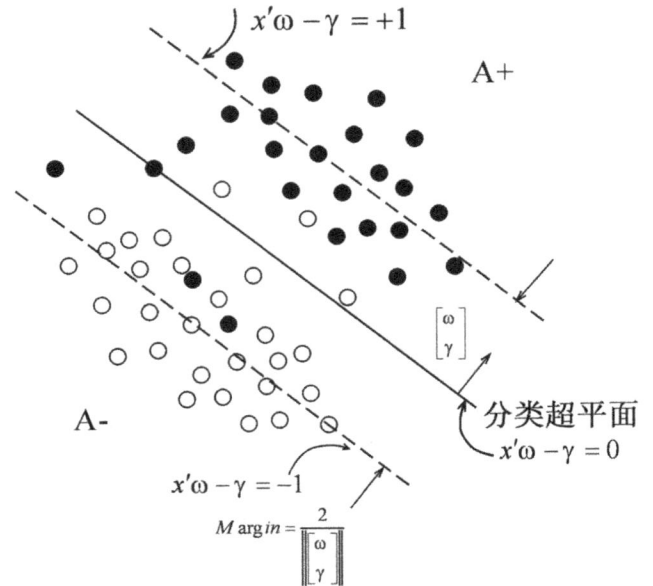

Figure 1. The Proximal Support Vector Machine Classifier: The planes $x'\omega-\gamma=\pm 1$ **around which points of the sets A+ and A- cluster and which are pushed apart by the optimization problem (1).**

$$\min_{(\omega,\gamma,y)\in R^{n+1+m}} \frac{c}{2}\|y\|^2 + \frac{1}{2}(\omega'\omega+\gamma^2)$$

$$s.t. \ D(A\omega-e\gamma)+y=e, \qquad (1)$$

where e is an m-dimensional vector of ones, and y is an error vector. When the two classes are strictly linearly separable, $y_i=0$ in (1) (which is not the case shown in Figure 1). As depicted in Figure 1, the variables (ω,γ) determine the orientation and location of the proximal planes:

$$x'\omega-\gamma=+1$$

$$x'\omega-\gamma=-1, \qquad (2)$$

around which the points of each class are clustered and which are pushed apart as far as possible by the term $(\omega'\omega+\gamma^2)$ in the objective function. Consequently, the plane:

$$x'\omega-\gamma=0, \qquad (3)$$

midway between and parallel to the proximal planes (2), is a separating plane that approximately separates A+ from A- as depicted in Figure 1. The distance $\frac{2}{\left\|\begin{bmatrix}\omega\\\gamma\end{bmatrix}\right\|}$ is called the "margin" (see Figure 1), and maximizing the margin enhances the generalization capability of a support vector machine [9,17].

The approximate separating plane (3) shown in Figure 1, acts as decision function as follows:

$$x'\omega - \gamma \begin{cases} > 0, \text{ then } x \in A+ \\ < 0, \text{ then } x \in A- \\ = 0, \text{ then } x \in A- \text{ or } x \in A- \end{cases} \quad (4)$$

Architecture of FNPSVM

In this paper, we will employ the following norms of a vector $x' \in R^n$ [17]:

$$L_1 \text{ norm of } x : = \|x\|_1 = \sum_{i=1}^{n} |x_i| \quad (5)$$

$$L_2 \text{ norm of } x : = \|x\|_2 = \left(\sum_{i=1}^{n} (x_i)^2 \right)^{1/2} \quad (6)$$

$$L_\infty \text{ norm of } x : = \|x\|_\infty = \max_{1 \le i \le n} (|x_i|) \quad (7)$$

The fuzzy nonlinear binary category proximal support vector machine. Generally, real data sets are corrupted with noises. And as a result, it's not always the case that one classifier obtained by training with noisy data would correctly classify some of the data samples. Since the optimal hyperplane only depends on a small part of the data points, it may become sensitive to noises or outliers in the training set [18,19]. We can associate each data point with a fuzzy membership that reflects their relative degrees as meaningful data, and accounts for the uncertainty in the class to which it belongs. Those noises or outliers are treated as less important and have lower fuzzy membership. This equips the classifier with the ability to train data that has noises or outliers. Such is done by setting lower fuzzy memberships to the data points that are considered to be noises or outliers with higher probability. A classifier that is able to use information regarding this fuzzy degree can improve its performance, and reduce the effects of noise or outliers. Thus we proposed the following the optimization problem in determining the classifier:

$$\min_{(\omega,\gamma,y) \in R^{n+1+m}} \frac{c}{2} \|Sy\|^2 + \frac{1}{2}(\omega'\omega + \gamma^2)$$

$$s.t. \quad D(A\omega - e\gamma) + y = e \quad (8)$$

where S denotes a diagonal matrix, i.e. $S = diag(s_1, s_2, \cdots, s_m)$, whose diagonal elements correspond to the membership values of the data samples belonging to A+ or A-; and e is the vector of plus ones. And $0 < \sigma \le s_i \le 1 (i = 1, 2, \cdots, m)$.

According to the objective function of (8), y can be replaced by ω and γ, so we then arrive at the following unconstrained minimization problem:

$$\min_{(\omega,\gamma) \in R^{n+1}} \frac{c}{2} \|S(D(A\omega - e\gamma) - e)\|^2 + \frac{1}{2}(\omega'\omega + \gamma^2) \quad (9)$$

To obtain fuzzy nonlinear proximal classifier, we modify formula

(9) as in [12,20] first by substituting the variable ω with its dual equivalent $\omega = A'Du$, and then by modifying the last term of the objective function to be the norm of the new dual variable u and γ. Now we obtain the following problem:

$$\min_{(u,\gamma) \in R^{m+1}} \frac{c}{2} \|S(D(AA'Du - e\gamma) - e)\|^2 + \frac{1}{2} \left\| \begin{bmatrix} u \\ \gamma \end{bmatrix} \right\|^2 \quad (10)$$

If we now replace the linear kernel AA' by a nonlinear kernel K(A, A)', we obtain:

$$\min_{(u,\gamma) \in R^{m+1}} \frac{c}{2} \|S(D(K(A,A')Du - e\gamma) - e)\|^2 + \frac{1}{2} \left\| \begin{bmatrix} u \\ \gamma \end{bmatrix} \right\|^2 \quad (11)$$

Let $F(u,\gamma) = \frac{c}{2} \|S(D(K(A,A')Du - e\gamma) - e)\|^2 + \frac{1}{2} \left\| \begin{bmatrix} u \\ \gamma \end{bmatrix} \right\|^2$, and setting one-order derivative of $F(u,\gamma)$ with respect to u and γ to zero, i.e. $\begin{cases} \partial F(u,\gamma)/\partial u = 0 \\ \partial F(u,\gamma)/\partial \gamma = 0 \end{cases}$, we arrive at the following formula:

$$\begin{cases} c(SDK(A,A')D)'S(D(K(A,A')Du - e\gamma) - e) + u = 0 \\ c(SDe)'S(D(-K(A,A')Du + e\gamma) + e) + \gamma = 0 \end{cases}, \quad (12)$$

where both D and S are diagonal matrices, and so that $D = D'$, $S = S'$ and $D^2 = I$. Further, we deal with the above formula (12), and obtain the equations with respect to u and γ:

$$\begin{cases} (c(SDK(A,A')D)'SDK(A,A')D + I)u - c(SDK(A,A')D)'SDe\gamma \\ \quad - c(SDK(A,A')D)'Se = 0 \\ \quad - c(SDe)'SDK(A,A')Du \\ \quad + (c(SDe)'SDe + 1)\gamma + c(SDe)'Se = 0 \end{cases} \quad (13)$$

Now let

$$M_1 = c(SDK(A,A')D)'SDK(A,A')D + I,$$
$$L_1 = -c(SDK(A,A')D)'SDe,$$
$$C_1 = -c(SDK(A,A')D)'Se,$$
$$M_2 = -c(SDe)'SDK(A,A')D,$$
$$L_2 = c(SDe)'SDe + 1,$$
$$C_2 = c(SDe)'Se$$

And thus formula (13) can be expressed by the following formula:

$$\begin{bmatrix} M_1 L_1 \\ M_2 L_2 \end{bmatrix} * \begin{bmatrix} u \\ \gamma \end{bmatrix} = \begin{bmatrix} -C_1 \\ -C_2 \end{bmatrix} \quad (14)$$

We can work out u and γ by solving formula (14), and hence the binary category nonlinear classifier can be written as follows:

$$K(x,A)Du - \gamma \begin{cases} > 0, \text{ then } x \in A+ \\ < 0, \text{ then } x \in A- \\ = 0, \text{ then } x \in A+ \text{ or } x \in A- \end{cases} \quad (15)$$

The fuzzy nonlinear proximal support vector machine. There are roughly four types of support vector machines that handle multi-class problems [21]. Two strategies have been proposed to adapt the SVM to N-class problems [22], namely the "one-against-one" strategy and the "one-against-rest" strategy. The "one-against-one" strategy is to construct a machine for each pair of classes, resulting in N (N- 1)/2 machines. When applied to a test pixel, each machine gives one vote to the winning class, and the pixel is labeled with the class having most votes. The "one-against-rest" strategy is to break the N-class case into N two-class cases, in each of which a machine is trained to classify one class against all others [4]. In this paper, we employed the above mentioned strategies.

♦ "One-against-one" strategy:

$$A = \left[A^1 \cdots A^k \right], A+ = A^r, A- = A^j,$$
$$r \in \{1, \cdots, k-1\}, j \in \{2, \cdots, k\}, r > j.$$

Here, k is the class number, while $A^r \in R^{m^r \times n}$ and $A^j \in R^{m^j \times n}$ represent the m^r and m^j points in class r and class j, respectively. Let $m = m^r + m^j$, and thus D is a $m \times m$ diagonal matrix as follows:

$$D_{ii} = 1 \, \text{for} \, A_i \in A^r, D_{ii} = -1 \, \text{for} \, A_i \in A^j,$$

From formula (14), the $k \times (k-1)/2$ unique u and γ can be obtained, and thus $k \times (k-1)/2$ proximal surfaces are generated:

$$K(x, A) D u^s - \gamma^s = 0, s = 1, \cdots, k \times (k-1)/2$$

A new given point $x \in R^n$ is assigned the ith class T_i $(i = 1, \cdots, k)$ timed by $k \times (k-1)/2$ proximal surfaces, and finally x is assigned i^{th} class in terms of the following formula:

$$K(x, A) D u^i - \gamma^i = \max T_i, i = 1, \cdots, k,$$

supposing the dataset is to be classified into M classes. Therefore, M binary SVM classifiers may be created where each classifier is trained to distinguish one class from the remaining M-1 classes. For example, class one binary classifier is designed to discriminate between class one data vectors and the data vectors of the remaining classes. Other SVM classifiers are constructed in the same manner. During the testing or application phase, data vectors are classified by finding margin from the linear separating hyperplane. The final output is the class that corresponds to the SVM with the largest margin [23].

♦ "One-against-rest" strategy:

$$A = \left[A^1 \cdots A^k \right], A+ = A^r, A- = \left[A^1 \cdots A^{r-1} A^{r+1} \cdots A^k \right],$$
$$r \in \{1, \cdots, k\}, A+ = A^r,$$

where k is the class number, $A^r \in R^{m^r \times n}$ represents the m^r points in class r. Letting $m = m^1 + m^2 + \cdots + m^k$, so that D is a $m \times m$ diagonal matrix as follows:

$$D_{ii} = 1 \, \text{for} \, A_i \in A^r, D_{ii} = -1 \, \text{for} \, A_i \notin A^r, r \in \{1, \cdots, k\}$$

From formula (14), the k unique u and γ can be obtained, and thus k proximal surfaces are generated:

$$K(x, A) D u^r - \gamma^r = 0, r = 1, \cdots, k$$

A new given point $x \in R^n$ is assigned class t, depending on which of the k nonlinear halfspaces generated by the k surfaces it lies deepest in, namely:

$$K(x, A) D u^t - \gamma^t = \max K(x, A) D u^r - \gamma^r, r = 1, \cdots, k.$$

In this method, SVM classifiers for all possible pairs of classes are created. Therefore, for M classes, there will be binary classifiers. The output from each classifier in the form of a class label is obtained. The class label that occurs most is assigned to that point in the data vector. In case of a tie, a tie-breaking strategy may be adopted. A common tie-breaking strategy is to randomly select one of the class labels that are tied [23].

Training algorithm of FNPSVM

Fuzzy membership model

In order to improve classification performance and to reduce the corruption of data samples from noises, we defined a fuzzy membership function to a given class, where a membership is assigned to each data point. It is written as:

$$f(x) = \begin{cases} 1 & 0 \leq x \leq t_1 \\ e^{-\frac{t_1 \cdot t_2}{t_2 - t_1} x} & t_1 \leq x \leq t_2 \\ 0 & t_2 \leq x \leq 1 \end{cases},$$

where x denotes the distance between the data sample and the center of the class that it belongs to. In addition, t_1 and t_2 that tune the fuzzy membership of each data point in the training are two user-defined constants, and they determine the range in which the data sample absolutely does or does not belong to a given class. On the other hand, they also control the figure of the curve (see Figure 2).

A reducing value of x would indicate that the distance between the data sample point and the center of the given class is smaller, and the probability of this sample belonging to this certain class is higher. When x is between 0 and t_1, the data sample point belongs to the given class with absolute certainty; and when x is between t_2 and 1, the data sample point doesn't belong to the given class. When the value of x is known, the values of t_1 and t_2 would influence the values of fuzzy memberships, and thus would also influence the ultimate classification result.

The distance x is the key of each training sample's fuzzy membership, and it can be obtained as follows:

$$M_t = \frac{1}{n} \sum_{i=1}^{n} VF_{ti}, t = 1, \cdots, p.$$

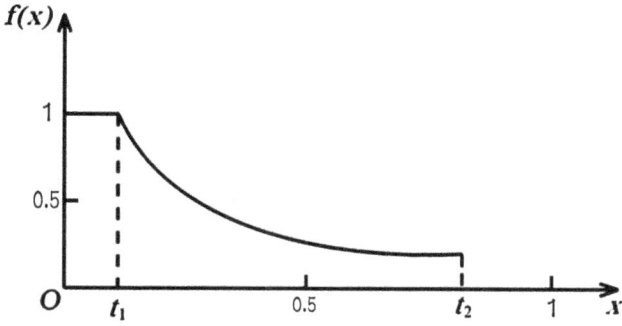

Figure 2. Figure of Fuzzy Membership Function: t_1 and t_2 that tune the fuzzy membership of each data point in the training are two user-defined constants, and they determine the range in which the data sample absolutely does or does not belong to a given class.

$$DM_t = \max|VF_{ti} - M_t|, i = 1, \cdots, n, t = 1, \cdots, p.$$

$$x_i = \frac{1}{p}\sum_{t=1}^{p} |VF_{ti} - M_t|/DM_t, i = 1, \cdots, n, t = 1, \cdots, p,$$

where n is the number of training samples to a given class, and p is the number of feature selected, with VF_{ti} representing the tth feature value of the ith sample. M_t is the mean value of tth feature of n samples to a given class; DM_t is the max value of the distances between all sample points and the center (M_t) of the tth feature to a given class; and x_i denotes the average distance between the ith sample and the centers of all features.

Sample Selection

The choice in sample size and sampling design affect the performance and reliability of a classifier. Sufficient samples are necessary. A previous study indicated that this factor alone could be more important than the selection of classification algorithms in obtaining accurate classifications [24].

Sample selection includes two parts, namely sample data size and selection method. Increases in sample data size generally will lead to improved performances, though at the same time resulting in a higher calculation cost. The sample size must be sufficient enough to provide a representatively meaningful basis for training of a classifier and for accuracy assessment. The basic sampling designs, such as simple random sampling, can be appropriate if the sample size is large [25] enough. The adoption of a simple sampling design is also valuable in helping to meet the requirements of a broad range of users [26]. In this paper, we apply simple random sampling design to collect training samples and testing samples.

Kernel Function Strategy

The concept of the kernel is introduced to extend SVM's ability in dealing with nonlinear classification. It can transform non-linear boundaries in low-dimensional space into linear ones in high-dimensional space by mapping feature vector into a high-dimensional space, and thus the training data can be classified in the high-dimensional space without knowing the specific form of the mapping function. A kernel function is a generalization of the distance metric that measures the distance between two data

points as the data points are mapped into a high dimensional space in which the data are more clearly separable [27,28].

Three kernel functions for nonlinear SVM, including the radial basis function (RBF), the polynomial, and the sigmoid are widely used. In this paper, we have adopted the Gaussian RBF kernel as the default kernel function model due to the fact that: (1) The RBF kernel can handle the case where the relation between class labels and attributes is nonlinear [29]; (2) The polynomial function spends a longer time in the training stage of SVM, and some previous studies [30–32] have reported that the RBF function would provide better performance compared to polynomial function. In addition, the polynomial kernel has more hyper parameters than RBF kernel does, and may approach infinity or zero while the degree is large [29]; (3) The sigmoid kernel behaves like the RBF under certain parameters; however, it is not valid under some parameters [9]; (4) When the size of sample data is quite large, convergent ability of RBF kernel is stronger than that of the other kernels above.

The Gaussian kernel function is expressed as:

$$K(A,B)_{ij} = e^{-\sigma\left\|A_i^{'} - B_j\right\|^2}, i = 1, \cdots, m, j = 1, \cdots, k$$

Here, the matrix $A \in R^{m \times n}$, and $B \in R^{n \times k}$; A_i is the ith row of A, which is a row vector in R^n, while B_j is the jth column of B; the kernel $K(A,B)$ maps $R^{m \times n} \times R^{n \times k}$ into $R^{m \times k}$. In particular, if x and y are column vectors in R^n, then $K(x',y)$ is a real number, $K(x',A')$ is a row vector in R^m, and $K(A,A')$ is a $m \times m$ matrix. The parameter σ of the RBF kernel is a user-defined positive constant regulating the width of the Gaussian kernel, which has an important impact on kernel performance. There is however little guidance in the literatures on the criteria of selecting kernel-specific parameters [33], hence we carried out lots of trials to acquire the optimal parameter σ.

Parameter Selection Method

Regardless of using a simple or a more complex classifier, the learning parameters have to be chosen carefully in order to yield a good classification performance. The FNPSVM algorithm proposed in this paper requires four given parameters, specifically c, σ, t_1 and t_2. Vapnik [9] discovered that varying kernel functions would slightly affect classification results of SVM, while the parameters of the kernel functions and penalty constant c would have a strong effect on the performance of SVM.

One such parameter $c > 0$ is an important quantity in determining a trade-off between the empirical error (number of wrongly classified inputs) and the complexity of the found solution. Normally large values for c lead to fewer training errors (and a narrower margin), all at the cost of more training time; whereas small values generate a larger margin, with more errors and more training points situated inside the margin. Since the number of training errors cannot be interpreted as an estimate of the true risk, this knowledge does not really help in choosing a suitable value for the parameter. The parameter σ of the Gaussian kernel affects the complexity of the decision boundary. Improper selection of these two parameters can cause over-fitting or under-fitting problems [29,34]. Nevertheless, there is little explicit guidance to solve the problem of choosing parameters for SVM. Recently, Hsu [35] suggested a method in determining parameters, namely grid-search and cross validation. For multi-category however, the cross validation method is not feasible. In this paper, we advanced his method and proposed an approach named the multi-layer grid search and random-validation.

The basic idea of random-validation is that we randomly divide the sample set into training set and test set of different size to each category. The test set is sequentially tested using the classifier trained on the training set, and the classification accuracy is derived. The above procedure is iteratively executed for n times during each cycle, and n accuracies are obtained. Finally, the random-validation accuracy is the mean of n accuracies.

We recommend the "multi-layer grid search" method on c and σ using n random-validation, in order to accurately find the optimal parameters while lowering computational cost. We first acquire the boundary of the parameters c and σ, and the 2-dimentional grid of pairs of (c_i, σ_j) is roughly constructed. Here, $i = 1, 2, \cdots, m$, and $j = 1, 2, \cdots, n$, thus $m \times n$ gird-plane and $m \times n$ pairs of (c_i, σ_j) are obtained. The FNPSVM algorithm uses each pair of (c_i, σ_j) to learn by basing on n random-validation, and obtains the classification accuracy. The corresponding $(c_i, \sigma_j)_{high}$ of the best accuracy is the optimal pair. If the best accuracy does not satisfy the requirement of classification, a new 2-dimensional grid-plane that's based on the center of the pair of $(c_i, \sigma_j)_{high}$ should be constructed, and the learning by using new pairs of (c, σ) in the new grid-plane is executed to acquire higher accuracy. The above procedure is performed iteratively to find the optimal parameters c and σ.

Although the multi-layer grid search and random-validation seem simple, it is actually practical because of the fact that: (1) For each parameter, a finite number of possible values is prescribed, and then all possible combinations of (c, σ) are considered to find one that yields the best result; (2) the computational time in finding good parameters through the approach isn't much more than that of advanced methods, since there are only two parameters (generally the complexity of grid search grows exponentially with the number of parameter); (3) The grid-search can be easily parallelized because each (c, σ) is independent, unlike some other advanced methods that require iterative processes.

Experiments and Discussion

All experiments were run on 1800 MHz ADM Sempron (tm) processor 3000^+ under Windows XP using Matlab 7.0 compiler. We have adopted the classification criterion of Chen [36]; saline-alkalized lands are classified into heavy saline-alkalized land, moderate saline-alkalized land, and light saline-alkalized land.

Classification Experiments Using ETM$^+$ Image

Experiment summary. We have selected Da'an, a city in northern China with a total area of 4,879 km^2 as our test area. Multi-spectral (Landsat-7 ETM$^+$) remote sensing data (30 m spatial resolution, UTM project) acquired on August 30th, 2000 was used to classify the image data into nine land cover types (heavy saline-alkalized land, moderate saline-alkalized land, light saline-alkalized land, water area, cropland, grassland, rural residential area, urban residential area, and sand land).

According to the topographic maps of Da'an city (1:100,000 scale), we implemented precise geometric correction and resampling of the image. Geometric correction of image was accomplished through two-order polynomial while resampling was achieved through cubic convolution with the error of matching less than one pixel. We selected 270 samples (90 for training and 180 for testing) for each class using a random sampling procedure from the image, totally 810 training samples and 1,620 test samples for nine classes. For each sample set, the test set was independent of the training set.

To demonstrate the effectiveness of the proposed method, both "one-against-one" and "one-against-rest" strategies that are based

on Gaussian RBF kernel in dealing with the n-class case were used, and the results (various accuracies, training speed, and classification speed) obtained using FNPSVM algorithm were compared with those derived from the four conventional classification methods including the maximum likelihood classifier (MLC), back propagation neural network (BPN), support vector machine (SVM), and proximal support vector machine (PSVM) under different training conditions (shown in Table 1).

Feature extraction and feature selection. (1) Feature extraction. Feature extraction has a strong impact on classification accuracy. In this paper, we extracted 14 features, including six bands of ETM$^+$ image, the first principle components of K-L transform and K-T transform, soil index, NDVI (normalized difference vegetation index), composition index, as well as H (hue), S (saturation), and I (intensity) color components of HSI color space. Some of the features can be obtained as follows:

$$\text{Soil index}: \quad SI = (B_5 - (255 - B_4))/(B_5 + (255 - B_4)) \text{ [37]}$$

$$\text{NDVI}: \quad VI = (B_4 - B_3)/(B_4 + B_3)$$

$$\text{Composition index}: \quad CI = (B_5 - B_1)/(B_5 + B_1) \text{ [37]}$$

Here B_1, B_3, B_4, B_5 represent the first band, third band, forth band, and fifth band of ETM$^+$ image, respectively.

In the field of digital image processing, a number of color models were proposed, such as RGB, HSI, CIE, etc. But selecting the most optimal color space is still a problem in color image segmentation [20].

The RGB color model is suitable for color display, but less so for color analysis because of its high correlation among R, G, and B color components [38]. In color image processing and analysis, we know that: (1) H and S components are closely correlated to the color sense of the eyes; (2) Hue information and intensity information are distinctly differentiated in HSI model; (3) By HSI model, computer program can easily process color information after the color sense of the eye has been transformed into specific values, so we extracted H, S, and I color components of HSI color space as three features of classification. False color image composite of bands 5, 4, and 2 were performed, after which the image was exported into RGB image. And finally the RGB model was transformed into HSI model according to the following formulas [39]:

$$H = \arccos\left\{ \frac{[(R-G)+(R-B)]/2}{[(R-G)^2+(R-B)(G-B)]^{1/2}} \right\}$$

$$S = 1 - \frac{3}{R+G+B}[\min(R,G,B)], I = \frac{1}{3}(R+G+B),$$

(2) Feature selection. Normally, the size of a real dataset is so large that learning might not work, and the running time of a learning algorithm might be drastically increased before removing these unwanted features. Thus we must select some features that are neither irrelevant nor redundant to the target concept.

Feature selection for classification is a well-researched problem, striving to improve the classifier's generalization ability, and to reduce the dimensionality and the computational complexity. It directly reduces the number of original features by selecting a

Table 1. Training data conditions under which the classification algorithms were tested.

Sample size		Number of features	Training case no.
Training sample number	Testing sample number		
60	210	4	A
		7	B
		10	C
		14	D
90	180	4	E
		7	F
		10	G
		14	H
120	150	4	I
		7	J
		10	K
		14	L

subset of them that still retains sufficient information for classification [40]. Feature selection attempts to select the minimally sized subset of features according to the following criterion. The criterion can be [41]:

1) The classification accuracy does not significantly decrease; and

2) The resulting class distribution when given only the values for the selected features, is as close as possible to the original class distribution when given all features.

For this paper, in terms of the above criterion, the data types and the characteristics of remote sensing image, we adopted traditional DB Index rules which used the methods of between-class scatter and within-class scatter to select classification features. DB Index rules are as follows [42]:

1)

$$S_i = \frac{1}{N_i} \sum_{x \in N_i} \|x - X_i\|,$$

where N_i denotes the number of samples of ith class; and X_i represents the center of the ith class.

2)

$$d_{ij} = \|X_i - X_j\|,$$

where d_{ij} is the distance between the centers of the two classes.

3) DB Index $DB_k = \frac{1}{k} \sum_{i=1}^{k} R_i$, $R_i = \max_{j=1,\cdots,k,j \neq i} \frac{S_i + S_j}{d_{ij}}$, where k is the number of classes.

The smaller the value of DB_k is, the better the performance of classification is. Based on the above rules and 270 sample points of each category, we obtained DB indices of fourteen features and their ranks (see Table 2).

Parameter setting. Due to the differing nature of the impacts that algorithm parameters have on different algorithms, it is impossible to account for such differences in evaluating the comparative performances of the algorithms [4]. To avoid this

problem, the corresponding parameters of the best performance of each algorithm were chosen for the purpose of comparison.

(1) Parameter setting of PSVM and FNPSVM. The performance of classification algorithms is affected by the parameter settings of those algorithms. As described in section 3.4, we searched for the optimal parameters t_1, t_2, c, and σ for FNPSVM classifier. In this procedure, we used two steps to find the best parameters. In the first step, we set the parameters $t_1 = 0.1$ and $t_2 = 0.8$, and searched for the kernel parameter σ and penalty constant c as described in section 3.5. In the second step, we set the parameters σ and c as found in the first step, and searched for the parameters t_1 and t_2 of the fuzzy membership mapping function. In the first step, we constructed the two-dimensional grid for the first layer. The values of c and σ were prescribed from 2^{-14} to 2^{14}, multiplied by 2^4. The grid-search using 5-time random-validation was executed, and we found that the optimal parameter pair (c,σ) was $(2^{10}, 2^{-10})$, having the highest overall classification accuracy (93.31%) and kappa

Table 2. DB indices of fourteen features and their ranks.

Rank	Feature	DB index
1	the 6th band of ETM image	2.0408
2	the 5th band of ETM image	4.2657
3	the 4th band of ETM image	5.0092
4	CI (Composition Index)	6.3319
5	the 1st component of K-L transform	7.0428
6	H component of HSI color space	7.6135
7	SI (Soil Index)	8.5819
8	NDVI (Normalized Difference Vegetation Index)	9.8511
9	the 1st component of K-T transform	10.8020
10	the 1st band of ETM image	14.8599
11	the 3rd band of ETM image	25.2807
12	the 2nd band of ETM image	26.7408
13	I component of HIS color space	29.8844
14	S component of HIS color space	153.2745

value (0.9248). Table 3 summarized the results of first-layer grid-search. Subsequently we constructed the second-layer grid based on the center $(2^{10}, 2^{-10})$; and the values of c and σ were chosen from 2^7 to 2^{13} and from 2^{-7} to 2^{-13}, multiplied by 2 respectively; and the grid-search using 5-time random-validation was implemented. As was shown in Table 4, $c = 2^{13}$ and $\sigma = 2^{-13}$ gave the best overall classification accuracy (93.56%) and kappa coefficient (0.9275). As the accuracies could fundamentally satisfy our classification demand, we began the next step, where we set the parameters $c = 2^{13}$ and $\sigma = 2^{-13}$, and searched for the parameters t_1 and t_2. Unfortunately, we couldn't find that the changes of parameters t_1 (0.05~0.2) and t_2 (0.7~0.9) to be able to significantly improve the performance of the FNPSVM, hence we set $t_1 = 0.1$, $t_2 = 0.8$.

(2) Parameter setting of BP neural network. There are many parameters associated with BP neural network, including neuron number, transfer function, learning rate, iteration time and so on. It is not easy to know beforehand which values of these parameters are the best for a problem. Consequently in this paper, in order to yield the optimal classification performances, the settings of some key parameters of BP neural network were achieved by repeated trials and some experiences from previous studying.

A BP neural network with a hidden layer can approximate with arbitrary precision an arbitrary non-linear function that's defined on a compact set of R^n [43,44]. We employed three-layer BP neural network including input layer, hidden layer and output layer. The number of neurons in the hidden layer is one of the primary parameters of BPN algorithm; currently however there is no authoritative rule to determine it. Larger number of hidden units leads to a poor generalization and increases training time, but too few neurons would cause the networks to unfit the training set and to prevent the correct mapping of inputs and outputs. In this paper, the number of neurons in the hidden layer was determined by the empirical formula [44] to be 20, thus the network structure became n-20-9 (n denotes the number of features).

We chose log-sigmoid function as the transfer functions from input layer, while setting the limit on the neural network's iteration number to be 1,000 times for each desired output. Levenberg-Marquard optimum algorithm (trainlm function in Matlab software) was utilized as the training function because it could greatly increase the training speed of the network by utilizing a lot of memory. Gradient descent with momentum weight and bias learning function was employed to calculate a given neuron's weight change from the neuron's input and error, the weight, learning rate, and the momentum constant according to the gradient descent with momentum. The other parameters of the network are chosen as follows: learning rate $\eta = 0.5$, momentum factor $\alpha = 0.8$, minimum gradient $\delta = 10^{-20}$, and minimum mean square error $\varepsilon = 10^{-6}$. Figure 3 shows the classification maps using the MLC, BPN, PSVM, and FNPSVM, all based on the settings of above parameters of various classifiers.

Performance assessments. Normally, settings of the various parameters on different algorithms affect the classification results, so it is difficult to evaluate the comparative performances of the algorithms because of the changing parameters. To address this problem, the best performance of each algorithm on each training case was listed in the following tables. The criterion for evaluating the performances of classification algorithms includes accuracy, speed, stability and comprehensibility, among others [4]. In this paper, we chose one group of criteria, consisting of classification accuracy, speed and stability to assess the performances of different algorithms. Table 5 gave overall accuracies and kappa coefficients using various multi-class strategies and classifiers with ETM$^+$ data on different cases. Using different classifiers under different training conditions, Table 6 gave training speed and classification speed of the entire data set. Means and standard deviations of the overall classification accuracies basing on different training samples, testing samples and features, were manifested in Table 7. Figure 4 shows the boxplots of the overall classification accuracies, developed by randomly selecting training samples and testing samples from the 270 samples of each class for six times.

(1) Classification accuracy. In this paper, classification accuracy, one of the most important criterions in evaluating the performance of the classifier, was measured using overall accuracies and kappa coefficients computed by the confusion or error matrix. The most widely used way to represent the classification accuracy of remote sensing data should be in the form of an error matrix, applicable for a variety of site-specific accuracy assessments. Numerous researchers have recommended using error matrix in representing accuracy in the past, and it has now become one of the standard conventions to adopt such practice. The effectiveness of the error matrix in representing accuracy can be seen from the fact that accuracies of each category are fundamentally described along with both the errors of inclusion and errors of exclusion present in the classification [25,45]. In order to accommodate the effects of chance agreement, some researchers suggest using kappa coefficient and adopting it as a standard measure of classification accuracy [46]. Foody [47] also pointed out that since many of the remote sensing data sets are dominated by mixed pixels, the standard accuracy assessment measures such as the kappa

Table 3. The overall accuracies (%) and kappa coefficients of the first layer grid-search using 5-time random-validation based on ETM$^+$ image.

c \ σ	2^{-14}	2^{-10}	2^{-6}	2^{-2}	2^2	2^6	2^{10}	2^{14}
2^{-14}	40.87/0.3348	69.15/0.6530	17.92/0.0766	11.11/0	11.11/0	11.08/0	11.11/0	11.11/0
2^{-10}	59.65/0.5461	74.25/0.7103	59.18/0.5408	12.45/0.0151	11.09/0	11.11/0	11.11/0	11.11/0
2^{-6}	64.00/0.5950	81.48/0.7916	75.83/0.7281	25.00/0.1563	11.43/0.0036	11.52/0.0046	11.34/0.0026	11.34/0.0026
2^{-2}	76.93/0.7405	88.02/0.8653	85.30/0.8346	42.13/0.3490	12.53/0.0160	11.57/0.0051	11.62/0.0058	11.49/0.0043
2^2	85.60/0.8380	90.71/0.8955	90.32/0.8911	47.95/0.4145	13.15/0.0230	11.55/0.0050	11.60/0.0055	11.54/0.0048
2^6	89.33/0.8800	92.78/0.9188	89.91/0.8865	46.20/0.3948	13.80/0.0303	11.42/0.0035	11.43/0.0036	11.61/0.0056
2^{10}	91.86/0.9085	93.31/0.9248	89.36/0.8803	48.60/0.4218	13.49/0.0268	11.57/0.0051	11.55/0.0050	11.70/0.0066
2^{14}	92.44/0.9150	93.08/0.9221	87.70/0.8616	45.14/0.3828	13.79/0.0301	11.71/0.0068	11.61/0.0056	11.67/0.0063

Table 4. The overall accuracies (%) and kappa coefficients of the second layer grid-search using 5-time random-validation based on ETM$^+$ image.

c σ	2^{-7}	2^{-8}	2^{-9}	2^{-10}	2^{-11}	2^{-12}	2^{-13}
2^7	91.20/0.9010	92.82/0.9193	92.34/0.9138	92.91/0.9203	92.71/0.9180	92.11/0.9113	91.00/0.8988
2^8	91.14/0.9003	92.45/0.9151	92.74/0.9183	92.71/0.9180	92.42/0.9148	92.07/0.9108	91.71/0.9068
2^9	91.34/0.9026	91.95/0.9095	92.94/0.9206	92.68/0.9176	92.75/0.9185	92.57/0.9165	92.10/0.9111
2^{10}	89.70/0.8841	92.45/0.9151	93.36/0.9253	92.48/0.9155	93.17/0.9231	92.91/0.9203	92.08/0.9110
2^{11}	90.99/0.8986	91.37/0.9030	92.37/0.9141	92.75/0.9185	92.99/0.9211	92.29/0.9133	92.57/0.9165
2^{12}	90.19/0.8896	91.49/0.9043	92.23/0.9126	93.08/0.9221	92.63/0.9171	92.96/0.9208	93.06/0.9220
2^{13}	90.16/0.8893	91.57/0.9051	92.42/0.9148	92.82/0.9193	93.05/0.9218	92.80/0.9190	93.56/0.9275

coefficient is often not suitable for accuracy assessment in remote sensing. Although its sensitivity to the density or frequency of the dynamic change in real world had some researchers arguing about its effect, the fact remains that the kappa coefficient has many intriguing features as an index of classification accuracy. More specifically, it offers some compensation for chance agreement, and a variance term could be calculated, enabling the statistical testing of the significance of the difference between two coefficients [25,48].

We also need to emphasize that the various measures of accuracy are to evaluate different components of accuracy and to make different assumptions on the data [49]. The fact is that the measurement and meaning of classification accuracy depend substantially on individual perspective and demands [49,50]. An accuracy assessment can be conducted for a variety of reasons, and many researchers have recommended that measures such as the kappa coefficient of agreement be adopted as a standard [25,46].

$$Overall\, accuracy = \sum_{k=1}^{q} n_{kk}/n \times 100\%$$

$$Kappa\, coefficient = \frac{n\sum_{k=1}^{q} n_{kk} - \sum_{k=1}^{q} n_{k+}n_{+k}}{n^2 - \sum_{k=1}^{q} n_{k+}n_{+k}}$$

In terms of the above parameters selected from different algorithms, and basing on the 270 samples of each category obtained through simple random sampling design, we obtained overall classification accuracies and kappa coefficients using various multi-class strategies and classifiers on 12 training cases with the ETM$^+$ dataset consisting of 4,037,099 points (see Table 5). Unfortunately, confronting such a large dataset, SVM failed on this problem because it required the more costly solution of a linear or quadratic program. Several patterns can be observed from Table 5 and Table 7, explained as follows:

1) As far as the multi-class classification strategies of PSVM and FNPSVM were concerned, the accuracies of "one-against-one" strategy in all training cases were about 1–2% higher than those of "one-against-rest" strategy. Also, through experiments, we found that compared to the classification speed of "one-against-rest" strategy, the classification speed of "one-against-one" strategy was at least two times higher, for both PSVM and FNPSVM (not listed in the following tables). So in this paper, we employed "one-

against-one" multi-class classification strategy of PSVM and FNPSVM for comparison with the other two classifiers.

2) The level of classification accuracies achieved by PSVM and FNPSVM was significantly higher than that produced by either the MLC or BPN classifier. In addition, they yielded significantly better results than the MLC or BPN classifier did in all 12 training cases (Table 5). The accuracy differences between the PSVM and FNPSVM were rather small, and quite the same as that between the MLC and BPN (Table 5). The mean overall accuracies of the PSVM and FNPSVM were remarkably higher than those of MLC and BPN, however the differences between MLC and BPN or between PSVM and FNPSVM were only slight (Table 7). This is expected because the PSVM and FNPSVM are designed to locate an optimal separating hyperplane, while the other two algorithms may not be able to locate this separating hyperplane. Statistically, the optimal separating hyperplanes located by the PSVM and FNPSVM should be generalized to unseen samples with the least errors among all separating hyperplanes. Generally, as the number of available features increases, the overall accuracies and kappa coefficients of PSVM and FNPSVM grow gradually. Unexpectedly however, the increase in the number of available features didn't always lead to an improvement of the accuracies of MLC and BP. On the contrary, MLC and BP showed better comparative performances on training cases with ten features than they did on training cases with fourteen features, which might be explained by the presence of a large number of irrelevant features that would hurt the classification performances. This again demonstrates the importance of feature selection. In terms of Table 5, it could be seen that the accuracies and kappa coefficients of the four algorithms improved with the increase in training data size, though not significantly.

3) The overall accuracy differences between MLC and BPN on the data set used in this study were generally small, and those between PSVM and FNPSVM were also not obvious. However, many of them were statistically significant.

(2) Training speed and classification speed. Training speed and classification speed are two important criterions in evaluating the performances of classification algorithms. Shown in Table 6, the training speed and classification speed of the four classifiers were substantially different. Generally, the training time and classification time rise with an increase in available features. The training speed of BPN was significantly lower than those of the other three classifiers because of its complex network structure. As far as classification speed was concerned, in all training cases, those of the PSVM and FNPSVM were remarkably lower than those of the MLC and BPN. The classification of the MLC and BPN in all training cases took from less than an hour to only a few minutes,

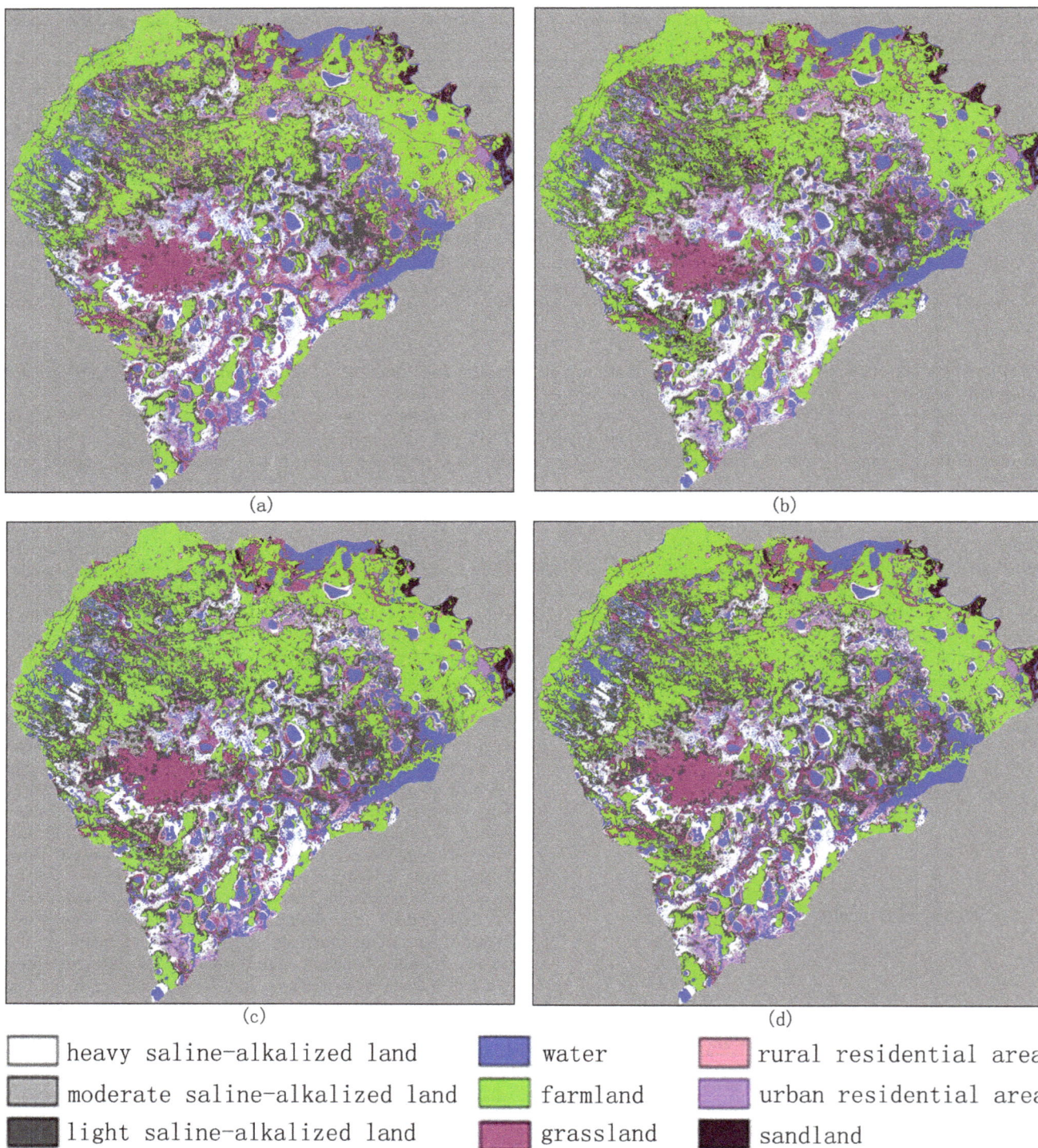

Figure 3. Classification maps for the test area in northern China using various classifiers under the same training case (90 training samples for each class, 10 features). (a) MLC algorithm. (b) BPN algorithm, $\eta = 0.5$, $\alpha = 0.8$, $\delta = 10-20$, $\varepsilon = 10-6$. (c) PSVM algorithm, c = 213, $\sigma = 2-13$. (d) FNPSVM, t1 = 0.1, t2 = 0.8, c = 213 and $\sigma = 2-13$.

while the PSVM and FNPSVM took more than several hours and ten hours, respectively. This was due to the fact that PSVM and FNPSVM involved large matrix calculation and reverse matrix operation during the process of classification. In addition, it should be noted that we have spent much time in searching for the

optimal key parameters including the kernel parameters σ and the constant c in the training process, therefore yielding a better performance. Compared with PSVM, the training speed and classification speed of FNPSVM were more than twice its counterparts. The reason was that in terms of the comparison

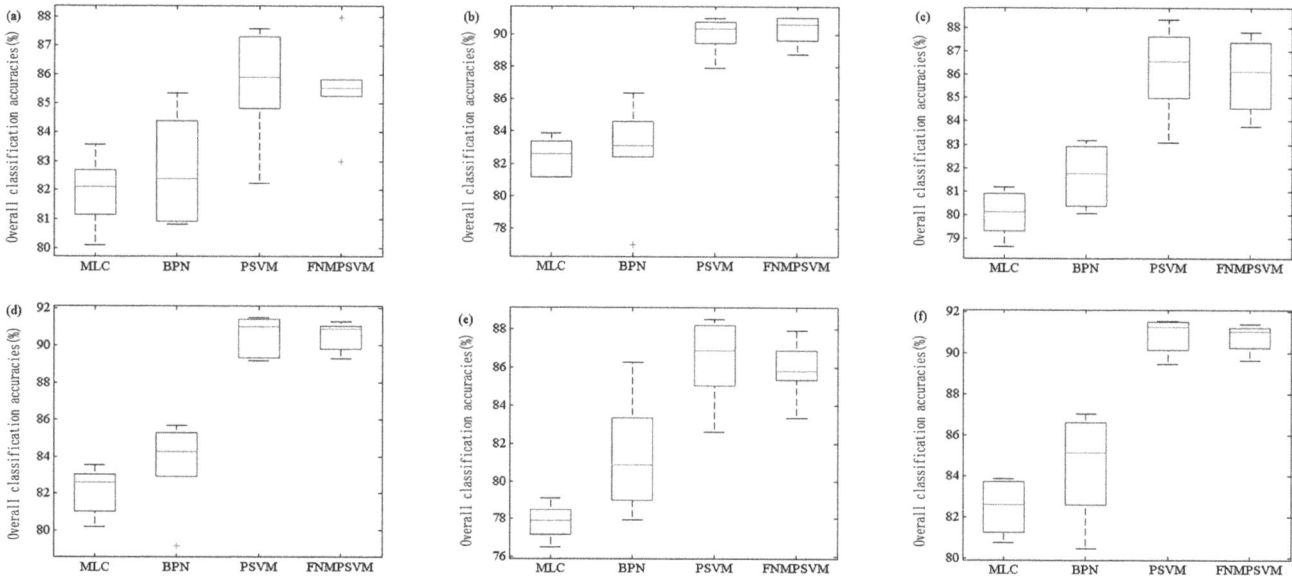

Figure 4. Boxplots of the overall classification accuracies developed by randomly selecting training samples and testing samples for six times from 270 samples of each class based on ETM$^+$ image. (a) Training samples = 60, testing samples = 210, number of features = 4. (b) Training samples = 60, testing samples = 210, number of features = 10. (c) Training samples = 90, testing samples = 180, number of features = 4. (d) Training samples = 90, testing samples = 180, number of features = 10. (e) Training samples = 120, testing samples = 150, number of features = 4. (f) Training samples = 120, testing samples = 150, number of features = 10.

between the FNPSVM algorithm in section 2.2.1 and the PSVM algorithm [12], it was easy to find that the PSVM algorithm dealt with the product of an n-dimensional (n being the number of features) row vector and a matrix (m being the number of training samples), thus requiring a high calculation cost; while the FNPSVM algorithm avoided such problem.

To summarize, the training speeds and classification speeds of the above four algorithms are affected by many factors, including numbers of training samples and features, the size of training data set, as well as algorithm parameter settings. The training speed and classification speed of BPN depend on network structure, momentum rate, learning rate and converging criteria; while those of the PSVM and FNPSVM were affected by the number of features, kernel function, key parameter settings, as well as class separability.

(3) Algorithm stability. Various accuracies in Table 5 were obtained by randomly selecting training samples and testing samples only once at each sample size level. In order to evaluate

Table 5. Overall accuracies (%) and kappa coefficients using various multi-class strategies and classifiers on different cases based on ETM$^+$ image.

Training case no.	MLC OA/KC	BPN OA/KC	PSVM One against one OA/KC	PSVM One against rest OA/KC	FNPSVM One against one OA/KC	FNPSVM One against rest OA/KC
A	81.28/0.7419	81.94/0.7923	87.82/0.8601	85.59/0.8347	87.97/0.8617	85.45/0.8332
B	81.33/0.7427	85.05/0.8291	89.36/0.8777	88.21/0.8645	89.23/0.8763	88.02/0.8624
C	82.34/0.7548	85.09/0.8288	91.19/0.8989	89.05/0.8739	91.14/0.8982	88.59/0.8687
D	80.15/0.7291	82.66/0.8006	91.20/0.8989	90.21/0.8875	91.21/0.8989	89.57/0.8802
E	81.85/0.7484	82.93/0.8034	87.52/0.8566	86.16/0.8411	87.87/0.8606	86.09/0.8405
F	82.35/0.7541	86.29/0.8430	89.84/0.8831	88.24/0.8648	89.88/0.8836	88.29/0.8655
G	84.18/0.7755	84.83/0.8258	91.05/0.8985	88.02/0.8635	91.92/0.9071	88.87/0.8719
H	81.41/0.7433	83.03/0.8048	91.85/0.9062	90.26/0.8880	92.81/0.9158	90.93/0.8943
I	82.02/0.7503	82.05/0.7935	87.76/0.8593	86.50/0.8449	88.00/0.8621	86.51/0.8451
J	82.39/0.7547	88.56/0.8685	89.91/0.8840	88.83/0.8715	90.73/0.8919	89.79/0.8811
K	84.85/0.7831	86.47/0.8442	91.99/0.9079	89.16/0.8751	91.71/0.9046	88.79/0.8709
L	81.72/0.7460	84.01/0.8157	92.12/0.9093	90.15/0.8868	92.94/0.9172	90.83/0.8931

Note: OA and KC denote overall accuracy (%) and kappa coefficient, respectively.

Table 6. Training time and classification time of whole data set (4,037,099 pixels) using various classifiers on different cases based on ETM$^+$ image unit:second.

Training condition	MLC		BPN		PSVM		FNPSVM	
	Training time	Classification time	Training time	Classification time	Training time	Classification time	Training time	Classification time
Training samples = 1080 Feature number = 4	13	408	556	775	54	62925	17	27968
Training samples = 1080 Feature number = 7	17	816	566	785	56	65389	19	29891
Training samples = 1080 Feature number = 10	20	1147	632	794	59	66426	22	31880
Training samples = 1080 Feature number = 14	21	1626	658	1135	65	68562	24	34193

the stabilities of the four classifiers and for the results to be statistically valid, we randomly selected training samples and testing samples for six times at three sample data size levels from the 270 samples of each class: 60 training samples and 210 testing samples, 90 training samples and 180 testing samples, as well as 120 training samples and 150 testing samples. Thus each classification algorithm was trained six times by various-sized training samples with four and ten features, respectively. Afterwards we calculated the means and standard deviations of the overall classification accuracies of each classifier (see Table 7).

The standard deviation of the overall accuracy of an algorithm estimated in cross validation is a quantitative measure of its relative stability [4]. Both Table 7 and Figure 4 revealed that the stabilities of the algorithms differed greatly and were affected by the training data size, testing data size, and the number of features. Generally, the overall classification accuracies of the algorithms became more robust when trained by using large-sized pixels than using small-sized pixels, especially when ten features were used (Figures 5 (b), (d), and (f)). Unexpectedly however, MLC showed higher reliability and lower mean overall accuracy when trained with only four out of a total 14 features (Table 7). This is probably due to the fact that MLC algorithm itself is sensitive to some relevant features, while some features that are partially or completely irrelevant to the classification target only increase the uncertainty of the classification results. On the other hand, according to Hughes effect [51], the effect of increasing dimensionality is thought to lower the reliability of the estimates of statistical parameters required for the computation of probabilities [10]. The FNPSVM showed more stable overall accuracies than the other three classifiers did when trained with ten features at different training sample size levels; however, when trained with four out of the total 14 features, the stability of FNPSVM was significantly lower than that of MLC, although clearly higher than those of the other two algorithms (Figure 4 (a), (c), and (e)). The likely cause of the stability of FNPSVM being lower than that of MLC on data with four features is that the applicability of the FNPSVM to non-linear decision boundaries depends on whether the decision boundaries can be transformed into linear ones by mapping the input data into a high-dimensional space. When the data contain very few features, the FNPSVM can't successfully transform non-linear decision boundaries in the original feature space into linear ones in a high-dimensional feature space, while the MLC algorithm is useful when there is a fair amount of randomness under which the data are generated. The theoretical statistical distribution allows the use of the MLC approach that is optimal in the sense that, using too many irrelevant features probably affects its stability; so that even when the data contain very few features, it has better comparative reliability performance over the FNPSVM.

Compared to PSVM, FNPSVM generated better reliability in all of the 6 training cases (Table 7), owing to automatically associating each data point with a fuzzy membership that can reflect their relative degrees as meaningful data, and FNPSVM becoming more applicable in reducing the effects of noises or outliers in the process of training. Of the four algorithms, the BPN gave overall accuracies in a wider range than the other three algorithms (Figure 4) did for all cases, and showed the worst reliability (Table 7) because of its complex network structure and lots of optional parameters that affect the classification performance.

Classification Experiments Using SPOT Image

Experiment summary. This study took the SPOT remote sensing image captured on September 12th, 2004 (scene number :

Table 7. Means and standard deviations (σ) of overall classification accuracies based on various samples and features using ETM$^+$ image.

Training condition	MLC		BPN		PSVM		FNPSVM	
	Mean	σ	Mean	σ	Mean	σ	Mean	σ
Training samples = 60 Testing samples = 210 Feature number = 4	80.62	1.2501	82.70	2.7309	85.62	2.6845	86.51	1.8390
Training samples = 60 Testing samples = 210 Feature number = 10	81.14	1.8666	82.76	3.6037	89.98	1.3334	90.29	1.0604
Training samples = 90 Testing samples = 180 Feature number = 4	82.05	1.2069	81.68	3.0933	86.18	2.4831	86.95	1.9367
Training samples = 90 Testing samples = 180 Feature number = 10	82.17	1.7292	83.61	2.8588	90.56	1.2268	90.54	0.9945
Training samples = 120 Testing samples = 150 Feature number = 4	80.83	1.1455	81.37	4.6063	84.36	2.4258	85.85	1.6640
Training samples = 120 Testing samples = 150 Feature number = 10	82.47	1.5977	84.50	3.3208	90.84	1.1358	91.74	1.0548

64002TH200409121049401018) as the data source, covering the western area of Da'an city in China and including near-infrared, red and green band. We cut 1,734*1,969 sized image from the SPOT image as test data set. The experiment area mainly contained several land types, namely heavy saline-alkalized land, moderate saline-alkalized land, light saline-alkalized land, water area and farmland. And then 120 samples (60 training samples and 60 test samples) were selected from each land type to train classification algorithm and to evaluate the accuracy of classification.

To evaluate the performance of FNPSVM algorithm on extracted saline-alkalized land using high spatial resolution (SPOT

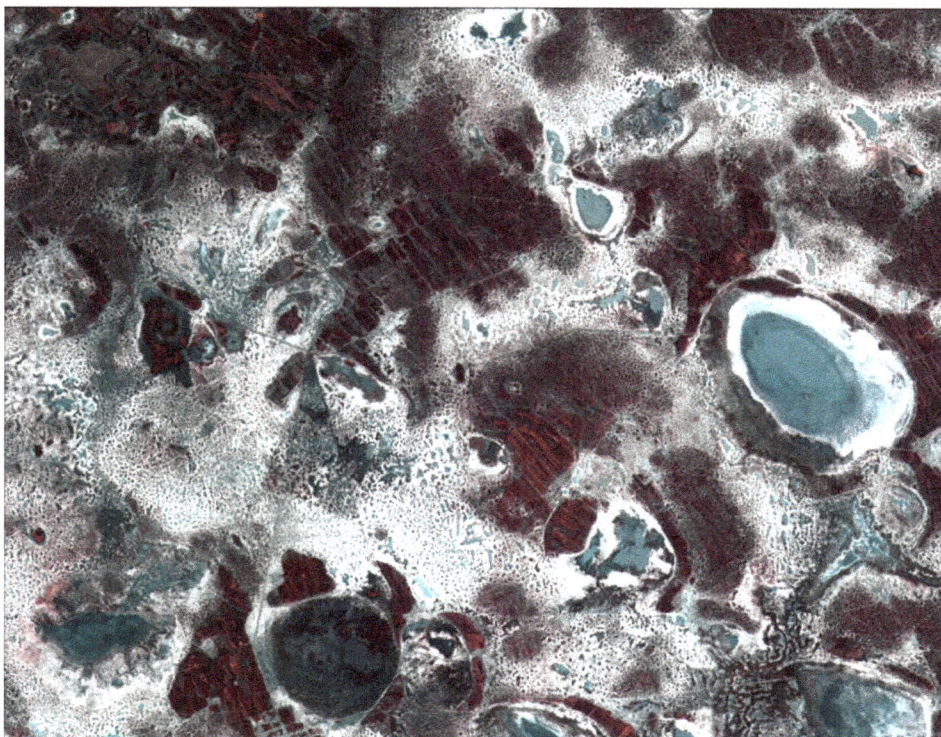

Figure 5. Original SPOT image in study area (composite of bands 3, 2 and 1).

with 20m resolution), we adopted the "one-against-one" strategy based on the Gaussian RBF kernel function, and compared with MLC, BPN, and PSVM methods in terms of classification accuracy and classification speed.

Feature extraction. The paper extracted 8 features from the SPOT image data, including near-infrared band, red band, green band, the 1st component of K-L transform, NDVI, and the H, S and I components of HSI color space. NDVI is expressed by the following formula:

$$NDVI = (B_3 - B_2)/(B_3 + B_2),$$

where B_2 and B_3 are red band and near-infrared band of SPOT, respectively. RGB image is acquired by basing on the false-color composite using the third, the second, and the first band of SPOT image. The RGB model is then transformed to HSI model by the following formulas:

$$I = \frac{1}{3}(R + G + B)$$

$$S = 1 - \frac{3}{R + G + B}[\min(R, G, B)]$$

$$H = \arccos\left\{\frac{[(R-G)+(R-B)]/2}{[(R-G)^2+(R-B)(G-B)]^{1/2}}\right\}.$$

It's not necessary to choose feature because of the limited features, so we extracted saline-alkalized land information based on the above eight features using various algorithms.

Key parameter setting. We used the method in section 3.4 to obtain the parameters of PSVM and FNPSVM classifier. The accuracy of FNPSVM classifier didn't change significantly when t_1 changed in the 0.05–0.2 range and t_2 changed in the 0.7–0.9 range, so the paper still set $t_1 = 0.1$ and $t_2 = 0.8$. Afterwards, the optimal parameters of c and σ were searched, and the results were shown in Table 8 and 9. The overall classification accuracy increased to a maximum of 97.26% at $c = 2^{11}$ and $\sigma = 2^{-13}$.

The parameter setting of BPN was as same as the BPN parameters of section 4.1.3.2 except for the neural network structure. We chose log-sigmoid function as the transfer functions

from input layer to output layer, and set the limit on the neural network's iteration number to 1,000 times for each desired output. Levenberg-Marquard optimum algorithm is used as the training function. The other parameters of the network were set as follows: learning rate $\eta = 0.3$, momentum factor $\alpha = 0.8$, minimum gradient $\delta = 10$–20, and minimum mean square error $\varepsilon = 10$-6. In this study, the number of neurons in the hidden layer is finally acquired through repeated experiments, eventually arriving at 12. Therefore the structure of neural network is 8-12-5.

Performance assessments. (1) Vision effect. Original SPOT image in study area is obtained by compositing bands 3, 2 and 1 (see Figure 5). Figure 6 shows the experiment classification result using MLC, BPN, PSVM and FNPSVM algorithms which were based on 8 feature vectors and 60 training samples per class. When comparing the classification result map of each algorithm with the original SPOT image, on the macro level the differences of various classification result maps are not clear; but in detail we can find that the spatial patterns of land cover classification from the PSVM and FNPSVM method are significantly better than the other two methods, while the spatial patterns of classification maps are quite similar between PSVM and FNPSVM. As seen in Figure 6, when using MLC and BPN classifier, not only were the patches fragmented, but the saline-alkalized land and water area were also mistakenly mixed; and a large number of saline-alkalized lands were wrongly classified as water area. But BPN, PSVM and FNPSVM classifiers all overcome the drawbacks of MLC classifier.

(2) Classification accuracy. We obtained the confusion matrix according to 60 test samples, and then calculated the overall accuracy and kappa coefficient of various classifiers (see Table 10).

Seen from Table 10, the overall accuracy and kappa coefficient of all classifiers were higher, except that the classification accuracy of MLC classifier was less than 90%. The accuracies of the other classifiers were all higher than 95%. The overall accuracy of FNPSVM was the highest (97.33%), meaning that the performance of FNPSVM was better than the others, which was mainly in accordance to the strict mathematical theory.

(3) Classification speed. Classification speed is one of the important indicators to evaluate the performance of the classifier. Table 10 gave the classification time based on the 8-dimensional feature vector. As can be seen from the table, the MLC classifier was the fastest, mainly because of the simplicity of the algorithm and the low amount of computation. This was followed by BPN classification by only a few minutes. The classification speed of PSVM and FNPSVM dropped substantially from that, with several hours lagging; the speed of FNPSVM was twice the such of

Table 8. The overall accuracies (%) and kappa coefficients of the first layer grid-search using 6-time random-validation based on SPOT image.

σ \ c	2^{-14}	2^{-10}	2^{-6}	2^{-2}	2^2	2^6	2^{10}	2^{14}
2^{-14}	66.13/0.57	80.13/0.75	85.40/0.81	94.80/0.93	95.73/0.94	96.33/0.95	92.80/0.91	93.40/0.91
2^{-10}	79.93/0.74	86.46/0.83	94.60/0.93	96.33/0.95	93.53/0.91	88.53/0.85	96.53/0.95	93.26/0.91
2^{-6}	28.00/0.10	48.93/0.36	72.40/0.65	76.06/0.70	89.53/0.86	88.46/0.85	88.93/0.86	84.73/0.81
2^{-2}	20.00/0	20.86/0.01	24.86/0.06	42.60/0.28	40.53/0.25	44.26/0.30	45.40/0.31	48.73/0.36
2^2	20.00/0	20.06/0	20.40/0	22.33/0.03	24.06/0.05	24.53/0.05	23.80/0.04	24.06/0.05
2^6	20.00/0	20.20/0	20.33/0	20.93/0.01	21.20/0.01	21.46/0.02	20.80/0.01	21.26/0.01
2^{10}	20.00/0	20.00/0	20.06/0	20.33/0	20.20/0	20.60/0.01	20.53/0.01	20.60/0.01
2^{14}	20.00/0	20.13/0	20.26/0	20.46/0	20.73/0.01	20.66/0.01	20.60/0.01	20.33/0

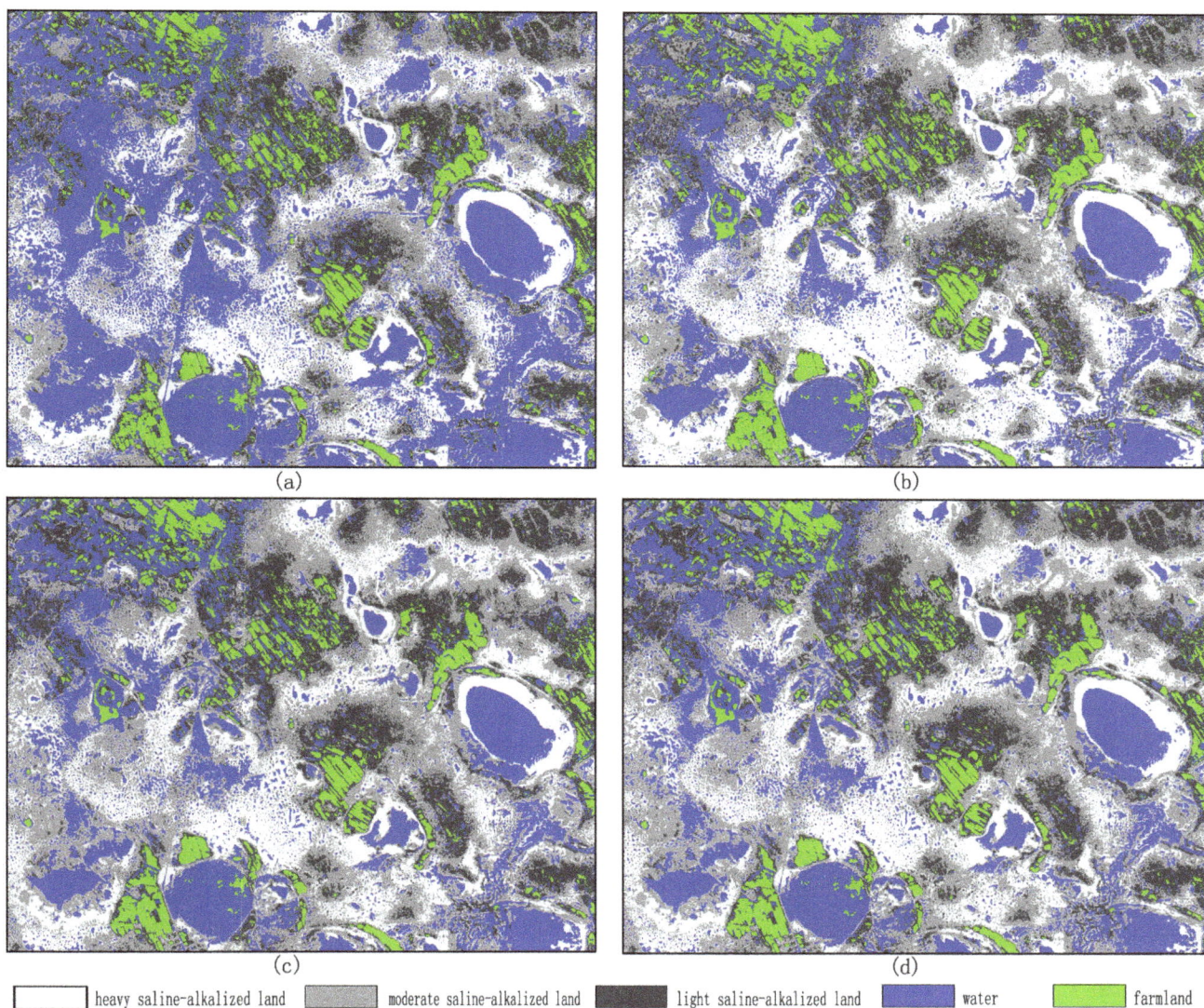

Figure 6. Classification maps for the western part of test area of Da'an city in China using various classifiers under the same training cases (120 training samples for each class, 8 features) based on SPOT image. (a) MLC algorithm. (b) BPN algorithm, $\eta = 0.3$, $\alpha = 0.8$, $\delta = 10^{-20}$, $\varepsilon = 10^{-6}$. (c) PSVM algorithm, $c = 2^{11}$, $\sigma = 2^{-13}$. (d) FNPSVM, t1 = 0.1, t2 = 0.8, c = 2^{11} and $\sigma = 2^{-13}$.

PSVM, for the same reason that explains ETM image classification speed, of which the paper will not discuss at this time.

(4) Algorithm stability. Based on the above experiment data, feature extraction and key parameter settings, we obtained the overall classification accuracies of various classifiers (see Table 10), and calculated the means and standard deviations (see Table 11).

Seen from Table 11, the stability of each algorithm under the same training condition is quite different. The stability of BPN classifier is the lowest because of its complex network structure and the many optional parameters affecting the classification performance. The FNPSVM showed more stable overall accuracies than the other three classifiers did, as by automatically associating each data point with a fuzzy membership in the process of training, FNPSVM could effectively reduce the effects of noises.

Conclusions

Considered as a kind of regularized least squares SVM, PSVM requires the solution of a single set of linear equations, and thus

can be considerably faster than conventional SVM. Jayadeva [16] extended the PSVM and proposed fuzzy linear proximal support vector machine. In order to increase nonlinear separability of real data set, we presented in this paper the fuzzy nonlinear proximal support vector machine (FNPSVM), and described the strategy for setting fuzzy membership in FNPSVM, therefore making FNPSVM more feasible in the application of reducing the effects of noises or outliers. Numerous experiments were performed to evaluate the comparative performances of this algorithm and three other popular classifiers, including the MLC, BPN and PSVM in saline-alkalized land classification. In addition, impacts of the key parameters of FNPSVM algorithm on its performance as well as the impacts of the selection of training data and features on all four classifiers were also evaluated.

The results of our experiments supported the use of "one-against-one" strategy for multi-class classification problems, and indicated that of the four algorithms evaluated, both the PSVM and FNPSVM achieved considerably higher levels in overall

Table 9. The overall accuracies (%) and kappa coefficients of the second layer grid-search using 6-time random-validation based on SPOT image.

σc	2^7	2^8	2^9	2^{10}	2^{11}	2^{12}	2^{13}
2^{-7}	94.33/ 0.93	92.46/ 0.91	91.86/ 0.89	94.53/ 0.93	87.73/ 0.84	83.66/ 0.79	91.60/ 0.89
2^{-8}	92.33/ 0.90	96.46/ 0.95	92.00/ 0.90	91.66/ 0.89	91.60/ 0.89	92.00/ 0.90	91.86/ 0.90
2^{-9}	97.60/ 0.97	93.20/ 0.91	93.13/ 0.91	96.73/ 0.95	92.73/ 0.91	96.6/ 0.95	96.53/ 0.95
2^{-10}	96.33/ 0.95	96.46/ 0.95	97.60/ 0.97	97.13/ 0.96	89.53/ 0.87	92.86/ 0.91	96.60/ 0.95
2^{-11}	92.80/ 0.91	92.46/ 0.91	97.06/ 0.96	97.00/ 0.96	96.86/ 0.96	97.13/ 0.96	96.73/ 0.96
2^{-12}	96.86/ 0.96	96.53/ 0.95	97.00/ 0.96	89.13/ 0.86	93.73/ 0.92	96.66/ 0.95	97.00/ 0.96
2^{-13}	93.00/ 0.91	96.86/ 0.96	89.66/ 0.87	96.73/ 0.95	97.26/ 0.96	88.46/ 0.85	92.33/ 0.90

accuracies and kappa coefficients than either the MLC or the BPN did, especially so in high-dimensional feature space; and comparatively, the accuracies and kappa coefficients of the MLC were lowest in all 12 training cases. The results should be attributed to the abilities of PSVM and FNPSVM in locating the optimal separating hyperplanes, as shown in Figure 1. Statistically, the optimal separating hyperplanes found by the PSVM and FNPSVM classifiers should be generalized as unlabeled samples with errors smaller than any other separating hyperplanes that might be located by other classifiers. In terms of the performances of PSVM and FNPSVM classifiers, the absolute differences of their overall accuracies were quite small. Many of the differences were however, statistically significant.

The stabilities of PSVM and FNPSVM are closely correlated to the features used in the classification. The PSVM and FNPSVM algorithms gave higher stability than either the MLC or BPN did when being trained with 10 features by different sizes of pixels. When reduced to only 4 features however, the MLC manifested comparatively better reliability. As far as the PSVM and FNPSVM classifiers were concerned, the stability of FNPSVM was significantly higher than that of the PSVM, because the application of fuzzy set approach reduced the effects of noises or outliers. With regard to classification speed, based on larger dataset consisting of 4,037,099 points, the MLC and BPN were much faster than the PSVM and FNPSVM, and the computational cost of the PSVM was more than twice the cost of the

Table 10. Overall accuracies (%), kappa coefficients and classification speed using various classifiers based on SPOT image.

	MLC	BPN	PSVM	FNPSVM
overall accuracy	89.67	95.33	96.33	97.33
kappa coefficients	0.87	0.94	0.95	0.97
classification speed	153	407	17167	8285

Table 11. Means and standard deviations (σ) of overall classification accuracies based on SPOT image.

	MLC	BPN	PSVM	FNPSVM
Mean	81.43	82.66	89.72	90.79
σ	1.9212	3.5050	1.3646	1.0211

FNPSVM. This indicated that the algorithm in this paper had predominant advantage in running speed compared to PSVM, and noted that both the PSVM and FNPSVM were affected by training data size, key parameter settings, class separability and so on.

When we adopted the multi-layer grid search and random-validation method in searching for the optimal parameters t_1 and t_2, it was found that with t_1 valued smaller than 0.2 and t_2 valued larger than 0.7, accuracy variation of FNPSVM classifier was rather small, and the classification accuracy was relatively high. Within this range, the sensitivity wasn't high. With t_1 valued between 0.2 and 0.5, or t_2 valued between 0.5 and 0.7, accuracy variation of FNPSVM classifier was relatively high, while the classification accuracy significantly reduced, showing obvious cases of misclassification. Within this range, the sensitivity was relatively high. This is mainly because t_1 and t_2 would determine the range of which this sample point absolutely does or doesn't belong to a given class. When t_1 has too high a value or t_2 too low a value, it would automatically lead to a higher probability in misclassification, at which point the classification result is relatively sensitive to the values of t_1 and t_2. This is also in consistency with our experience.

Both the selection and the number of training samples and testing samples affect the performances of all four classifiers. It is impractical and even impossible to determine the minimum number of samples needed for the sufficient training of an algorithm according to the results from these experiments. Fuzzy classification technique can reduce the corruption that the noises have on the data samples, and consequently, classification robustness is improved. To a greater extent, feature extraction and feature selection exert a strong impact on the substantial increases in accuracy, as some irrelevant features probably lead to a declining classification performance for some algorithms. As SVM is becoming a popular learning machine for object classification, the principal contribution of this paper is the presentation of a fast, simple and efficient classification algorithm in the research field of SVM, which is important as it significantly reduces the running time of classification and improves the stability performance.

Currently, due to China's increasing population and the far from perfect land management system, the conflict in more people having less land is becoming increasingly apparent. Drastically reduced farmland and land degradation have drawn the public's attention. The FNPSVM method proposed in this paper can rapidly and accurately extract information in land use and dynamic changes. As a foundation in establishing general planning for land utilization, protecting basic farmland, and ensuring sustainable development of the land, the method is also capable of providing decision supports for the authorities in land utilization and management. The algorithm has been successfully used in lands extraction with remote sensing image, and with the development in constructing the digital city, the extraction of land use information in cities becomes increasingly important.

This algorithm can be used to classify patches of cities, and it is devoted to the construction of a digital city.

Acknowledgments

We are grateful to the undergraduate students and staff of the Laboratory of Forest Management and "3S" technology at Beijing Forestry University.

Author Contributions

Conceived and designed the experiments: XZ JL HD. Performed the experiments: JL SD GW. Analyzed the data: XZ JL HD GW ZZ. Contributed reagents/materials/analysis tools: YJ YW ZZ LW FY. Wrote the paper: XZ JL.

References

1. Lin J, Zhong Y, Peng X (2010) Region multi-center method for land use classification of multispectral RS imagery,Journal of Remote Sensing 14 (1): 173–179.
2. Dwivedi RS, Sreenivas K (1998) Image transforms as a tool for the study of soil salinity and alkalinity dynamics. International Journal of Remote Sensing 19 (4) : 605–619.
3. Zhu J, Li J, Ye J (2011) Land Use Information Extraction from Remote Sensing Data Based on Decision Tree Tool. Geomatics and Information Science of Wuhan University 36 (3): 301–305.
4. Huang C, Davis LS, Townshend JRG (2002) An assessment of support vector machines for land cover classification. International Journal of Remote Sensing 23 (4) : 725–749.
5. Guo H, Wang W (2009) Research on SVM learning algorithms based on neural networks. Computer Engineering and Applications 45 (2): 51–54.
6. Yi J, Wang Q, Zhao D, Wen J (2007) BP neural network prediction-based variable-period sampling approach for networked control systems. Applied Mathematics and Computation 185(2): 976–988.
7. Hecht Nielsen R(1989) Neural Networks IJCNN. In:International Joint Conference.pp.593–605.
8. Liu H, Yin S, Liu D (2010) FPGA implementation of dynamic neural network for support vector machines. Journal of Shanhai Jiaotong University 44 (7): 962–967.
9. Vapnik VN (1995) The Nature of Statistical Learning Theory. New York: Springer-Verlag Press.
10. Pal M, Mather PM (2005) Support vector machines for classification in remote sensing. International Journal of Remote Sensing 26 (5): 1007–1011.
11. Wang C, Luo J, Zhou C, Ming D, Chen Q, et al. (2005) Extraction of road network from high resolution remote sensed imagery with the combination of gaussian markov random field texture model and support vector machine. Journal of Remote Sensing 9 (3): 271–276.
12. Fung G, Mangasarian OL (2001) Proximal support vector machine classifiers.In:Proceeding KDD '01 Proceedings of the seventh ACM SIGKDD international conference on Knowledge discovery and data mining (ACM New York, NY, USA).pp.77–86.
13. Lin C, Wang S (2004) Training algorithms for fuzzy support vector machines with noisy data. Pattern Recognition Letters 25: 1647–1656.
14. Li Q, Gao D, Yang B (2009) Journal of Jilin University (Engineering and Technology Edition) 39 (2): 131–134.
15. Jayadeva R, Suresh C (2004) Fast and robust learning through fuzzy linear proximal support vector machines. Neurocomputing 61: 401–411.
16. Jayadeva R, Suresh C (2005) Fuzzy linear proximal support vector machines for multi-category data classification. Neurocomputing 67: 426–435.
17. Golub G, Van Loan C(1996) Matrix Computations (3rd Edition).Maryland: The Johns Hopkins University Press.
18. Boser BE, Guyon I, Vapnik V (1992) A training algorithm for optimal margin classifiers. In:Fifth Annual Workshop on Computational Learning Theory.pp.144–152.
19. Zhang X (1999) Using class-center vectors to build support vector machines. In:Proceedings of 1999 IEEE Signal Processing Society Workshop on Neural Networks for Signal Processing 4.pp.3–11.
20. Mangasarian OL (2000) Generalized support vector machines. In: A.Smola, P.Bartlett, B.Scholkopf, D.Schuurmans, editors. Advances in Large Margin Classifiers. 135–146.
21. Daisuke T, Shigeo A (2003) Fuzzy least squares support vector machines for multiclass problems. Neural Networks 16: 785–792.
22. Gualtieri JA, Cromp RF (1998)Support vector machines for hyperspectral remote sensing classification. In:Proceedings of the 27th AIPR Workshop: Advances in Computer Assisted Recognition. 221–232.
23. Mahesh P (2005) Multiclass Approaches for Support Vector Machine Based Land Cover Classification. In: Proceeding of Computer Vision and Pattern Recognition: Neural and Evolutionary Computing, MapIndia, arXiv: 0802.2411.
24. Hixson M, Scholz D, Fuhs N, Akiyama T(1980) Evaluation of several schemes for classification of remotely sensed data. Photogrammetric Engineering and Remote Sensing 46 (12): 1547–1553.
25. Foody GM (2002) Status of land cover classification accuracy assessment. Remote Sensing of Environment 80: 185–201.
26. Stehman SV, Czaplewski RL (1998) Design and analysis for thematic map accuracy assessment: fundamental principles. Remote Sensing of Environment 64: 331–344.
27. Girolami M (2002) Mercer kernel-based clustering in feature space. IEEE Trans. Neural Networks 13 (3): 780–784.
28. Kim DW, Lee K, Lee D, Lee K (2005) A kernel-based subtractive clustering method. Pattern Recognition Letters 26: 879–891.
29. Min J, Lee Y (2005)Bankruptcy prediction using support vector machine with optimal choice of kernel function parameters. Expert Systems with Applications 28: 603–614.
30. Mei J, Duan S, Qin Q (2004)Design and implementation of high precision map symbol library based on GDI+. Geomatics and Information Science of Wuhan University 29 (10): 912–915.
31. Huang Z, Chen H, Hsu CJ, Chen WH, Wu S (2004) Credit rating analysis with support vector machine and neural networks: A market comparative study. Decision Support Systems 37: 543–558.
32. Kim KJ (2003) Financial time series forecasting using support vector machines. Neurocomputing 55: 307–319.
33. Eitrich T, Lang B (2006) Efficient optimization of support vector machine learning parameters for unbalanced datasets. Journal of Computational and Applied Mathematics 196: 425–436.
34. Tay FEH, Cao L (2001) Application of support vector machines in financial time series forecasting. Omega 29: 309–317.
35. Hsu CW, Chang CC, Lin CJ (2004) A practical guide to support vector classification. Technical Report, Department of Computer Science and Information Engineering, National Taiwan University.
36. Chen J, Zhang S, Chen J, Tang J (2003) Soil salinization detection with remote sensing and dynamic analysis in DaQing City. Journal of Arid Land Resources and Environment 17 (4): 101–107.
37. Luo Y, Chen H (2001)Grading salinization soil based on TM image with the support of GIS.Remote sensing information 4: 12–15.
38. Shih F, Cheng S (2005) Automatic seeded region growing for color image segmentation. Image and Vision Computing 23 (10) : 877–866.
39. Zhan Y (2004) Image processing and analysis. Beijing: Tsinghua University Press.
40. Liu Y, Zheng Y (2006) FS_SFS: A novel feature selection method for support vector machines. Pattern Recognition 39: 1333–1345.
41. Dash M, Liu H (1997)Feature selection for classification. Intelligent Data Analysis 1: 131–156.
42. Theodoridis S, Koutroumbas K (2003) Pattern Recognition (2th Edition).Beijing: China Machine Press.
43. Bai Y, Jin Z (2005) Prediction of SARS epidemic by BP neural networks with online prediction strategy.Chaos, Solitons and Fractals 26: 559–569.
44. Zenren Y (1999) The Applications of Artificial Neural Network.Beijing:Tsinghua University Press.
45. Congalton R (1991)A review of assessing the accuracy of classifications of remotely sensed data. Remote Sensing of Environment 37: 35–46.
46. Smits PC, Dellepiane SG, Schowengerdt RA (1999) Quality assessment of image classification algorithms for land-cover mapping: A review and proposal for a cost-based approach. International Journal of Remote Sensing 20: 1461–1486.
47. Foody GM (1996) Approaches for the production and evaluation of fuzzy land cover classification from remotely-sensed data. International Journal of Remote Sensing 17: 1317–1340.
48. Rosenfield GH, Fitzpatrick-Lins K (1986) A coefficient of agreement as a measure of thematic classification accuracy. Photogrammetric Engineering and Remote Sensing 52: 223–227.
49. Stehman SV (1999) Comparing thematic maps based on map value. International Journal of Remote Sensing 20: 2347–2366.
50. Campbell JB (1996) Introduction to remote sensing (2nd ed.). London: Taylor and Francis.
51. Hughes GF (1968) On the mean accuracy of statistical pattern recognizers. IEEE Transactions on Information Theory 14: 55–63.

PERMISSIONS

LIST OF CONTRIBUTORS

Seth Stapleton and Todd Atwood
United States Geological Survey, Alaska Science Center, Anchorage, Alaska, United States of America

David Garshelis
Department of Fisheries, Wildlife and Conservation Biology, University of Minnesota, St. Paul, Minnesota, United States of America
Minnesota Department of Natural Resources, Grand Rapids, Minnesota, United States of America

Michelle LaRue and Claire Porter
epartment of Earth Sciences, University of Minnesota, Minneapolis, Minnesota, United States of America

Nicolas Lecomte and Stephen Atkinson
Department of Environment, Government of Nunavut, Igloolik, Nunavut, Canada
Minnesota Department of Natural Resources, Grand Rapids, Minnesota, United States of America

Mei-chen Feng, Lu-jie Xiao, Mei-jun Zhang and Wu-de Yang
Institute of Dryland Farming Engineer, Shanxi Agricultural University, Taigu, People's Republic of China

Guang-wei Ding
Department of Chemistry, Northern State University, Aberdeen, South Dakota, United States of America

Xiaomei Yang and Chenghu Zhou
Institute of Geographic Sciences and Natural Resources Research, Chinese Academy of Sciences, Beijing, China

Fan Meng
Institute of Geographic Sciences and Natural Resources Research, Chinese Academy of Sciences, Beijing, China
University of Chinese Academy of Sciences, Beijing, China

Dong Jiang, Yaohuan Huang, Dafang Zhuang, Yunqiang Zhu, Xinliang Xu and Hongyan Ren
State Key Lab of Resources and Environmental Information System, Institute of Geographical Sciences and Natural Resources Research, Chinese Academy of Sciences, Beijing, China

Kathy L. Murray, James A. Y. Moore, George Shedrawi and Barton G. Huntley
Department of Environment and Conservation, Kensington, Western Australia, Australia

Stuart N. Field and Richard D. Evans
Department of Environment and Conservation, Kensington, Western Australia, Australia
Oceans Institute, University of Western Australia, Crawley, Western Australia, Australia

Peter Fearns, Mark Broomhall, Lachlan I. W. McKinna and Daniel Marrable
Remote Sensing and Satellite Research Group, Department of Imaging and Applied Physics, Curtin University, Bentley, Western Australia, Australia

Dehua Zhao, Hao Jiang, Ying Cai and Shuqing An
Department of Biological Science and Technology, Nanjing University, Nanjing, China

Minchen Zhu and Binghan Liu
College of Mathematics and Computer Science, Fuzhou University, Fuzhou, Fujian, China

Weizhi Wang
College of Civil Engineering, Fuzhou University, Fuzhou, Fujian, China

Jingshan Huang
School of Computing, University of South Alabama, Mobile, Alabama, United States of America

Wei Wang
School of Land Sciences and Technology, China University of Geosciences, Beijing, China

Tao Yuan, Xinqi Zheng, Xuan Hu and Wei Zhou
School of Land Sciences and Technology, China University of Geosciences, Beijing, China
Key Laboratory of Land Regulation, Ministry of Land and Resources, Beijing, China

Huanyu Xu, Quansen Sun, Nan Luo, Guo Cao and Deshen Xia
School of Computer Science and Technology, Nanjing University of Science and Technology, Nanjing, Jiangsu, China

Mitch Bryson, Richard J. Murphy and Daniel Bongiorno
Australian Centre for Field Robotics, The University of Sydney, Sydney, NSW, Australia

Matthew Johnson-Roberson
The Department of Naval Architecture and Marine Engineering, University of Michigan, Ann Arbor, Michigan, United States of America

Qin Dai, JianBo Liu, ShiBin Liu and Jin Yang
Institute of Remote Sensing and Digital Earth, Chinese Academy of Sciences, Beijinng, China

Yang Liu
Institute of Remote Sensing and Digital Earth, Chinese Academy of Sciences, Beijinng, China
University of Chinese Academy of Sciences, Beijing, China

Dehua Zhao, Meiting Lv, Hao Jiang, Ying Cai, Delin Xu and Shuqing An
Department of Biological Science and Technology, Nanjing University, Nanjing, P R China

Cheng He and Siyu Zhang
Nanjing Forest Police College, Nanjing, China

Matteo Convertino
School of Public Health-Division of Environmental Health Sciences, University of Minnesota Twin-Cities, Minnesota, United States of America
Institute on the Environment, University of Minnesota Twin-Cities, Minnesota, United States of America

Zhongke Feng
Institute of GIS, RS&GPS, College of Forestry, Beijing Forestry University, Beijing, China

Abbey Rosso and Peter Neitlich
National Park Service, Winthrop, Washington, United States of America

Robert J. Smith
Department of Botany and Plant Pathology, Oregon State University, Corvallis, Oregon, United States of America

Heng Sun and Jian Weng
Department of Computer Science, College of Information Science and Technology, Jinan University, Guangzhou, People's Republic of China

Guangchuang Yu
Key Laboratory of Functional Protein Research of Guangdong Higher Education Institutes, Institute of Life and Health Engineering, College of Life Science and Technology, Jinan University, Guangzhou, People's Republic of China

Richard H. Massawe
International School, Jinan University, Guangzhou, People's Republic of China

Yong Wang, Dong Jiang, Dafang Zhuang, Yaohuan Huang and Xinfang Yu
State Key Laboratory of Resources and Environmental Information System, Institute of Geographical Sciences and Natural Resources Research, Chinese Academy of Sciences, Beijing, China

Wei Wang
School of Computer and Information Engineering, Beijing Technology and Business University, Beijing, China

Xiaomei Zhong
Tianjin Chengjian University, Tianjin, China

Jianping Li, Huacheng Dou, Shijun Deng, Guofei Wang, Yu Jiang, Yongjie Wang, Zebing Zhou and Li Wang
Tianjin Institute of Geotechnical Investigation and Surveying, Tianjin, China
Tianjin StarGIS Information Engineering Company Limited, Tianjin, China

Fei Yan
Beijing Forestry University, Beijing, China

Index